ELEMENTARY ALGEBRA for COLLEGE

ELEMENTARY ALGEBRA for COLLEGE

HARRY LEWIS
Jersey City State College

D. VAN NOSTRAND COMPANY
New York Cincinnati Toronto London Melbourne

D. Van Nostrand Company Regional Offices:
New York   Cincinnati

D. Van Nostrand Company International Offices:
London   Toronto   Melbourne

Copyright © 1978 by Litton Educational Publishing, Inc.
Library of Congress Catalog Card Number: 77-99181

ISBN: 0-442-24767-2

All rights reserved. No part of this work covered by the copyright hereon may be reproduced or used in any form or by any means—graphic, electronic, or mechanical, including photocopying, recording, taping, or information storage and retrieval systems—without written permission of the publisher. Manufactured in the United States of America.

Published by D. Van Nostrand Company
135 West 50th Street, New York, N.Y. 10020

10 9 8 7 6 5 4 3

Preface

This book is designed for either a one-semester or one-quarter terminal course in mathematics or in helping establish the foundation for further study of mathematics. Its basic goal is to ease the student through the first steps beyond the elementary algorithms of arithmetic to an understanding of basic algebra. Every effort has been made to keep the writing "light" to encourage student reading. Readers of the manuscript were almost unanimous in their comment that the student who is willing to read the material will have little trouble in following the development of the mathematical concepts.

Specifically, these are a few of the distinguishing features of this book:

1. All concepts are developed so that the student will understand exactly *why* he is called upon to perform whatever operations are needed at the moment.

2. The development of the course flows from topic to topic so that the student can comprehend immediately the relationship between the topic under examination and prior topics.

3. Every topic is introduced by justifying why that particular topic is examined at that particular point of the course.

4. Each chapter closes with a review of all topics covered in that chapter.

5. The text is divided into chapters that, in turn, are subdivided into sections. Usually, each section is devoted to the development of *one* concept. At the close of each section there are enough exercises to provide the student with ample practice to consolidate the understanding of that concept.

6. The style of writing and the readability of the text lend themselves to self-instruction.

7. The style of writing gives the student the feeling that the instructor is working along with him making certain that each point is fully understood before moving on to the next point.

8. The exercises are abundant and very carefully graded. Even if some of them are used as illustrative material, sufficient exercises will remain for student practice.

9. The illustrative examples that appear prior to each set of exercises cover the background needed to complete the exercises successfully. The instructor can make assignments without fear of the inevitable "surprise" of exercises that are in no way related to the explanatory material.

10. Every illustrative example is accompanied by an "Explanation" to show the need for each step in the "Solution" that follows. Each "Solution" presents a neat orderly format as a guide to the students.

11. Definitions and "principles" are presented only as an outgrowth of explanations that lay a thorough foundation for their need. The statements of both are in a language befitting the mathematical maturity of the student just being introduced to this subject. It was felt that the rigor of mathematical "proof" and sophisticated mathematical terminology had best be kept at a minimum in this course.

Answers to the odd-numbered exercises appear at the end of the book. Answers to even-numbered exercises appear in the *Instructor's Manual.* As a further aid, the *Instructor's Manual* contains a separate test for each chapter in addition to a midsemester examination and a final examination.

To the following people the author extends his thanks for their comments and suggestions: Anita Kitchens, Appalachian State University; Albert W. Liberi, Westchester Community College; David Russell, Prince George's Community College; Richard D. Semmler, Northern Virginia Community College; and Ara B. Sullenberger, Tarrant County Junior College.

Harry Lewis

Contents

1 THE SYSTEM OF REAL NUMBERS — 1

    1.1   The Number Line — 1
    1.2   The Extended Number Line — 5
    1.3   The Mathematical Sentence — 8
    1.4   The Exponent — 13
    1.5   The Solution Set of an Equality — 18
    1.6   The Solution Set and Graph of an Inequality — 22
    Chapter Review — 26

2 THE OPERATIONS OF ADDITION AND SUBTRACTION — 29

    2.1   Addition of Real Numbers — 29
    2.2   Addition of Monomials — 40
    2.3   Addition of Polynomials — 49
    2.4   Subtraction of Real Numbers — 53
    2.5   Subtraction of Polynomials — 59
    2.6   Simplifying Expressions Involving Parentheses — 62
    Chapter Review — 67

3 THE OPERATIONS OF MULTIPLICATION AND DIVISION 69

3.1	Multiplication of Real Numbers	69
3.2	Multiplication of Monomials	76
3.3	Multiplication of Polynomials	83
3.4	Special Products	91
3.5	Division of Real Numbers	99
3.6	Division of Monomials	102
3.7	Division of a Polynomial by a Monomial	107
3.8	Division of Polynomials	110
	Chapter Review	117

4 THE SOLUTION OF AN OPEN SENTENCE IN ONE VARIABLE 119

4.1	Solution of a Linear Equation by the Addition Principle	119
4.2	Solution of a Linear Equation by the Multiplication Principle	125
4.3	Solution of a Linear Equation Using Both the Addition and Multiplication Principles	130
4.4	Solution of an Equation Involving Absolute Values	143
4.5	Solution of an Inequality	149
4.6	Solution of an Inequality Involving Absolute Values	153
	Chapter Review	158

5 PROBLEM SOLVING IN MATHEMATICS 161

5.1	The Mathematical Phrase	161
5.2	Number Problems	165
5.3	Coin Problems	175
5.4	Percent Problems	181
	Chapter Review	191

6 FACTORING — 193

6.1	Factors of a Number	194
6.2	Factoring for the Common Factor	198
6.3	Factoring the Difference of Two Squares	210
6.4	Factoring a Trinomial	216
6.5	Summary of Factoring	230
6.6	Solution of the Quadratic Equation by Factoring	231
6.7	Problem Solving Involving Quadratic Equations	236
	Chapter Review	246

7 THE OPERATIONS WITH FRACTIONS — 249

7.1	Reducing Fractions to Lowest Terms	251
7.2	Multiplication of Fractions	258
7.3	Division of Fractions	262
7.4	Addition and Subtraction of Fractions	267
7.5	Solution of Equations Containing Fractions	285
7.6	Solution of Inequalities Containing Fractions	291
7.7	Solution of Fractional Equations	293
7.8	Solution of Literal Equations	298
7.9	Problem Solving Involving Fractions	304
	Chapter Review	314

8 THE GRAPH OF AN OPEN SENTENCE IN TWO VARIABLES — 317

8.1	Naming Points in a Plane	318
8.2	The Graph of a Linear Equation in Two Variables	330
8.3	The Slope of a Line	352
8.4	The Graph of a Linear Equation in Two Variables Where Absolute Values are Involved	365
8.5	The Graphs of Linear Equations in One Variable	370
8.6	The Graph of a Linear Inequality	373
	Chapter Review	381

9 SYSTEMS OF OPEN SENTENCES IN TWO VARIABLES — 385

9.1	Solving a Linear System of Equations by Graphing	386
9.2	Solving a Linear System of Equations by Substitution	395
9.3	Solving a Linear System of Equations by Addition or Subtraction	402
9.4	Solving a Linear System of Inequalities by Graphing	417
9.5	Problem Solving Involving Two Variables	420
	Chapter Review	442

10 REAL NUMBERS REVISITED — 445

10.1	The Solution of the Pure Quadratic Equation	445
10.2	Solution of the Quadratic Equation by the Pure Quadratic Method	449
10.3	Solution of the Quadratic Equation by Completing the Square	452
10.4	Solution of the Quadratic Equation by the Quadratic Formula	461
10.5	The Theorem of Pythagoras	471
10.6	The Rational Number	477
10.7	The Irrational Number	482
10.8	The Product of the Square Roots of Two Quantities	485
10.9	The Quotient of the Square Roots of Two Quantities	491
10.10	The Sum of the Square Roots of Quantities	496
10.11	Rationalizing the Denominator of a Fraction	499
10.12	Radical Equations	504
	Chapter Review	509

ANSWERS TO ODD-NUMBERED PROBLEMS — A1

INDEX — I1

ELEMENTARY ALGEBRA for COLLEGE

1
The System of Real Numbers

1.1 THE NUMBER LINE

Basically, the study of mathematics is concerned with an understanding of numbers and their relationships to one another. And just as we often resort to photographs or drawings to give people a better idea of what we are talking about, so, too, does the mathematician use a "drawing" in order to present a visual picture of a number. To do this, he draws a line and then proceeds to name each of the points on that line with a number.

Thus, in Figure 1-1 the mathematician begins by selecting any point on a line and naming it with the number "0." He then has but one more decision

FIGURE 1-1

to make, that is, how far from the zero point should he select the point that he will name with the number "1." Once he has made that second decision, then the names for all other points on the line are locked into place.

FIGURE 1-2

1

For instance, the point named by the number 2 in Figure 1-2 must be exactly the *same distance* from the point named 1 as the point named 1 is from the starting point 0. Similarly, the point named by the number 3 must be the same distance from the point named 2 as the point named 2 is from the point 1.

This pattern will continue in the same manner where any two consecutive whole numbers such as 4 and 5, or 8 and 9, or 23 and 24, or any two others will always name two points on this line which are the same distance apart as the points named by the first two numbers, 0 and 1.

FIGURE 1-3

Notice that in Figure 1-3 the points named by the numbers 0, 1, 2, 3, and 4 are spread much further apart than the points named by these same numbers in Figure 1-2. This came about because the point selected to be named 1 in Figure 1-3 is much further to the right of the 0 point than it is in Figure 1-2. This, in turn, affected the name for every other point in Figure 1-3.

There are many, many points on this line that have not been named. For instance, as shown in Figure 1-4, the name for the point that falls halfway between the points 2 and 3 should apparently have the name $2\frac{1}{2}$ or possibly

FIGURE 1-4

2.5. Either name is fine for both represent the same number. Similarly, the point that lies at $\frac{3}{4}$ of the distance from 1 and 2 is named either as $1\frac{3}{4}$ or 1.75. In fact, there are no numbers with which we are familiar that do not name some point on this line. Can you locate the point named by the number $2\frac{1}{4}$; the number 4.75; the number $\frac{1}{2}$; the number $1\frac{1}{3}$?

In spite of the fact that we are familiar with more numbers than anyone can possibly count, there are still points on this line that have no names in terms of the numbers that we know are present. Part of our study in this course will concern the discovery of new numbers that will name some of these presently unnamed points.

A line where each of its points is named by some number in the manner just described is called a **number line**. And every number that names a point on this line is called a **real number**. The reason the word *real* is used goes back many centuries to the time when mathematicians thought that these numbers were the only ones that made any sense in the *real* world around us. Now, though, mathematicians, engineers, and scientists in general constantly use numbers other than those that name the points on the number line and use them for very practical computations. But since those numbers do not name points on the number line they are not *real* numbers.

Example 1

Draw a number line and locate the points named by each of the following numbers.

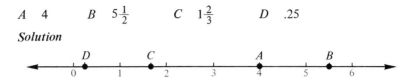

Solution

Explanation After drawing a line locate the starting point 0 and the point to be named 1. The point 2 is then placed the same distance from 1 as 1 is from 0. And the same applies to the points 3, 4, 5, and 6. The point named by 4 is darkened for that is the first point we were asked to find and that was labeled A. The point named by $5\frac{1}{2}$ is halfway between 5 and 6; it is labeled B. The point named by $1\frac{2}{3}$ is $\frac{2}{3}$ of the distance from 1 to 2; it is labeled C. Lastly, the point named by .25 is $\frac{1}{4}$ of the distance from 0 to 1 and is labeled D.

Example 2

Draw a number line and locate the points named by all the numbers that are greater than 5 and less than 9.

Solution

Explanation We indicate all the points named by numbers greater than 5 and less than 9 by drawing a heavy line from the point 5 to the point 9. To indicate that neither the point 5 nor the point 9 is to be included, we draw a circle about each of these points. Had we wanted to include the point named 5 but not the one named 9 we would do this by filling in the circle that we drew around the point 5 but leaving the circle around the point 9 as an "empty" circle.

Example 3

What are the names for all the points covered by the heavy line drawn on the number line below?

Solution The names for the points are the numbers from 3 through 11, including 11 but not 3.

Explanation The number 11 is included since the circle around that point is filled in. This is not the case with the point 3; hence, that point is not named.

Notice that in the explanation above we refer to the "point 3." This is often used as an abbreviation for the statement "the point named by the number 3."

EXERCISES 1.1

1. Draw a number line and locate each of the following points on that number line.
   - a. 3
   - b. 7
   - c. 12
   - d. 0
   - e. $4\frac{1}{2}$
   - f. $8\frac{3}{4}$
   - g. $2\frac{1}{3}$
   - h. $\frac{1}{2}$
   - i. $9\frac{1}{4}$
   - j. $\frac{3}{4}$
   - k. 6.75
   - l. 7.25

2. Draw a number line for each of the following exercises and locate the points named by the numbers in each of these exercises.
   - a. The numbers greater than 2 and less than 6.
   - b. The numbers greater than 4 and less than 10.
   - c. The numbers greater than 7 and less than 8.
   - d. The numbers greater than $3\frac{1}{2}$ and less than $9\frac{1}{2}$.
   - e. The numbers from 1 to 5, including 1 and 5.
   - f. The numbers from 0 to 8, including 0 and 8.
   - g. The numbers from $2\frac{1}{2}$ to 11, including $2\frac{1}{2}$ and 11.
   - h. The numbers from 7 to 10, including 7 but not 10.
   - i. The numbers from 9 to 12, including 9 but not 12.
   - j. The numbers from 0 to 10, including 10 but not 0.
   - k. The numbers from 3 to 4, not including either 3 or 4.

3. What are the names of all the points covered by the heavy line drawn on the number line in each of the exercises below?

   a.

   b.

   c.

   d.

   e.

   f.

   g.

   h.

4. Draw a number line for each of the following exercises and locate the points named by the numbers in each of these situations.

   a. The multiples of 4 that are less than 13 (the multiples of 4 are the numbers in the "4 times table." They are the numbers 4, 8, 12, 16, . . .).
   b. The multiples of 3 that are less than 16.
   c. The multiples of 5 that are greater than 7 and less than 22.
   d. The multiples of 10 that are less than 8.
   e. The multiples of 6 that are greater than 9 and less than 15.

1.2 THE EXTENDED NUMBER LINE

Perhaps it seemed odd to you that after selecting the point 0 on the number line only the points to the right of the starting point are named by numbers. Actually, the reason for this is that we knew no numbers by which to name the points to the left of the zero point. Our goal now is to invent those numbers.

You may recall looking at an outdoor thermometer on a very cold day when the temperature fell below zero. To indicate that the temperature registered 15 degrees below zero the manufacturer of the thermometer had very likely printed the number as,

$$-15°$$

and we may have read this as "minus 15 degrees." Actually, this is read as,

$$\text{negative } 15°.$$

In general, all numbers such as -7° (negative 7 degrees), -26° (negative 26 degrees), and others having the negative sign in front of them imply that the temperature is *below* zero.

In the same way, a movement of

$$-6 \text{ yards}$$

by a football team during a certain play implies that the team has *lost* 6 yards during the play.

Following this same scheme of things the points to the *left* of the starting point are named by placing a negative sign before the numbers. Thus, the number -1 will name the point that is the same distance to the left of the 0 point as 1 is to the right of the 0 point. Similarly -2 names a point that is as

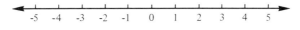

FIGURE 1-5

far to the left of 0 as 2 is to the right. And as we can see in Figure 1-5 the negative numbers continue off to the left following the same pattern as the numbers to the right.

In view of the fact that we begin or *orginate* the naming of the points on the number line from the 0 point, we call this point the **origin**. And since the points to the left of the origin are named by having a negative sign in front of them, we frequently will name the points to the right of the origin by placing a **positive** sign before each of their names. Thus, the point named 1 is also named +1 (positive 1). *This implies that the numbers of arithmetic are in reality the positive numbers which name the points on the number line that are to the right of the origin.* The number line with the new names for the points appears as in Figure 1-6.

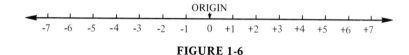

FIGURE 1-6

The + (positive) sign tells us that the point is located to the *right* of the origin while the - (negative) sign implies it is to the *left* of the origin. Since these signs indicate *direction* from the origin, they are called **signs of direction**. Although they are exactly the same as the symbols for addition (+) and subtraction (-), which are called **signs of operation**, we will have very little trouble mistaking the one for the other because of the manner in which they are used.

Were we asked which of the following two people is the wealthier,

"A person with -2 dollars"   or   "A person with 0 dollars"

our guess would probably be that the "-2 dollars" means that the person is $2 in debt. Hence, the person with the $0 is really the wealthier of the two. This reasoning is actually quite correct.

There is, though, another way in which this can be examined. Were we to compare the size of the number 7 with that of 1 (Figure 1-7) we would say that 7 is greater than 1. Now, when we locate the points on the number line that are named by these numbers we discover that 7 (+7) is *further to the right* on the number line than 1 (+1). Similarly, we notice that 0 is further to the right than -2 (Figure 1-8) and it was our guess that 0 should be the larger of these two numbers. Thus it would seem that,

---

When comparing the size of two numbers, the number that names the point the further to the right on the number line will be the larger number.

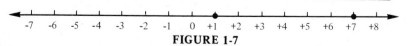

FIGURE 1-7

1.2/The Extended Number Line

FIGURE 1-8

Example 1
Interpret the statement, "The change in the value of a share of the Fisk Company on Tuesday was -$8."

Explanation The negative sign implies that the stock fell in value.

Solution "The value of a share of stock of the Fisk Company fell $8 on Tuesday."

Example 2
Using the number line, determine which of the two numbers, -7 or +2, is the greater.

Solution
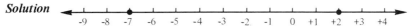

Explanation Since +2 names a point further to the right on the number line than does -7, then +2 is the greater number.

Example 3
Draw a number line and locate the points named by all the numbers that are greater than -6 and less than -2.

Solution

Explanation Since the numbers are greater than -6, the points will have to be to the right of the point -6. However, the points named must not be to the right of -2 as the numbers must be less than -2. Since neither -6 nor -2 are included, empty circles are drawn around the points.

EXERCISES 1.2

1. Interpret each of the following statements.
   a. The temperature reading is -5 degrees.
   b. The temperature reading is +10 degrees.
   c. The temperature reading is 35 degrees.
   d. The football team moved +5 yards on the last play.
   e. The quarterback ran the ball -10 yards during the third down.
   f. During the day's trading the value of the stock changed +$2.50.
   g. Over the past month the value of the stock changed -$9.75.
   h. The elevation of this town is +1,520 feet (the elevation is measured in relation to sea level).
   i. The lake is at an elevation of -240 feet.

2. Which number in each of the following pairs of numbers is the larger? Using the number line, justify your answer in each exercise.

   a. +5 or +7   e. -4 or -7   i. +7 or -3   m. -7 or +7
   b. 0 or -6   f. -2 or +2   j. -10 or +1   n. -6 or 0
   c. 0 or -5   g. -5 or -1   k. -1 or -9   o. +6 or -8
   d. -5 or -3   h. +3 or -1   l. +1 or +9

8 1/The System of Real Numbers

3. Draw a number line for each of the following exercises and locate the points named by the numbers in each of these exercises.

  a. The numbers greater than +1 and less than +7.
  b. The numbers greater than 0 and less than +5.
  c. The numbers greater than −9 and less than −5.
  d. The numbers greater than −7 and less than −6.
  e. The numbers greater than −4 and less than 0.
  f. The numbers greater than −3 and less than +3.
  g. The numbers greater than −2 and less than 4.
  h. The numbers from −10 to −4, including −10 and −4.
  i. The numbers from −5 to +1, including −5 but not +1.
  j. The numbers from −4 to +5, including +5 but not −4.

4. What are the names of all the points covered by the heavy line drawn on the number line in each of the exercises below?

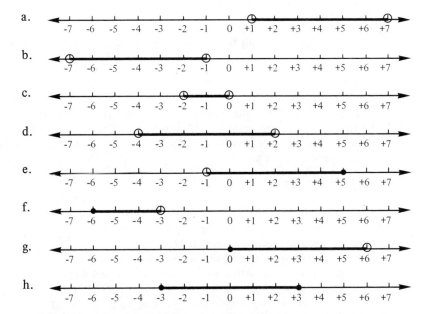

1.3 THE MATHEMATICAL SENTENCE

You may have noticed that some of the exercises we examined earlier were stated in a rather lengthy form. For instance, in one situation we were asked to locate the points on the number line that were named by the numbers from 7 through 10, including 7 but not 10. Over the many, many centuries in the development of mathematics many attempts have been made to express statements such as this in a more compact form. In order to do this, symbols were designed to replace not only single words but even groups of words.

1.3/The Mathematical Sentence

One of the most important symbols that the mathematician has invented is the use of a letter of the alphabet to hold the place for either a single number or even an entire collection of numbers. As an illustration, the statement

$$\text{The numbers are greater than seven.} \qquad (1)$$

can be expressed in a simpler way as,

$$x \text{ is greater than } 7. \qquad (2)$$

By doing this, we are using the letter x to *hold the place* for all those numbers that are greater than 7. Some of these numbers are 8, 9, 17, 26, $7\frac{1}{2}$, 11.25, and so on. However, x is *not* holding the place for a number such as 6, 5, or 2.34, for each of these numbers is less than 7. What are some other numbers that x can represent? Name some numbers that x cannot represent.

Although statement (2) above is somewhat shorter in length than (1), it is not simplified enough to satisfy the mathematicians. They have designed the symbol ">" to replace the words "is greater than" thus enabling us to rewrite (2) as,

$$x > 7. \qquad (3)$$

These few symbols replace the statement,

$$\text{The numbers are greater than seven.} \qquad (1)$$

Rather than use the symbol x we could just as well have used a, c, y, or any other letter of the alphabet. When letters of the alphabet are used in mathematics to *hold the place* of numbers they are called **placeholders** or **variables**. The name *variable* comes from the fact that the numbers for which the letters stand may vary. In the example above this can be seen from the fact that the replacement for x can be any of the numbers 8, 9, 17, etc., for each is greater than 7.

Just as there is a symbol replacing the words "is greater than," so, too, is there a symbol replacing the words "is less than." Rather than have the arrowhead point to the right, we reverse the direction and have it point to the left.

$$x < 7 \qquad \text{means} \qquad \text{"The numbers are less than seven."}$$

A device used to recall whether the symbol should be read as "is greater than" or "is less than" is to keep in mind the fact that *the arrowhead always points to the smaller number*. Whether the statement is written as,

$$x < 7 \quad \text{or} \quad 7 > x$$

it would be well at this stage of our work to read these sentences by starting with the variable no matter where that variable appears in the sentence. In the first of these we will read the sentence from left to right while in the second since the x appears at the far right, we will read it from right to left. Also, as the arrowhead points to the x we know that the x will be smaller than the 7. Hence, this sentence is read as,

$$x \text{ is less than } 7.$$

The numbers are less than seven.

Should we want to indicate that the numbers are equal to seven we can do this by using the equality sign (=). Thus, the sentence,

$$x = 7$$

is the symbol form of the much lengthier sentence,

The numbers are equal to seven.

As there is only one number that is equal to seven, then the only number that can replace the variable x in this situation is the number 7.

We are now in a position to express some rather complex statements with very few symbols. As an example,

"The numbers are either greater than nine or also equal to nine." (4)

can be rewritten with symbols as,

$$x \geqslant 9. \tag{5}$$

What we have really done here is to take the two sentences

$$x > 9 \quad \text{and} \quad x = 9 \tag{6}$$

and expressed them in the single sentence shown in (5). However, rather than use both lines of the equality sign (=), we have used only one of the lines. The sentence (5) is read as,

$$x \text{ is either greater than or equal to } 9.$$

Similarly, the sentence,

$$x \leqslant -3 \tag{7}$$

is read as,

$$x \text{ is either less than or equal to negative } 3. \tag{8}$$

1.3 / The Mathematical Sentence

And this simply means

The numbers are either less than or equal to negative three.   (9)

Some of the numbers whose place x is holding are -5, -7, -26, and so forth for each of the numbers is less than -3. The numbers -5, -7, -26 all name points that are to the left of -3 on the number line.

Sentences such as (7) in which *symbols* are used to represent not only numbers but also one of these three relations,

equals   or   greater than   or   less than

are called **mathematical sentences**. Sentences such as (9) in which the *words* are used rather than the symbols are called **English sentences**.

Example 1

a. Rewrite the following mathematical sentence as an English sentence: $y \geqslant -5$.
b. Locate the points on the number line that are named by the variable in the sentence: $y \geqslant -5$.

Solution   a. The numbers are either greater than or equal to negative five.

b.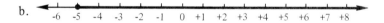

Explanation   Since the numbers have to be greater than -5, these numbers will name points to the right of the point -5 on the number line. The circle around the point -5 is filled in because the information stated that the numbers also include the number -5. The arrowhead is drawn at the right end of the heavy line to indicate that the points yet to be named continue on and on off to the right on the number line.

Example 2

a. State the English sentence that names the points covered by the heavy line on the number line below.
b. Rewrite the English sentence as a mathematical sentence.

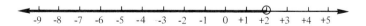

Solution   a. The numbers are less than positive two.
          b. $b < +2$.

Explanation   In view of the fact that the heavy line covers points that are to the left of $+2$, then the numbers naming these points must be less than $+2$.

1/The System of Real Numbers

Since the circle around the point +2 is empty, this implies that this point is not included among the points covered by the heavy line. Finally, since the arrowhead points to the left we know that there are many points in that direction still to be covered by the heavy line.

EXERCISES 1.3

1. Rewrite each of the following English sentences as a mathematical sentence.
    a. The numbers are greater than twelve.
    b. The numbers are less than fourteen.
    c. The numbers are less than positive eleven.
    d. The numbers are greater than negative five.
    e. The numbers are either greater than positive four or equal to positive four.
    f. The numbers are either less than zero or equal to zero.
    g. The numbers are either greater than negative ten or equal to negative ten.
    h. The numbers are equal to six.

2. Rewrite each of the following mathematical sentences as an English sentence.

    a. $x > 9$
    b. $y > +7$
    c. $z < 20$
    d. $w > -5$
    e. $a < +4$

    f. $d = 6$
    g. $v \geqslant 8$
    h. $m \leqslant +14$
    i. $n \geqslant -6$
    j. $p \geqslant +50$

    k. $5 < x$*
    l. $+7 < y$
    m. $-16 > b$
    n. $-15 \geqslant w$
    o. $-1 \leqslant w$

    *Remember to read this from right to left.

3. Draw a separate number line for each of the exercises below and locate the points named by the mathematical sentence.

    a. $y > 1$
    b. $w < 5$
    c. $a > +2$
    d. $b < -3$
    e. $z = +8$

    f. $w = -7$
    g. $x = 0$
    h. $y \geqslant -4$
    i. $z \leqslant -5$
    j. $c \leqslant +2$

    k. $+9 < w$
    l. $-2 > d$
    m. $-7 \leqslant x$
    n. $11 = y$
    o. $+5 \geqslant m$

4. State the mathematical sentence that names the points covered by the heavy line in each of the exercises below.

a.

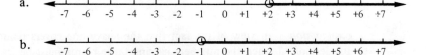

b.

1.4/The Exponent 13

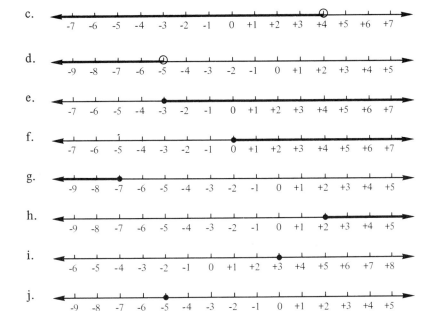

5. Name three numbers that can be used as replacements for the variable in each of the following mathematical sentences.

a. $x > +5$
b. $y < -2$
c. $z > -10$
d. $w < 0$
e. $m > 30$
f. $12 < a$
g. $-6 < b$
h. $+9 > c$
i. $-6 > d$

6. Name the smallest or largest number that can be used as a replacement for the variable in each of the following mathematical sentences. State whether it is the "smallest" or the "largest."

a. $w \leq 10$
b. $z \geq -9$
c. $m \leq +4$
d. $f \leq -6$
e. $n \geq 17$
f. $p \geq 0$

1.4 THE EXPONENT

Having decided upon using the letters of the alphabet as variables, we have placed ourselves in the position where we can no longer use the letter "x" to indicate the operation of multiplication. For instance, upon seeing an expression such as,

$$a \text{ x } b$$

doubt would arise in our minds as to whether the letter "x" implied that the variable a were to be multiplied by the variable b or whether the three variables $a, x,$ and b had simply been written next to one another.

In view of this, the "x" symbol for multiplication is replaced by any one of the following forms:

1. $4 \cdot 3$ 2. $(4)(3)$ 3. $4(3)$ 4. $(4)3$ 5. $[4][3]$ 6. $4[3]$ 7. $[4]3$

Each of the seven expressions above is read as,

$$4 \text{ times } 3.$$

In the first of these, the dot resembles a decimal point. However, whereas the decimal point is placed at the same level as the lower part of the numbers, the dot implying the operation of multiplication is placed at a level that is halfway between the top and bottom of the two numbers.

The first three of the expressions above are the ones most often used when writing 4 times 3. The symbols ( ) are called **parentheses** and the symbols [ ] are called **brackets**.

Indicating the product of two variables such as a and b is done in a manner even shorter than any of the ways above. When the a is written next to the b without any symbol between them, this will imply that the operation of multiplication must be performed between the variables a and b.

$$ab \qquad \text{means} \qquad a \text{ times } b$$

Were we to try to express the product of 4 and 3 in the same manner as we have just written a times b, that is, ab, we find that it appears as 43 which we recognize as the number **forty-three**. Hence, when writing the product of two Arabic numerals we must place some symbol between the numerals.

Just one more piece of information before leaving this topic. Were we involved in writing the product of two or more numbers in which all of these numbers are *exactly the same*, we can resort to a rather short way of doing this. Thus,

$$7 \cdot 7 \cdot 7 \cdot 7 \cdot 7$$

can be expressed simply as,

$$7^5$$

The small number 5 written at the upper right corner of the number 7 tells us that the number 7 appears 5 times when writing the product.

Similarly,

$$4 \cdot 4 \cdot 4 \qquad \text{can be written as} \qquad 4^3$$

in which the small 3 at the upper right corner implies that the 4 appears 3 times when writing the product.

1.4/The Exponent

The reverse of this situation, of course, is also true. That is,

$$6^4 \quad \text{means} \quad 6 \cdot 6 \cdot 6 \cdot 6.$$

The small 4 again implies that the number 6 is to appear 4 times in the product.

The small number at the upper right corner is called an **exponent**. It merely indicates the number of times that the larger number written to its left will appear in the product. The larger number such as the 6, or the 4, or the 7, which we used in the three situations above, is called the **base**.

The expression 6^4 is read either as,

6 to the fourth power   or   6 with an exponent of 4.

If the exponent happens to be a 2, there is yet another way in which that quantity can be read. For instance,

$$6^2 \quad \text{can be read as} \quad 6 \text{ square.}$$

A comparable situation exists when the exponent is 3, for the expression

$$6^3 \quad \text{can be read as} \quad 6 \text{ cube.}$$

Example 1

a. How do you read 5^3?
b. What is the meaning of 5^3?
c. What is the value of 5^3?

Solution a. 5^3 is read as "five cube" or "five to the third power."
b. 5^3 means "5 times 5 times 5."
c. The value of 5^3 is found by multiplying 5 by 5, then taking that product and multiplying it by 5. Thus, $5 \cdot 5 \cdot 5 = 125$.

Example 2

Find the value of $3a^2$ where $a = 4$.

Solution $3a^2 = 3 \cdot 4^2$
$= 3 \cdot 16$
$= 48$

Explanation There are several things that have to be clarified in the solution above. The first is the fact that there is no symbol of operation between the 3 and the a^2. From what we have learned this implies that the operation is understood to be *multiplication*. Notice also that we must first square the 4, for it is the number 16 that must be multiplied by 3.

In an expression such as,

$$5a + 2a^2b - 4b^3 + \frac{2a}{5b} \qquad (1)$$

each of the quantities,

$$5a, 2a^2b, 4b^3, \frac{2a}{5b}$$

are called **terms**. As can be seen here, a term is simply a combination of numbers and variables that are separated from the rest of the expression either by an addition sign or a subtraction sign or possibly both of these signs. In (1) above, the $5a$ is separated from the rest of the expression by the addition sign that follows it. The $2a^2b$ is separated at the left by the addition sign while at the right we find the subtraction sign.

There is an important point that must be made concerning a term. There will be occasions when we will know the replacement values for the variables in expressions such as (1) above. When this occurs each term must be considered as a *single number.* Our task will be to determine the single number for each term before we can add or subtract the numbers as the signs indicate. The following two illustrations should help clarify this concept.

Example 3

Find the value of $2a + 5b - ac$ where $a = 3, b = 8$, and $c = 4$.

Solution $2a + 5b - ac = 2 \cdot 3 + 5 \cdot 8 - 3 \cdot 4$
$= 6 + 40 - 12$
$= 34$

Explanation Notice that when the variables are replaced by numbers, it is necessary to use the "dot" to indicate the operation of multiplication. The original expression consists of three terms: $2a, 5b,$ and ac. Each of these terms must eventually be converted into a single number. Thus, after replacing the a with 2 in the first term, that term becomes $2 \cdot 3$ and the single number that it represents is 6. Similarly, after replacing b with 8, the single number that represents the $5 \cdot 8$ is 40. And finally, the single number representing $3 \cdot 4$ is 12. The computation is then completed by following the directions given by the addition and subtraction signs between the numbers 6, 40, and 12.

Example 4

Find the value of $a^3 + a^2b^4 - 8b^2$ where $a = 5$ and $b = 3$.

Solution $a^3 + a^2b^4 - 8b^2 = 5^3 + 5^2 \cdot 3^4 - 8 \cdot 3^2$
$= 125 + 25 \cdot 81 - 8 \cdot 9$
$= 125 + 2{,}205 - 72$
$= 2{,}078$

1.4/The Exponent

Explanation After replacing each of the variables with the numbers they represent, our first step is to find the value of each of the quantities having exponents. That is, 5^3 has to be changed to 125; 5^2 to 25; 3^4 to 81; and 3^2 to 9. Once this is done we can complete the computation in exactly the same manner as is done in Example 3.

Example 5

Find the value of a^2 where $a = \frac{4}{5}$.

Solution $a^2 = \left(\frac{4}{5}\right)^2$

$= \frac{16}{25}$

Explanation This example is called to your attention to point out the need for placing the parentheses around the fraction $\frac{4}{5}$. The original expression, a^2, calls for the need to square the number that is the replacement for a. Were we to write this as $\frac{4^2}{5}$ it would imply that only the numerator is to be squared and not the entire fraction. The answer $\frac{16}{25}$ is arrived at by multiplying $\frac{4}{5}$ by $\frac{4}{5}$.

EXERCISES 1.4

1. Write each of the following expressions in a shorter form.

   a. $6 \cdot 6 \cdot 6$
   b. $5 \cdot 5$
   c. $9 \cdot 9 \cdot 9 \cdot 9$
   d. $14 \cdot 14 \cdot 14$
   e. $11 \cdot 11 \cdot 11 \cdot 11 \cdot 11$
   f. $a \cdot a \cdot a$
   g. $bbbbb$
   h. $\frac{2}{3} \cdot \frac{2}{3}$
   i. $\frac{5}{6} \cdot \frac{5}{6} \cdot \frac{5}{6} \cdot \frac{5}{6}$

2. What is the meaning of each of the following expressions?

   a. 5^3
   b. 3^5
   c. 6^8
   d. a^4
   e. b^2
   f. $\left(\frac{3}{4}\right)^2$
   g. $\left(\frac{2}{3}\right)^4$
   h. $\left(\frac{1}{2}\right)^5$
   i. $a^3 b^2$

3. What is the value of each of the following expressions?

   a. 3^2
   b. 7^2
   c. 9^2
   d. 1^2
   e. 2^3
   f. 4^3
   g. 15^2
   h. 23^2
   i. 6^3
   j. 1^3
   k. 2^4
   l. 5^4
   m. 3^5
   n. 2^7
   o. 1^9
   p. $.1^2$
   q. $.02^2$
   r. $.02^3$
   s. $.5^3$
   t. $\left(\frac{1}{2}\right)^3$
   u. $\left(\frac{2}{5}\right)^4$

4. Find the value of each of the following expressions under the conditions that $x = 2$ and $y = 3$.

a. $x + y$
b. $3x + 4y$
c. $2y - x$
d. x^3
e. y^4
f. $4x^2$
g. $5y^2$
h. xy^2
i. yx^2

j. $2x^2 + y^2$
k. $x^5 + 4y$
l. $y^4 - 5x$
m. $4xy$
n. $3xy^2$
o. $5x + 6y - xy$
p. x^2y^3
q. x^3y^2
r. $4x^4y - 2y^3 + xy^2$

5. Find the value of each of the following expressions under the conditions stated.

a. a^2 where $a = \frac{2}{3}$.

b. b^2 where $b = \frac{5}{9}$.

c. a^3 where $a = \frac{3}{4}$.

d. a^2 where $a = \frac{1}{2}$.

e. c^{15} where $c = 1$.

f. b^2 where $b = .01$.

g. c^3 where $c = .03$.

h. $a^2 + b^2$ where $a = .2$ and $b = .3$.

i. $a^2 + b^2$ where $a = .2$ and $b = .03$.

j. $a^3 + b^3$ where $a = 2$ and $b = .4$.

k. $2a^4 - ab$ where $a = 1$ and $b = .1$.

1.5 THE SOLUTION SET OF AN EQUALITY

Earlier in our work we translated the mathematical sentence,

$$x \leq -3$$

into the English sentence,

    The numbers are either less than or equal to negative three.

At this time we would like to examine this same mathematical sentence but from a somewhat different point of view.

1.5 / The Solution Set of an Equality

In reality, a mathematical sentence that involves variables is much the same as a completion sentence. Thus, in the statement,

_____ was a president of the United States,

there are many names that can be used as replacements for the blank space. Names such as Benjamin Franklin, John Paul Jones, Thomas Edison, George Washington, John Adams, and Abraham Lincoln are all equally fine replacements for the blank space so long as we are not limited to the truth or falsity of this statement. However, our selection of names is usually limited to those names that will make the sentence true. Under this condition, only the last three names of the six listed plus the name of any other former president are acceptable.

The mathematical sentence,

$$x \leq -3 \tag{1}$$

is also a completion sentence which can be written as

$$_____ \leq -3.$$

Now we are called upon to replace the blank space with numbers. However, here, again, as in the English completion sentence, our choice of numbers is limited to *only those that will make the sentence true.* Thus, the variable x is holding the place for those numbers that are less than or equal to -3.

Those numbers that can replace the variable in a mathematical sentence so as to make that sentence *true* are called **elements in the solution set** of that sentence. Some of the elements in the solution set for the mathematical sentence $x \leq -3$ are -5, -12, -26, -3, for each of these is less than -3 or equal to -3 and hence make this sentence true. However, -2, 0, 5, 29 are not members of the solution set for these numbers are greater than -3 and therefore would make the sentence false.

A statement in which we are asked to find the solution set of a mathematical sentence is simply interpreted as asking to name all the numbers that will make the sentence true. The word set is probably not new to us for we have very likely encountered it many times in our study of mathematics prior to now. You may recall that a set is simply a collection of things called either *elements* or *members*. In mathematics, of course, those elements are numbers.

Our concern now is to seek out some way by which we can determine the solution set of mathematical sentences that are somewhat different from those examined in an earlier section. Thus, the sentence

$$x + 4 = 6.$$

can be written as,

$$_____ + 4 = 6.$$

We would read this as,

The number plus 4 equals 6.

It is evident that there is only one number that will make this sentence true. That number is 2. This number is the solution set of the sentence $x + 4 = 6$. We can express this with symbols by writing,

$$x = 2$$

or by writing,

The solution set of the sentence is: $\{2\}$.

The symbols $\{\ \}$ are called **braces**. They are used to enclose the elements of a set. In this situation they enclose the number 2 which is the only element in the solution set of the sentence $x + 4 = 6$. A mathematical sentence that contains an = symbol is called an **equation**.

Example 1

Find the solution set of the equation $x - 5 = 6$.

Solution $x - 5 = 6$.
The translation of this equation is, "The number minus 5 is equal to 6."

$$x = 11 \quad \text{or} \quad \text{"The solution set is: } \{11\} \text{ ."}$$

Explanation Based on the English translation, the only number from which 5 can be subtracted and leave a difference of 6 must be 11. No other number but 11 will make the equation true.

Example 2

Find the solution set of the equation $15 = 3a$.

Solution $15 = 3a$.
The translation of this equation is, "Fifteen is equal to 3 times the number."

$$x = 5 \quad \text{or} \quad \text{"The solution set is: } \{5\} \text{ ."}$$

Explanation There is only one number that when multiplied by 3 will yield 15 as the product. That number is 5; hence 5 is the only number in the solution set.

Example 3

Find the solution set of the equation $3 + 2x = 15$.

1.5/The Solution Set of an Equality

Solution $3 + 2x = 15$.

The translation of this equation is, "Three added to 2 times the number is equal to 15."

$$x = 6 \quad \text{or} \quad \text{"The solution set is:} \{6\} \text{."}$$

Explanation This equation requires a bit more thought than the earlier ones. However, after testing a few numbers we soon discover that the number must be 6. For instance, had we chosen 5 as the replacement for x, then by adding 3 to 2 times 5 we find the sum to be 13 and not 15. Hence, we simply try a larger number. Had we tried 8, then 3 added to 2 times 8 would have given us 19, which would imply that the number we had chosen is too large and that we should try a smaller number.

EXERCISES 1.5

1. Translate each of the following English sentences to an equation.

    a. The number added to five is equal to seven.
    b. Seventeen added to the number is equal to twenty-three.
    c. The number minus six is equal to two.
    d. Twelve minus the number is one.
    e. Fifteen is equal to the number plus eleven.
    f. Thirty-six is equal to fifty-two minus the number.
    g. Three times the number is equal to twenty-seven.
    h. The number times six is equal to thirty.
    i. The number multiplied by itself will give eighty-one as a product.
    j. Sixty-four is equal to the number squared.
    k. The cube of the number is equal to sixty-four.
    l. Six added to two times the number is equal to twenty-seven.
    m. Four times the number minus five is equal to thirty-one.

2. Translate each of the following equations to an English sentence.

    a. $x + 4 = 7$
    b. $5 + x = 9$
    c. $x - 16 = 4$
    d. $12 - x = 7$
    e. $x + 11 = 19$
    f. $3x = 12$
    g. $x^2 = 16$
    h. $x^3 = 27$
    i. $2x + 5 = 17$
    j. $12 - 3x = 3$

    k. $15 = x + 6$
    l. $11 = 4 + x$
    m. $25 = x - 5$
    n. $32 = 40 - x$
    o. $56 = 8x$
    p. $41 = x + 21$
    q. $49 = x^2$
    r. $32 = x^2$
    s. $19 = 3x + 1$
    t. $27 = 29 - 2x$

3. State whether each of the following mathematical sentences is true or false.

a. $4 + 2 = 6$
b. $5 - 1 > 2$
c. $3 + 7 < 8$
d. $2 \cdot 9 > 15$
e. $16 > 8 + 2$
f. $27 < 3 \cdot 9$
g. $2 \cdot 4 + 3 < 16$
h. $18 + 2 > 24 - 6$
i. $14 - 5 > 10 - 3$
j. $18 < 3 \cdot 5 + 10$
k. $16 \div 2 < 8$
l. $14 + 2^3 > 20$
m. $5 + 2 \cdot 3 = 11$
n. $9 \cdot 2 - 5 < 2 \cdot 3 + 8$
o. $3^2 + 4^2 > 20$
p. $4 \cdot 5 < 6 + 2^3$
q. $7^2 - 2^2 > 5^2$
r. $1^4 + 2^3 < 4 + 3^2$
s. $3^3 + 4 > 2^5$
t. $4 \cdot 2^4 > 3 \cdot 4 \cdot 5$

4. Find the solution set for each of the following equations.

a. $x + 2 = 3$
b. $x + 9 = 16$
c. $x + 27 = 50$
d. $x - 5 = 3$
e. $x - 12 = 2$
f. $14 - x = 11$
g. $3x = 18$
h. $5x = 35$
i. $9x = 54$
j. $17 = x + 15$
k. $23 = x + 19$
l. $16 = x - 6$
m. $27 = x - 10$
n. $15 = 23 - x$
o. $46 = 60 - x$
p. $21 = 7x$
q. $48 = 6x$
r. $4 = x \div 2$

5. Find the solution set for each of the following equations.

a. $x + 4 = 6\frac{1}{2}$
b. $5 + x = 9\frac{1}{2}$
c. $x - 7 = 2\frac{3}{4}$
d. $12\frac{1}{4} = x + 7$
e. $3x = 7$
f. $5x = 12$
g. $16 = 7x$
h. $11 = 9x$

6. Find the solution set for each of the following equations.

a. $2x + 1 = 9$
b. $2x - 1 = 15$
c. $3 + 2x = 19$
d. $7 + 3x = 22$
e. $15 - 3x = 0$
f. $5x - 1 = 29$
g. $17 = 3x - 4$
h. $25 = 4x - 3$
i. $31 = 10x + 1$
j. $19 = 10 + 3x$
k. $42 = 8x - 6$
l. $x^3 = 125$

1.6 THE SOLUTION SET AND GRAPH OF AN INEQUALITY

Far more interesting than finding the solution set of an equation is the situation in which we are asked to find the solution set of a mathematical sentence that

1.6/The Solution Set and Graph of an Inequality

involves either the relation of "greater than" or "less than." Mathematical sentences concerned with either "greater than" or "less than" are called **inequalities**. Whereas the solution set of those equations we have just examined involve only one number we will find that the solution set of an inequality involves **infinitely many numbers**. This term will be understood to mean that there are more numbers than any person or even any group of persons can ever possibly count.

Consider, as an example, the inequality

$$x > +7$$

The translation of this sentence into an English sentence is,

The numbers are greater than seven.

In answer to this, those numbers that will make this sentence true are numbers such as 8, 9, 23, $7\frac{1}{2}$, 8.34, 56.92, and on and on. In fact, as was just pointed out, there are far more numbers in the solution set than anyone can possibly list.

Rather than listing these numbers, which is impossible to do, the mathematician usually prefers to picture the solution set by resorting to the number line. Hence, for the solution set of the inequality

$$x > +7$$

the mathematician would draw the number line as we did earlier and fill in all points to the right of the point named by the number +7. This is done in Figure 1-9.

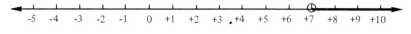

FIGURE 1-9

The picture in Figure 1-9 is called the **graph** of the inequality "x is greater than +7." In general, we say that

> A graph of a mathematical sentence is the set of points on the number line that are named by numbers that make the mathematical sentence true.

For instance, in Figure 1-9, the numbers 8, 11, $7\frac{1}{2}$, $9\frac{1}{4}$, in fact, all the numbers greater than +7 make the sentence $x > +7$ a true sentence. Hence, the points named by these numbers make up the graph of the sentence $x > +7$.

Before leaving this, it is interesting to note that the graphs of the equalities we examined in the previous section will consist of but one point. This is so because each of those equalities had but one number that will make the sentence true. For instance, when earlier we examined the equation,

$$x + 4 = 6$$

we found that the only number in the solution set is 2. Hence, were we to draw the graph of this equation we would mark only one point; the point named by the number 2. Figure 1-10 shows the graph of the equation $x + 4 = 6$.

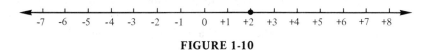

FIGURE 1-10

Example 1

Draw the graph of the equation $x - 5 = 4$.

Explanation After translating the mathematical sentence "$x - 5 = 4$" into the English sentence, "A number minus five is equal to four," we realize that the only number that will make the sentence true is the number 9. Hence the number 9 is the only number in the solution set.

Solution Solution set: $\{9\}$.

Graph:

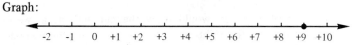

Example 2

Draw the graph of the inequality $y \geq -5$.

Explanation Notice in the graph below that the circle around the point -5 is filled in since the number -5 is included in the solution set.

Solution

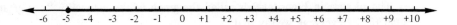

Example 3

Draw the graph of the inequality $x + 2 > 7$.

Explanation Our first objective is to translate this sentence into an English sentence. This is, "Each number plus 2 is greater than 7." After reading this we immediately realize that if "each number added to 2 must be greater than 7," then each of these numbers alone must be greater than 5. Were any of these numbers less than 5, then when added to 2 the sum would be less than 7 and this

1.6/The Solution Set and Graph of an Inequality

would not make the sentence true. Hence, the solution set consists of numbers that are greater than 5. Therefore, the graph will consist of all points to the right of the point +5.

Solution

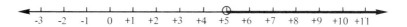

Example 4

Draw the graph of the inequality $6 < 3x$.

Explanation As in our earlier work we must read this sentence starting with the term that contains the variable; hence we must read this sentence from right to left: "Three times each number is greater than 6." We immediately realize that if "3 times each number is greater than 6," then each number alone must be greater than 2. Were any of the numbers less than 2 then when they were multiplied by 3 their product would be less than 6 and this would not make the sentence true. Hence, the solution set consists of the numbers that are greater than 2 and the graph consists of all the points to the right of the point +2.

Solution

EXERCISES 1.6

1. Translate each of the following English sentences to an inequality.
    a. Each number is less than seventeen.
    b. Each number is greater than twelve.
    c. Two times each number is less than twenty-four.
    d. Each number added to four is greater than ten.
    e. Five times each number is greater than fifty.
    f. Sixteen minus each number is greater than ten.
    g. Twenty-seven is greater than nine times each number.
    h. Thirty is less than each number plus one.
    i. Eight is less than two added to each number.
    j. Three times each number added to five is greater than eleven.
2. Translate each of the following inequalities to an English sentence.
    a. $x + 2 > 5$
    b. $x - 7 > 4$
    c. $3x > 12$
    d. $5x < 30$
    e. $16 - x > 7$
    f. $24 + x > 29$
    g. $18 < x + 5$
    h. $12 < 17 - x$
    i. $32 > 4x$
    j. $9 > x - 5$
    k. $45 < 9x$
    l. $19 < 2x + 1$

3. Draw the graph of each of the following equations.

   a. $x + 2 = 6$
   b. $x - 5 = 7$
   c. $6x = 18$
   d. $17 = x + 15$
   e. $4 = x - 3$
   f. $32 = 16x$
   g. $12 = x + 7$
   h. $2x + 1 = 13$
   i. $3x - 4 = 5$
   j. $22 = 5x + 2$

4. Draw the graph of each of the following inequalities.

   a. $x > +5$
   b. $x < -3$
   c. $x \geqslant +4$
   d. $x \leqslant 0$
   e. $x > 9$
   f. $x \leqslant 11$
   g. $x \geqslant -6$
   h. $x \leqslant -2$

5. Draw the graph of each of the following inequalities.

   a. $x + 3 > 7$
   b. $x + 6 > 14$
   c. $x - 1 > 5$
   d. $x - 4 > 1$
   e. $2x > 16$
   f. $3x > 24$
   g. $9 < x + 2$
   h. $14 < x + 10$
   i. $24 < 6x$
   j. $27 < x^3$
   k. $3x + 1 > 28$
   l. $2x - 1 > 17$

CHAPTER REVIEW

1. Draw a number line and locate each of the following points on that number line.

   a. 5
   b. +12
   c. -6
   d. 0
   e. -1
   f. +3
   g. $+4\frac{1}{2}$
   h. $-2\frac{1}{4}$

2. Draw a number line and locate the points named by the numbers in each of these exercises.

   a. The numbers greater than 5 and less than 9.
   b. The numbers greater than +4 and less than +11.
   c. The numbers greater than -3 and less than +7.
   d. The numbers less than -5 and greater than -10.
   e. The numbers from -6 to +2, including -6 but not including +2.

3. What is the English sentence that names all the points covered by the line drawn on the number line in each of the exercises below?

   a.
   b.
   c.

4. Rewrite each of the following English sentences as a mathematical sentence.

   a. The numbers are greater than positive six.
   b. The numbers are less than a negative two.

5. Rewrite each of the following mathematical sentences as an English sentence.

   a. $y > 5$
   b. $x < -3$
   c. $w = 4$
   d. $m \geq +4$
   e. $-6 \geq a$
   f. $+2 \leq b$

6. State the mathematical sentence that names the points covered by the bold line on the number line below.

7. Write each of the following expressions in a shorter form.

   a. $4 \cdot 4 \cdot 4 \cdot 4 \cdot 4$
   b. $y \cdot y$
   c. $\frac{3}{5} \cdot \frac{3}{5} \cdot \frac{3}{5}$

8. What is the meaning of each of the following expressions?

   a. x^3
   b. 7^6
   c. $(\frac{3}{4})^2$

9. What is the value of each of the following expressions?

   a. 2^4
   b. 3^2
   c. 1^7
   d. $(\frac{4}{7})^3$

10. Find the value of each of the following expressions under the conditions stated.

    a. $x + 2y$ where $x = 5$ and $y = 7$
    b. $x^2 + xy$ where $x = 4$ and $y = 9$
    c. $5x^2 y$ where $x = 3$ and $y = 6$

11. Translate each of the following equations to an English sentence.

    a. $x + 6 = 17$
    b. $15 - x = 4$
    c. $x^3 = 1$
    d. $12 = 9 + x$

12. Find the solution set for each of the following equations.

    a. $x + 2 = 9$
    b. $12 = 15 - x$
    c. $6x = 30$
    d. $x - 9 = 10$
    e. $3x + 1 = 7$
    f. $22 = 5x - 3$

13. Translate each of the following English sentences to an inequality.

   a. Each number is more than negative five.
   b. Two times each number is greater than fourteen.
   c. Twenty is less than four times each number.

14. Draw the graph of each of the following equations.

   a. $x + 3 = 10$ b. $9x = 36$

15. Draw the graph of each of the following inequalities.

   a. $x \leq +5$ b. $x - 2 > 6$

The Operations of Addition and Subtraction

Having extended our numbers to include both positive and negative numbers, we have placed ourselves in exactly the same position as we had during our study of arithmetic. Thus, each time we invented a new number such as the fraction, mixed number, or decimal, it was necessary for us to develop a method for performing the fundamental operations—addition, subtraction, multiplication, and division—with these numbers. Now we face exactly the same situation; that is, in what way can we determine a sum, difference, product, or quotient with these new numbers that we invented by extending the number line to the left. In this chapter we will learn both addition and subtraction of these numbers while in the next chapter we will discover how it is possible to multiply and divide.

2.1 ADDITION OF REAL NUMBERS

PART 1

The addition of two numbers is merely an operation by which we assign a single number to a pair of numbers. The manner in which this number is assigned is usually through a sequence of movements on the number line. These movements are sometimes referred to as translations or displacements.

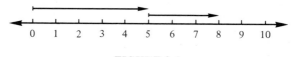

FIGURE 2-1

The number to be assigned to 5 and 3 by addition is found by considering 5 + 3 as a set of instructions related to the number line. These instructions are interpreted by reading 5 + 3 from left to right (Figure 2-1).

The first of these instructions informs us to move 5 units to the right of the origin on the number line, while the second instruction informs us to move another 3 units to the right beyond the point where we had just stopped. The first movement will bring us to the point named by the number 5, while the second movement carries us from the point 5 to the point three units beyond 5. The number that names this final point, which in this case is 8, is called "the sum of 5 and 3."

The fact that real numbers consist of both positive and negative numbers involves us in the need to extend our instructions just slightly. Should the sign of direction for the number be *positive* it will indicate that the direction of movement on the number line will be to the *right*. On the other hand, should the sign of direction of the number be *negative*, then the direction of movement on the number line will be to the *left*.

As an illustration, consider the situation in which we are asked to find the sum for

$$(+4) + (+3).$$

It is important that we realize that the sign within the parentheses immediately before the 4 is a *sign of direction* and the same is true for the sign in the parentheses immediately before the 3. However, the sign between the two pairs of parentheses is a *sign of operation*. That sign informs us that these two numbers are to be added.

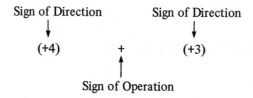

The very first instruction is the positive sign before the 4. This implies that we are to move *to the right* while the number 4 implies that we are to move four units. This brings us to the point +4. The addition sign between the two

2.1 / Addition of Real Numbers

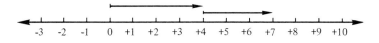

FIGURE 2-2

numbers informs us that we are now to start at the point +4 and continue our movement as the instructions indicate. The positive sign before the 3 again implies a movement to the right while the number 3 implies that the movement is to cover three units. Thus, we come to rest on the point named by the number +7. Hence, we say that the sum of +4 and +3 is a +7, as illustrated by Figure 2-2.

$$(+4) + (+3) = +7$$

Let us examine the addition of two numbers where one of these numbers is negative while the other is positive, for example,

$$(-4) + (+3).$$

Here we find that the first instruction is the negative sign before the 4. In view of this, the first movement will be to the *left* and it will be a movement covering 4 units. This brings us to the point -4. As before, the addition sign between the two numbers informs us that we are to continue our movement from point -4. However, the *positive* sign before the 3 informs us that the movement is now going to be to the *right* and that it will cover 3 units. This brings us to the point -1, thus the sum of -4 and +3 is -1, as shown in Figure 2-3.

$$(-4) + (+3) = -1$$

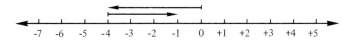

FIGURE 2-3

Example

Determine the sum in the exercise below.

$$(+8) + (-10)$$

Solution

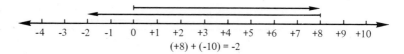

$(+8) + (-10) = -2$

Explanation The first movement is 8 units to the right, bringing us to the point +8. We are than told to move 10 units to the left, which brings us to the point -2. Hence, the sum of +8 and -10 is -2.

EXERCISES 2.1 (Part 1)

A

Use the number line to find the sum in each of the following exercises.

1. (+6) + (+4)
2. (+2) + (+7)
3. (+4) + (+8)
4. (+8) + (+4)
5. (+10) + (+5)
6. (-6) + (-2)
7. (-1) + (-9)
8. (-4) + (-8)
9. (-10) + (-7)
10. (-2) + (-14)
11. (+10) + (-3)
12. (+6) + (-4)
13. (+8) + (-11)
14. (+4) + (-4)
15. (+3) + (-9)
16. (+1) + (-11)
17. (+5) + (0)
18. (0) + (+6)
19. (0) + (-9)
20. (+8) + (-8)
21. (-6) + (+3)
22. (-10) + (+9)
23. (-4) + (+6)
24. (-5) + (+12)
25. (-7) + (+7)
26. (-4) + (0)
27. (-9) + (+9)
28. (-1) + (+7)
29. (-20) + (+3)
30. (-3) + (+20)

B

Use the number line to find the sum in each of the following exercises.

1. (+2) + (+3) + (+1)
2. (+5) + (+4) + (+2)
3. (-1) + (-4) + (-3)
4. (-4) + (-5) + (-3)
5. (+2) + (+6) + (-3)
6. (+5) + (-8) + (+3)
7. (+7) + (-3) + (-6)
8. (-2) + (+5) + (-3)
9. (-4) + (-6) + (+13)
10. (-5) + (+10) + (-8)

C

Each of the following exercises shows the gain or loss made by a football team on four successive downs. Using signed numbers and the number line,

2.1 / Addition of Real Numbers

determine the total number of yards won or lost during these four downs.

	First Down	*Second Down*	*Third Down*	*Fourth Down*
1.	4 yds loss	10 yds won	2 yds loss	5 yds loss
2.	6 yds won	2 yds won	12 yds loss	15 yds won
3.	3 yds won	5 yds loss	14 yds won	6 yds won
4.	9 yds loss	6 yds loss	2 yds loss	17 yds won
5.	8 yds won	8 yds loss	10 yds loss	12 yds won

PART 2

It is probably quite evident to you by now that we would not want to continue to find the sum of two real numbers by constantly having to refer to the number line. Our objective at this time is to try to discover some method that will enable us to determine the sum of two *signed numbers* without the need of resorting to the number line.

Let us examine the numbers 8 and 2 and attach to these numbers all possible combinations of signs of direction. Thus, when attaching the positive and negative signs to 8 we come up with +8 and -8. Similarly, when attaching these same signs to 2, the numbers become +2 and -2. Now, the various addition situations that involve the four numbers are:

Column One	*Column Two*
1. (+8) + (+2) =	(+2) + (+8) =
2. (-8) + (-2) =	(-2) + (-8) =
3. (+8) + (-2) =	(-2) + (+8) =
4. (-8) + (+2) =	(+2) + (-8) =

By referring to the number line we can find the sum in each of these eight situations. This is done in Figure 2-4.

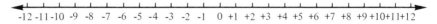

-12 -11 -10 -9 -8 -7 -6 -5 -4 -3 -2 -1 0 +1 +2 +3 +4 +5 +6 +7 +8 +9 +10 +11 +12

FIGURE 2-4

Column One	*Column Two*
1. (+8) + (+2) = +10	(+2) + (+8) = +10
2. (-8) + (-2) = -10	(-2) + (-8) = -10
3. (+8) + (-2) = +6	(-2) + (+8) = +6
4. (-8) + (+2) = -6	(+2) + (-8) = -6

The very first thing that catches our eye is the fact that each sum in column two is exactly the same number as the sum for the corresponding situation in column one. That is, when the two numbers are interchanged, the sum is in no way affected. We should have realized that this will be the case for when we add 5 and 4 we obtain the same sum as when we add 4 and 5. And we would not want our new signed numbers to act differently than our old numbers of arithmetic had.

The mathematician refers to the fact that 5 + 4 will give the same sum as 4 + 5 by saying that

---
Addition is commutative.

Hence we need only examine column one to determine whatever principles we can that will enable us to find the sum of two signed numbers. In situations (1) and (2) we notice that if *we ignore the signs of direction* the sums turn out to be the number 10 which is exactly the same number we would arrive at had

$$\underbrace{(+8) \quad + \quad \overbrace{(+2)}^{\text{Sum of 8 and 2}} \quad = \quad +10}_{} \qquad (1)$$

$$\underbrace{(-8) \quad + \quad \overbrace{(-2)}^{\text{Sum of 8 and 2}} \quad = \quad -10}_{} \qquad (2)$$

we added the two numbers 8 and 2 as we had in arithmetic. We also discover that in both of these situations the signs of direction of the two numbers are identically the same. Thus, in situation (1) both signs of direction are positive, while in situation (2) both signs of direction are negative. Hence, it appears that

Signs are both +.
$$(+8) \quad + \quad (+2) \quad = \quad +10 \qquad (1)$$

Signs are both −.
$$(-8) \quad + \quad (-2) \quad = \quad -10 \qquad (2)$$

when *adding two signed numbers if the signs of direction are the same we will find the sum by adding one number to the other as we had in arithmetic.*

By looking at the sums and ignoring the signs of direction in situations (3) and (4) we again discover that they are exactly the same number. Now, though,

2.1/Addition of Real Numbers

they are the number 6 and this is the number we would arrive at had we subtracted the 2 from the 8 as we had in arithmetic. Closer examination of these

$$\overbrace{(+8) \quad + \quad (-2)}^{\text{Subtract 2 from 8}} \quad = \quad +6 \qquad (3)$$

$$\overbrace{(-8) \quad + \quad (+2)}^{\text{Subtract 2 from 8}} \quad = \quad -6 \qquad (4)$$

two situations shows that the signs of direction of the two numbers are not the same in either situation (3) and (4). Hence, it appears that when *adding two signed numbers, if the signs of direction are different we will find the sum by subtracting one of the numbers from the other as we had in arithmetic.*

$$\overbrace{(+8) \quad + \quad (-2)}^{\text{Different Signs}} \quad = \quad +6 \qquad (3)$$

$$\overbrace{(-8) \quad + \quad (+2)}^{\text{Different Signs}} \quad = \quad -6 \qquad (4)$$

We are now faced with the problem of determining what the sign of direction of each answer will be. Although, to determine the *sum* in both (3) and (4) it is necessary to subtract one number from the other as we had in arithmetic, in (3) the sign of the direction turns out to be positive while in (4) it is negative. Oddly enough, the sign of 8 in each of these situations is the very same as the sign of the sum. Our first impulse is to say that the sign of the answer is the same as the sign of the larger of the two numbers.

$$(+8) \quad + \quad (+2) \quad = \quad +10 \qquad (1)$$

$$(-8) \quad + \quad (-2) \quad = \quad -10 \qquad (2)$$

$$(+8) \quad + \quad (-2) \quad = \quad +6 \qquad (3)$$

$$(-8) \quad + \quad (+2) \quad = \quad -6 \qquad (4)$$

Unfortunately, -8 is not larger than +2, for we compare the size of two numbers by determining which of them names a point on the number line farther to the right. The point named by +2 is much farther to the right than the one named by -8. This, of course, makes -8 smaller than +2 and hence implies that the principle stated at the end of the previous paragraph is not true.

There is, though, a feature about -8 relative to size wherein -8 is greater than +2. The feature to which we refer is that -8 is a greater distance from the origin on the number line than +2 is; see Figure 2-5.

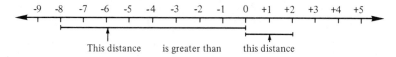

This distance   is greater than   this distance

FIGURE 2-5

The distance from a point named by a number to the origin is called the absolute value of the number.

Thus, we are now in a position where we can say that the sign of the sum in situations (3) and (4) is the same as the sign of the number having the larger *absolute value*. This statement is also true for situations (1) and (2). In (1) the sign of the 8 is positive and so is the sum a positive number (+10); while in (2) the sign of the 8 is negative and so, too, is the sign of the sum negative (-10).

In view of this we can now state the following principle.

Addition Principle

When finding the sum of *two* signed numbers,

a. The sign of direction of the sum is the same as the sign of direction of the number having the *larger* absolute value.
b. If the signs of direction of the two numbers are the same, then the *sum* is found by adding the two numbers as is done in arithmetic.
c. If the signs of direction of the two numbers are different, then the *sum* is found by subtracting one number from the other as is done in arithmetic.

Example 1

Determine the sum in the exercise below.

$$(+5) + (-6)$$

2.1 / Addition of Real Numbers

Explanation Since the signs of direction are not the same, one number is subtracted from the other in the same manner as is done in arithmetic. The sign of direction of the sum will be negative for the number −6 has a larger absolute value than +5.

Solution $\qquad (+5) + (-6) = -1$

Example 2

Determine the sum of the two numbers below.

$$\begin{array}{r} -9 \\ \underline{12} \end{array}$$

Explanation We have learned earlier that the numbers of arithmetic are in reality the positive numbers. Hence, at this stage of the work it might be well to write a positive sign immediately before any number where the sign of direction is not shown. In this example we should write the positive sign before the 12.

Solution
$$\begin{array}{r} -9 \\ \underline{+12} \\ +3 \end{array}$$

Example 3

Determine the sum of the three numbers below.

$$(-4) + (+6) + (-5)$$

Explanation The first step is to find the sum of the first two numbers. In this situation the sum of these two numbers is +2. This number is then added to the third number. Thus, a positive 2 added to a negative 5 gives a sum of negative 3.

Solution $\qquad (-4) + (+6) + (-5) = -3$

There are two important points about Example 3 that must be emphasized before going on. Notice that although the sign of the number having the *largest* absolute value in this example is *positive*, the sign of direction of the sum of the three numbers is *negative*. When we established the principle for adding signed numbers, this principle was based on finding the sum of but *two* numbers at one time and does *not* apply to three or more numbers.

The second point concerning this example is the fact that the sum of these three numbers can be found by grouping the numbers differently from the way the numbers are grouped in the explanation above. In that explanation the sum was found by initially grouping the *first number with the second* and then adding that answer to the third number. This can be illustrated by using brackets in the manner shown below.

$$(-4) + (+6) + (-5)$$
$$= [(-4) + (+6)] + (-5)$$
$$= \quad (+2) \quad + (-5)$$
$$= -3$$

However, the same sum can be found by grouping the *second number with the third* and then adding that answer to the first number. Thus,

$$(-4) + (+6) + (-5)$$
$$= (-4) + [(+6) + (-5)]$$
$$= (-4) + \quad (+1)$$
$$= -3$$

In view of this it would appear that no matter how the numbers are associated in addition, the sum will always be the same number. This principle is called the **associative principle of addition.**

EXERCISES 2.1 (Part 2)

A

Find the sum in each of the following exercises.

1. (+2) + (+5)
2. (+9) + (+7)
3. (6) + (+5)
4. (10) + (+8)
5. (+7) + (-3)
6. (+12) + (-5)
7. (+2) + (-8)
8. (+8) + (-10)
9. (+5) + (-5)
10. (+2) + (-20)
11. (-5) + (-2)
12. (-8) + (-6)
13. (-8) + (0)
14. (0) + (-12)
15. (-1) + (+1)
16. (-5) + (+8)
17. (-12) + (+22)
18. (-9) + (4)
19. (-8) + (15)
20. (6) + (-9)
21. (+10) + (-24)
22. (-14) + (-16)
23. (18) + (-30)
24. (-20) + (34)
25. (-16) + (0)
26. (46) + (+14)
27. (-51) + (84)
28. (46) + (-75)
29. (+127) + (-139)
30. (-127) + (139)

2.1 / Addition of Real Numbers

B

Find the sum in each of the following exercises.

1.	+6 +9	2.	+8 −3	3.	+10 −1	4.	+7 −12
5.	−8 +6	6.	−15 +12	7.	−7 −6	8.	−9 +9
9.	+5 −5	10.	6 −6	11.	−8 8	12.	12 12
13.	−18 +7	14.	+24 −15	15.	−19 21	16.	−17 −14
17.	+16 +23	18.	−25 +25	19.	+36 −36	20.	42 −18
21.	−24 +30	22.	−46 48	23.	56 −60	24.	+146 −215

C

Find the sum in each of the following exercises.

1. $(+2) + (+3) + (+8)$
2. $(+5) + (+7) + (-2)$
3. $(+6) + (+8) + (-10)$
4. $(+8) + (-2) + (+3)$
5. $(+9) + (-11) + (+5)$
6. $(+2) + (-8) + (+10)$
7. $(+15) + (-4) + (-6)$
8. $(+18) + (-3) + (-15)$
9. $(-4) + (-3) + (-7)$
10. $(-2) + (+6) + (-8)$
11. $(-4) + (-7) + (-10)$
12. $(-10) + (-5) + (-17)$
13. $(-8) + (7) + (-2)$
14. $(6) + (-10) + (14)$
15. $(-16) + (0) + (20)$
16. $(+25) + (+15) + (-12)$
17. $(-16) + (+9) + (20)$
18. $(+41) + (-53) + (-7)$
19. $(-34) + (-41) + (+60)$
20. $(+57) + (-105) + (+43)$

D

Find the sum in each of the following exercises.

1. $(+2) + (+4) + (-5) + (+8)$
2. $(-4) + (-7) + (+6) + (+1)$
3. $(+2) + (-5) + (+10) + (-5)$
4. $(-9) + (+3) + (+12) + (-12)$
5. $(7) + (-6) + (+18) + (1)$
6. $(-15) + (+24) + (-40) + (-24)$
7. $(+49) + (-52) + (+52) + (-49)$
8. $(+16) + (0) + (-20) + (-16)$

2.2 ADDITION OF MONOMIALS

PART 1

Having discovered how it is possible to determine the sum of two real numbers, we will now turn our attention to try to devise some means whereby we might add terms containing variables. You may recall that a term was a combination of numbers and variables that is separated from the rest of an expression by either an addition sign or a subtraction sign, or possibly both. In an expression such as the one below there are four terms.

$$4a + 2a^2 b - 4b^3 + \frac{2a}{5b} \tag{1}$$

Should the expression be such that no variable appears in the denominator of any fraction, then that expression is called by the special name, **polynomial**. The expression (1) above is *not* a polynomial for the fourth term contains the variable *b* in the denominator of the fraction. The expression (2) below is a polynomial, for although there are fractions in some of the terms, none of the terms has a variable in the denominator.

$$4a^3 + \tfrac{1}{2}a^2 b + \tfrac{3}{5}ab^2 + b^3 \tag{2}$$

There are a variety of names that are used with reference to polynomials. These names relate to either the number of variables that are found in the polynomial *or* the number of terms in the polynomial. For instance, the polynomial (2) above has *two* variables—the variable *a* and the variable *b*. In view of this, it is called *a polynomial in two variables*. If it had but *one* variable, such as in the one below

$$5a^4 + 8a^3 + 4a^2 - 7a + 2, \tag{3}$$

it would be called a polynomial in *one* variable.

On the other hand, if the polynomial contains but one term it is called a *monomial*.

Example of a monomial: $4a^2$

In the same way, the special polynomial that has but two terms is called a *binomial*.

Example of a binomial: $5a + 3$

Finally, the special polynomial that has three terms is called a *trinomial*.

Example of a trinomial: $5a^2 + \tfrac{2}{3}ab + 7b^2$

2.2 / Addition of Monomials

All other polynomials have no special names. They are simply called polynomials of four terms or polynomials of five terms, and so forth, depending on the number of terms in the polynomial. Example of a polynomial of 5 terms:

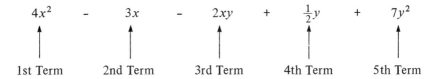

$$4x^2 \quad - \quad 3x \quad - \quad 2xy \quad + \quad \tfrac{1}{2}y \quad + \quad 7y^2$$

1st Term   2nd Term   3rd Term   4th Term   5th Term

Our objective in this section is to determine a method for adding certain monomials. In order to do this, let's first consider the situation in which we purchase

4 apples at 6¢ each.

We then go to a second store where we now buy

5 apples at 6¢ each.

Were we asked to determine the total price for all the apples we could determine the price by either of two ways. By one method we would find the cost of the apples in the first store, that is, 4 times 6¢, then find the cost of the apples in the second store, that is, 5 times 6¢, and then add these two answers to determine the total cost. This method is shown below:

$$4 \cdot 6 + 5 \cdot 6 \qquad (4)$$
$$= 24 + 30$$
$$= 54$$

However, the total cost can also be found by first determining the total number of apples purchased, which is 4 + 5, and then multiplying that number by the cost of each apple. This will again give a total price of 54¢. This method of finding the cost can be written as,

$$(4 + 5) \cdot 6 \qquad (5)$$
$$= 9 \cdot 6$$
$$= 54$$

It appears then that expressions (4) and (5) give exactly the same result,

$$4 \cdot 6 + 5 \cdot 6 = (4 + 5) \cdot 6 \qquad (6)$$

Mathematicians name the principle that we have just stated in (6) as the **distributive principle of multiplication over addition** and they usually represent this with variables, as

$$ba + ca = (b + c)a. \qquad (7)$$

This is the principle we must use to find the sum of two terms.

Suppose that we would like to find the sum of $3a$ and $4a$. This can be written in the form of

$$3a + 4a. \qquad (8)$$

By applying the distributive principle, we rewrite (8) as

$$(3 + 4)a. \qquad (9)$$

But, $3 + 4$ is equal to 7, hence (9) can be restated in the form of

$$7a. \qquad (10)$$

Actually, what we have just shown is that

$$3a + 4a = 7a. \qquad (11)$$

An examination of (11) suggests that we can move directly from the $3a + 4a$ to the $7a$ by merely adding the 3 and the 4 and simply "bringing along" the variable. This is much the same as if we were asked to find the sum for

$$3 \text{ apples} + 4 \text{ apples}. \qquad (12)$$

Here, too, we would simply add the 3 and the 4 and merely "bring along" the word "apples."

The number 3 in the expression "3 apples" indicates *how many* apples are under consideration and the word "apples" indicates the *kind of quantity* under consideration.

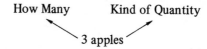

In the same way, it might be somewhat clearer to us were we to think of the term

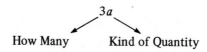

2.2/Addition of Monomials

where the 3 informs us of "how many" a's are under consideration. A number that appears before a variable, such as the 3 above, is called a **coefficient**. Thus, a coefficient tells us "how many" of that variable are under consideration.

Hence, it is possible for us to think of the earlier illustration of

$$3a + 4a = 7a \qquad (11)$$

as one in which the sum $7a$ is arrived at merely by adding three a's to four a's, thus yielding a sum of seven a's.

Example

Find the sum in the following exercise.

$$5y + 11y$$

Solution $\qquad 5y + 11y = 16y$

Explanation In this example we are called upon to add five y's and eleven y's. This is a total of sixteen y's.

EXERCISES 2.2 (Part 1)

A

Identify each of the following expressions as to whether each is a polynomial or not a polynomial. Justify each of your answers.

1. $3x^2 + 5x - 4$
2. $\frac{2}{3} x^2 y - 17xy^3$
3. $\frac{5x}{2y} + 4y^2$
4. $\frac{3xy}{2} + 5x + 4y$
5. $\frac{1}{4}x^2 + \frac{5}{7}y^2 + \frac{9}{10}xy$
6. $7x^2 y^3 + \frac{2}{x} + 4x$

B

Identify each of the following polynomials as to whether it is a monomial, binomial, or trinomial.

1. $4x^2 + 2x$
2. $\frac{2}{3}x^2 + 5x + 6$
3. $7x^5 - 4x^2 y + y^2$
4. $5x^2 y^3 z$
5. $\frac{1}{3}x^5 - \frac{2}{3}y^5$
6. $7x^4 y^3 + 6x^3 y^2 - 2x^2 y$
7. $x^3 y^3$
8. $x^3 - y^3$

C

Find the sum in each of the following expressions.

1. $2a + 7a$
2. $9x + 3x$
3. $5a + 4a$
4. $10c + 6c$
5. $3ab + 4ab$
6. $5xy + 1xy$
7. $2cd + 12cd$
8. $15ac + 23ac$

PART 2

Were we asked to find the sum of $5a$ and $2p$, that is,

$$5a + 2p, \qquad (1)$$

it would not be possible to rewrite this expression as we had earlier, for the variable in the first term is not the same as the variable in the second term. Hence, it is *not* possible to add these two terms. This is much the same as being asked to add 5 apples and 2 pears. Since the "kind" of fruit is different the sum cannot be determined. We can think of the first term in (1) as representing $5a$'s, while the second term represents $2p$'s and since each represents a *different* "kind" of quantity, their sum cannot be determined.

Now let's turn to a situation in which the variables involve exponents, such as

$$5x^3 + 4x^3. \qquad (2)$$

Again applying the distributive principle we arrive at

$$(5 + 4)x^3. \qquad (3)$$

This expression can be rewritten as,

$$9x^3. \qquad (4)$$

Hence, it appears that

$$5x^3 + 4x^3 = 9x^3. \qquad (5)$$

Notice here that the variables are the same and the exponents are also the same. When this is so, the sum is found by adding the coefficients of the two terms and leaving both the variables and the exponents exactly as they were. We should have suspected that the sum in (5) would turn out as it had for we can consider this situation as one in which we are asked to add five x-cubes to four x-cubes. Hence, the sum must be nine x-cubes.

2.2 / Addition of Monomials

Lastly, what will the situation be if the variables are the same but the exponents differ? For instance, can we find the sum of

$$7y^2 + 5y^3 \; ? \tag{6}$$

This is not possible, for the distributive principle can be applied only if the variables are *exactly* the same. The variables in $7y^2$ and $5y^3$ are not of the same power. Thus, we cannot apply this principle. Or, by thinking of the first term as involving *y-squares* while the second involves *y-cubes*, we again come to the conclusion that they are different "kinds" of quantities and hence their sum cannot be found.

Terms in which the variables and exponents are identically the same are called *like* terms or *similar terms*.

An example of like terms is: $7x^5$ and $4x^5$

Principle for Addition of Like Terms

1. Add the coefficients in the same manner as you would any two real numbers.
2. The variables in the sum will be *identical* to those of the terms.
3. The exponents in the sum will be *identical* to those of the terms.

Example 1
Determine the sum in the exercise below.

$$(-5x^3) + (+9x^3)$$

Explanation These terms are like terms, for the variables are the same and the exponents are the same. We can now proceed by adding the coefficients of −5 and +9 giving us a sum of +4. The variable x and the exponent remain the same in the sum as they are in the two terms.

Solution $\quad (-5x^3) + (+9x^3) = +4x^3$

Example 2
Determine the sum in the exercise below.

$$(x^2y) + (-3x^2y)$$

Explanation The only difficulty here is the fact that there is no coefficient stated in the first term. We know that 1 multiplied by any number leaves that number unchanged. Hence, 1 multiplied by x^2y is the same as x^2y. Therefore, we can rewrite x^2y as $1x^2y$. Also, from what we learned earlier, since no sign of direction appears before this term, the sign of direction is understood to be positive.

Solution $\qquad (+1x^2y) + (-3x^2y) = -2x^2y$

Let us summarize the two points made in the explanation above:

If a coefficient is not written before a term, it is understood to be the number 1.
If a sign of direction is not written before a term, it is understood to be the positive sign.

Example 3

Determine the sum in the exercise below.

$$(-6a^3b^4) + (-a^3b^4)$$

Explanation Since the coefficient is not written in the second term, it is possible for us to insert the number 1 as the coefficient.

Solution $\qquad (-6a^3b^4) + (-1a^3b^4) = -7a^3b^4$

You may have noticed that writing the sum of two monomials in the form

$$(-5x^3) + (+9x^3)$$

seems to be somewhat lengthy. The mathematicians felt the same way about this and hence they shortened this expression to the form,

$$-5x^3 + 9x^3.$$

When doing this, it is important that we realize that the "$-$" sign before the $5x^3$ is the sign of direction for the 5, while the "$+$" sign before the $9x^3$ is also the sign of direction for the 9.

Henceforth, any expression written in a form such as

$$-3a^2 + 4a^2 + 5a^2 - a^2$$

is to be understood as the abbreviated form for

$$(-3a^2) + (+4a^2) + (+5a^2) + (-a^2).$$

2.2 / Addition of Monomials

Example 4
Find the sum in the exercise below.

$$-6x^2 + 5x^2 + x^2$$

Solution $\quad -6x^2 + 5x^2 + 1x^2 = 0$

Explanation After writing the coefficient 1 for the third term, we add the coefficients as we have in the past. However, we do not write the sum as $0x^2$. Since 0 multiplied by any number is 0, that is 0 times 5 is 0, 0 times 27 is 0, so too 0 times x^2 equals 0. Hence, the sum is simply written as 0 rather than $0x^2$.

EXERCISES 2.2 (Part 2)

A

Find the sum in each of the following exercises. If it is not possible to add the two monomials in an exercise, write the words "not possible" as your answer to that exercise.

1. $(3a) + (6a)$
2. $(5b) + (12b)$
3. $(6x) + (x)$
4. $(+2y) + (-6y)$
5. $(-8w) + (-7w)$
6. $(-3c) + (-c)$
7. $(+2x) + (-3y)$
8. $(+5a^2) + (+2a^2)$
9. $(-7x^3) + (+5x^3)$
10. $(+8b^4) + (-10b^4)$
11. $(4x^3) + (-9x^2)$
12. $(-x^2) + (-x^2)$
13. $(+6b^2) + (-6b^2)$
14. $(-5x^5) + (-5x^5)$
15. $(-4a^3) + (4a^3)$
16. $(5ab) + (3ab)$
17. $(ab) + (2ab)$
18. $(-xy) + (3xy)$
19. $(+7x^2y) + (-12x^2y)$
20. $(-6a^3b^2) + (-6a^3b^2)$
21. $(5ab^4) + (-5ab^4)$
22. $(+3a^2b) + (7ab^2)$
23. $(+12xyz) + (+8xyz)$
24. $(-14x^2y^3) + (+10x^2y^3)$
25. $(a^3b) + (-7a^3b)$
26. $(-6ac) + (7ab)$
27. $(+x^2y) + (-x^2y)$
28. $(18ab^3) + (-25ab^3)$
29. $(-15x^3y^3) + (-20x^3y^3)$
30. $(-a^2b^3c) + (a^2b^3c)$

B

Find the sum in each of the following exercises. If it is not possible to add the two monomials in an exercise, write the words "not possible" as your answer to that exercise.

1. $7x$
    $4x$

2. $+9x^2$
    $+7x^2$

3. $-5a$
    $+3a$

4. $+6a$
    $-2b$

5. $9xy$
    xy

6. $-3a^2$
    $+2a$

7. $-ab$
    $+5ab$

8. $-xy$
    $-xy$

9. $-7x^2y$
    $18x^2y$

10. $5xy^2$
    $-26xy^2$

11. $-2a$
    $+2b$

12. $-7a^3$
    $-24a^3$

13. $+4ab^3$
    $16ab^3$

14. $8ab$
    $-8ab$

15. $-9ab^2c$
    $13ab^2c$

16. $42a^2b^2$
    $-54a^2b^2$

C

Determine the sum in each of the following exercises.

1. $3a + 5a$
2. $6a - 8a$
3. $-7a - 9a$
4. $a - 12a$
5. $11x^2 + 5x^2$
6. $-8x^2y - 7x^2y$
7. $-2ab^2 + 2ab^2$
8. $-14xy^3 + 16xy^3$
9. $-a^3b - a^3b$
10. $-ab^2c + ab^2c$
11. $7x - 2x - 9x$
12. $-3a^2 + 5a^2 + 7a^2$
13. $-2ab - 3ab - ab$
14. $ab^2 - ab^2 + ab^2$
15. $-2xy + xy + xy$
16. $9a^3b - a^3b + 2a^3b$
17. $5xyz + xyz - 7xyz$
18. $-10xy^2 + 2xy^2 + 10xy^2$
19. $-5x^2 + 4x^2 + 3x^2 - 9x^2$
20. $a^3b - a^3b + 2a^3b - a^3b$

D

Rewrite each of the following expressions in a shorter way and determine the sum in each exercise.

1. $(+5x) + (-3x)$
2. $(-4ab) + (+2ab)$
3. $(-7x^2) + (-16x^2)$
4. $(-3x) + (-x)$
5. $(2x^2) + (-5x^2)$
6. $(+x) + (+x) + (+x)$
7. $(-a) + (-a) + (-a)$
8. $(-3a^2) + (a^2) + (-a^2)$
9. $(2x^4y) + (-x^4y) + (5x^4y)$
10. $(-w^2) + (-w^2) + (w^2)$

2.3 ADDITION OF POLYNOMIALS

It is but one short step from addition of monomials to addition of polynomials. Given a polynomial of the form

$$3x^2 + 5x - 6,$$

we know that we cannot add the $+3x^2$ to $+5x$ nor can we add the -6 to either of these terms for the three terms are not like terms. However, were we required to add the polynomial $3x^2 + 5x - 6$ to the polynomial $-4x^2 + 6x - 9$, it would simply be a matter of writing the second polynomial below the first one, making certain that the like terms are placed in the same column. Thus,

$$\begin{array}{r} 3x^2 + 5x - 6 \\ \underline{-4x^2 + 6x - 9} \\ -1x^2 + 11x - 15 \end{array}$$

Example 1

Find the sum of the following polynomials.

$$\begin{array}{r} 4y^2 - 6y + 1 \\ -3y^2 - 5y - 7 \\ \underline{2y^2 + y + 9} \end{array}$$

Explanation After checking to make certain that like terms appear in the same column, we complete the exercise as though we were faced with three separate addition exercises in which in the first we had to find the sum of $4y^2$, $-3y^2$, $2y^2$; in the second the monomials to be added are $-6y$, $-5y$, $+y$; and in the third the monomials are $+1, -7$, and $+9$.

Solution
$$\begin{array}{r} 4y^2 - 6y + 1 \\ -3y^2 - 5y - 7 \\ \underline{2y^2 + y + 9} \\ +3y^2 - 10y + 3 \end{array}$$

Example 2

Find the sum of the polynomials in the following expression.

$$(3x - 4y - 2z) + (x - 4z) + (4y - 3x)$$

Explanation The first objective is to place the polynomials under one another so that the like terms are correctly aligned. When we examine the third polynomial, we notice that the $-3x$ term will have to be interchanged with the $4y$ term. Make certain to write the "+" sign before the $4y$. If this is not done then the terms $-3x$ and $4y$ will appear as

$$-3x4y.$$

This implies that $-3x$ is to be multiplied by $4y$, which is not so.

Solution
$$\begin{array}{l} +3x - 4y - 2z \\ +1x - 4z \\ \underline{-3x + 4y } \\ +1x - 6z \end{array}$$

Explanation continued Notice that the 0 for the sum of the $-4y$ and $+4y$ is not written in the answer. The fact that the sum of $-4y$ and $+4y$ is 0 may be indicated by drawing a line through the two terms.

As the course develops we will occasionally encounter expressions of the form

$$4a - 7b - 9a.$$

this implies that we are to find the sum of the three terms,

$$4a, -7b, \text{ and } -9a.$$

However, it is not possible to add the terms $4a$ and $-7b$, for they are not like terms. But, if we change the positions of the $-7b$ and the $-9a$, the expression will be

$$4a - 9a - 7b.$$

Under this arrangement, we can find the sum of $4a$ and $-9a$. This is $-5a$. The sum of $-5a$ and $-7b$ is then expressed in the usual manner as

$$-5a - 7b$$

and the answer is left in that form until such time that we are told the replacement values for a and b.

This process of rearranging the terms so that the like terms will be together and then adding these terms is often referred to as **collecting like terms**. However, rather than actually rearranging the terms and writing them as we have above, we merely move our eyes from term to term spotting those terms that are like terms and adding mentally as we move along.

Example 3

Collect like terms in the following expression.

$$5x + 6y - 3x - 7x + y - 2w$$
(with check marks over $5x$, $-3x$, $-7x$)

Explanation The initial step is to find another term that is similar to $5x$. The first of these that we come to as we move from left to right is the term $-3x$. We find the sum of this term and $5x$ to be $+2x$. After having checked off both the $5x$ and the $-3x$ we keep the sum of $+2x$ in mind and continue our examination from left to right until we find another x term. The next one we come to is $-7x$. When this is added to the previous sum of $+2x$, the new sum is $-5x$. We place a check mark over the $-7x$ and continue the process of examining the remaining terms for other x terms. Since there are none, we write down the first term in the answer as $-5x$. This same process is then repeated with the y terms and finally with the w terms. When the answer is determined, there should be a check mark over every term. If there is a term that has no check mark over it, it will probably imply that this term was overlooked.

Solution
$$5x + 6y - 3x - 7x + y - 2w$$
$$= -5x + 7y - 2w$$

EXERCISES 2.3

A

Find the sum of the polynomials in each of the following exercises.

1. $2x + 3y - 4w$
    $-5x + 2y - 6w$

2. $-4a + 3b - 2c$
    $-5a + b - c$

3. $7x^2 - 3x - 5$
    $2x^2 - 3x + 7$

4. $y^2 - 5y + 9$
    $-3y^2 - 5y - 9$

5. $a^2 - a + 1$
    $-a^2 - a + 1$

6. $x^2 + xy + y^2$
    $-x^2 + xy - y^2$

7. $3x - 5y + 2w$
    $-7x + 3w$

8. $-2a + 4c$
    $-3a - 5b + c$

9. $4a - 2b$
    $-5a + 2b - 6c$

10. $-5x^2$
    $7x^2 - 2xy + y^2$

11. $3a^2 - 4ab + b^2$
    $a^2 + 5ab$

12. $- 3b - 5c$
    $2a + 9b + 5c$

13. $6x + 4y - 2w$
    $-2x + 3y + 5w$
    $5x + 2y + 3w$

14. $7a^2 - 2a + 4$
    $5a^2 - 6a - 2$
    $-2a^2 - 3a - 5$

15. $x^2 - 3x + 1$
    $2x^2 + 5x - 1$
    $4x^2 - 7x + 6$

16. $-4y^2 - y + 3$
$y^2 + y - 7$
$5y^2 - 2y + 6$

17. $x^2 - xy + y^2$
$-2x^2 - xy + 3y^2$
$x^2 - xy - 5y^2$

18. $a^2 + 2ab + b^2$
$a^2 - 5ab - 6b^2$
$-3a^2 + 3ab - b^2$

19. $4x - 2y + 3w$
$5y + 7w$
$3x - 4w$

20. $3x^2 - 5x$
$ 2x + 7$
$-7x^2 - 7$

21. $-4x^2 + 9$
$4x^2 + 3x$
$ - 3x - 5$

B

Rearrange the terms properly in each of the following exercises and then find the sum of the polynomials.

1. $3x + 4y - 5w$
$-2y + 5x - 2w$

2. $a - 2b + c$
$c - 2a + 3b$

3. $5x - 4y + 1$
$3 + 5y - 2x$

4. $8x^2 - 5x + 4$
$3x - 5$

5. $x^2 - xy + y^2$
$y^2 - xy - x^2$

6. $a^2 - 3ab$
$3ab - b^2 + a^2$

7. $3a - 2b + 4c$
$2b - 3c + 4a$
$-2c + 5a - 7b$

8. $3x^2 - 5x + 6$
$4x - 6$
$x^2 - 5$

9. $x^2 - xy + y^2$
$y^2 + xy - x^2$
$xy - y^2$

C

In each of the expressions below, rewrite the polynomials in column form and then find their sum.

1. $(4x - 2y + 3w) + (5x + 7y - 7w)$
2. $(2x^2 - 5x + 9w) + (-3x^2 + 2x - 6w)$
3. $(2a - b + 3c) + (5a + 4b - 5c) + (a + b - c)$
4. $(6x^2 - 5x + 3) + (x^2 - x - 4) + (-x^2 - x - 1)$
5. $(5x^2 - 6xy + y^2) + (2x^2 - xy - y^2) + (x^2 + xy + 3y^2)$
6. $(3a + b - c) + (2a - 3c + 4b) + (-6c - 7b - 2a)$
7. $(2x - y + 3w) + (x - 2y - 4w) + (5w - x + 2y)$
8. $(5x^2 - 2xy + y^2) + (3y^2 + 4xy - x^2) + (xy + y^2 + x^2)$
9. $(7a - 2b) + (4a - 3c) + (5b + 7c)$
10. $(6a^2 - 4) + (7a - 5) + (-5a^2 - 7a)$
11. $(2xy - y^2) + (3x^2 - xy) + (x^2 - 2xy + y^2)$
12. $(+5x^2) + (4x^2 - xy + y^2) + (-3y^2 + 2x^2)$

2.4/Subtraction of Real Numbers

D

Collect like terms in each of the exercises below.

1. $3a - b + 5b$
2. $5x - 2y + 3x$
3. $4a + 3b - 5a - 6a$
4. $2x - 7y + 4 + 5x + 6y$
5. $x^2 + 3x - 5x + 2x^2$
6. $3x^2 - x - 2x^2 + 4x - x^2$
7. $8a^2 + a^2 - 3a - 10a^2 - 2a$
8. $2x^2 - x - 4x^2 - 5 + 2x - 3$
9. $x^2 - x + x^2 + x + 3x$
10. $2a^2 - 3ab - 4a^2 + 5ab - b^2$
11. $-a^2 - b^2 + 2ab - 2ab - 3a^2 + 4b^2$
12. $3x - 5 + 4y + 2 - 3x + 6 + y$
13. $x^2 - 5x + x^2 + y^2 + 6x - y^2 - x^2$
14. $10 - x + 4x - 6y - 12 - 2x - 5y$
15. $a + b - a + b + a - b - a$

2.4 SUBTRACTION OF REAL NUMBERS

PART 1

It is quite evident that having explored the operation of addition with real numbers and with polynomials, we would move on to examine the operation of subtraction.

Although the great majority of us learned the operation of subtraction by the "take away" method, there are many of us who were taught this operation by what is called the *additive method of subtraction*. As an example, consider the situation here where 5 is to be subtracted from 8. The person who

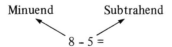

was taught subtraction by the "additive method" will ask himself, "5 plus what number will give a sum of 8?" Or, in general, the question will be, "The subtrahend plus what number will give the minuend?"

Mathematicians tend to use the meaning of subtraction as given by the "additive method" not only when dealing with real numbers but also with other new numbers they have invented over the years. In the example above, the question now arises as to exactly what we can do to determine the number that must be added to 5 so that the sum of these two numbers will be 8.

You may recall that the sum of two numbers was originally found through a succession of two movements on the number line. Thus, to find the sum of 4 and 2, the first movement carries us from the origin to the point named by the number 4 while the second movement carries us 2 units further to the point 6 which is the sum of 4 and 2 (see Figure 2-6).

FIGURE 2-6

We will now apply this method to find the difference in the following exercise.

$$8 - 5 = ?$$

When we rewrite this exercise in terms of the additive method it becomes,

5 added to what number will give 8?

Hence, the first step is to make a movement of 5 units from the origin (Figure 2-7).

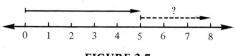

FIGURE 2-7

This brings us to the point 5. We now ask ourselves, "What second movement will bring us to the point 8?" This can be determined by counting the number of units from 5 to 8. Since there are 3 units, we say that the difference is 3.

$$8 - 5 = 3$$

This very same method is applied when finding the difference between any two signed numbers. As an example, consider the situation,

$$(+4) - (-2) = ?$$

Where before we asked, "5 added to what number will give 8?," now the question is, "-2 added to what number will give +4?" This implies that the

2.4/Subtraction of Real Numbers

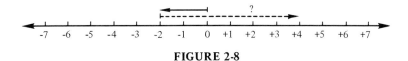

FIGURE 2-8

first movement will carry us from the origin to the point -2. Then, to determine the second movement that will carry us from -2 to +4, we simply count the number of units from -2 to +4. This turns out to be 6. However, since that movement is one in which we *travel to the right* (see the arrowhead on the dotted line in Figure 2-8) we attach a *positive* sign to the 6 and say that the difference between +4 and -2 is a +6.

$$(+4) - (-2) = +6$$

If the second movement is *to the left*, then the sign of direction of the difference is *negative*.

Example
Determine the difference in the exercise below.

$$(5) - (+8)$$

Explanation The initial step is to write the sign of direction before the 5. Since none is there, we understand the direction to be positive. The first movement takes us from the origin to the point named by the *subtrahend*. In this exercise, it brings us to the point +8. Now we ask, "What movement will carry us from +8 to +5?" Since that movement is to the left, the sign of direction

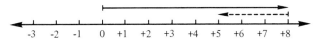

of the difference will be negative and since the number of units we travel from +8 to +5 is 3, then the difference between +5 and +8 is -3.

Solution $(+5) - (+8) = -3$

EXERCISES 2.4 (Part 1)

Use the number line to find the difference in each of the following exercises.

1. $(+4) - (+1)$
2. $(+8) - (+3)$
3. $(+10) - (+2)$
4. $(+9) - (+5)$
5. $(+12) - (+4)$
6. $(-6) - (-8)$
7. $(-2) - (-7)$
8. $(-1) - (-4)$
9. $(0) - (-6)$

10. (0) - (-2)   17. (-8) - (+1)   24. (+2) - (+7)
11. (+2) - (-3)   18. (0) - (+6)   25. (+5) - (+8)
12. (+5) - (-1)   19. (-6) - (0)   26. (+1) - (+12)
13. (+7) - (-7)   20. (+5) - (0)   27. (6) - (8)
14. (+1) - (-5)   21. (+6) - (+6)   28. (-5) - (7)
15. (-3) - (+2)   22. (-12) - (-14)   29. (9) - (-2)
16. (-5) - (+7)   23. (-14) - (-12)   30. (3) - (15)

PART 2

As you might have suspected, we would not want to continue to use the number line to find the difference between two numbers, for the process is rather lengthy. Let us examine the difference between the two numbers 8 and 2 where all possible combinations of signs of direction are attached to those numbers and all possible arrangements are made of the two numbers. Each of these is listed below.

Column One   *Column Two*

1. (+8) - (+2) =   5. (+2) - (+8) =
2. (-8) - (-2) =   6. (-2) - (-8) =
3. (+8) - (-2) =   7. (-2) - (+8) =
4. (-8) - (+2) =   8. (+2) - (-8) =

Notice that each of the subtraction exercises in column two comes from the one immediately to its left in column one by merely interchanging the positions of the two numbers. Now, by referring to the number line we can find the difference in each of these eight situations. This is done in Figure 2-9.

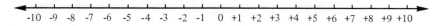

FIGURE 2-9

Column One   *Column Two*

1. (+8) - (+2) = +6   5. (+2) - (+8) = -6
2. (-8) - (-2) = -6   6. (-2) - (-8) = +6
3. (+8) - (-2) = +10   7. (-2) - (+8) = -10
4. (-8) - (+2) = -10   8. (+2) - (-8) = +10

The very first thing that should catch our eye is the fact that when the two numbers are interchanged as they are in situations (1) and (5), the differences are *not* the same. In (1) it is +6, while in (5) it is -6. Hence, whereas we discovered earlier that addition of real numbers is commutative, subtraction is *not* commutative.

2.4/Subtraction of Real Numbers

However, more important than this is the discovery that if we change the sign of direction of each of the subtrahends, every answer can be found by performing the operation of addition *as it pertains to signed numbers*. As an example, consider the first case in column one,

$$(+8) - (+2) = +6. \tag{1}$$

If we change the sign of direction of +2 to -2 and then add -2 to +8, the sum will be +6. Thus,

$$(+8) + (-2) = +6.$$

The same will be true for each of the other situations. As another example, consider exercise (4),

$$(-8) - (+2) = -10. \tag{4}$$

By changing the sign of direction of the subtrahend +2 to -2 and changing the operation from subtraction to addition the exercise becomes

$$(-8) + (-2).$$

The sum of -8 and -2 is -10 and this is exactly what we found earlier when using the number line to determine the answer to

$$(-8) - (+2). \tag{4}$$

Two numbers, such as +2 and -2, whose sum is 0, are called the **additive inverse** of each other. For instance, +7 is the *additive inverse* of -7, for the sum of +7 and -7 is 0. And in the same way, -15 is the *additive inverse* of +15 for the sum of these two numbers is 0.

The term *additive inverse* is introduced so that we might use it in stating the principle for subtraction of real numbers. Rather than say that we will change the sign of direction of the subtrahend we prefer to say that the subtrahend will be replaced by its additive inverse.

Subtraction Principle

When finding the difference between two signed numbers,

a. Replace the subtrahend with its additive inverse.
b. Add the two numbers as you would normally add two signed numbers.

Example 1

Determine the difference in the exercise below.

$$(-8) - (-12)$$

Explanation The first step is to replace the -12 with its additive inverse +12. This is followed by changing the operation from subtraction to addition.

Solution
$$(-8) - (-12)$$
$$= (-8) + (+12)$$
$$= +4$$

Example 2

Determine the difference between the two numbers, (5) and (-6), in the exercise below.

$$\begin{array}{r} 5 \\ -6 \\ \hline \end{array}$$

Explanation As usual, we must write the sign of direction before the 5; this is +. Rather than actually replacing the -6 with its additive inverse +6, we do this replacement mentally and think of +6 as being added to +5 thus giving an answer of +11.

Solution
$$\begin{array}{r} +5 \\ -6 \\ \hline +11 \end{array} \longleftarrow \text{Difference}$$

EXERCISES 2.4 (Part 2)

A

Find the difference in each of the following exercises.

1. (+10) - (+4)
2. (+8) - (+6)
3. (+12) - (+3)
4. (16) - (+7)
5. (9) - (+1)
6. (+6) - (3)
7. (-8) - (-10)
8. (-5) - (-9)
9. (-4) - (-12)
10. (0) - (-6)
11. (+3) - (-5)
12. (+4) - (-9)
13. (12) - (-2)
14. (14) - (-3)
15. (-4) - (-2)
16. (-9) - (-8)
17. (-7) - (-7)
18. (-6) - (+6)
19. (+6) - (-6)
20. (-3) - (5)
21. (7) - (-5)
22. (4) - (-8)
23. (-6) - (2)
24. (-5) - (10)
25. (7) - (2)
26. (+8) - (1)
27. (-9) - (+16)
28. (0) - (+7)
29. (+6) - (0)
30. (-5) - (0)

2.5/Subtraction of Polynomials

B

Find the difference in each of the following exercises.

1.	+10 +2	2.	+7 -3	3.	+5 -8	4.	-3 +4
5.	-2 +6	6.	+9 0	7.	+4 +6	8.	+8 +12
9.	-5 +6	10.	+9 +15	11.	-17 -12	12.	-13 -6
13.	-5 4	14.	1 -10	15.	6 9	16.	0 -8
17.	-5 0	18.	6 0	19.	0 12	20.	+15 +23
21.	-17 -26	22.	+23 -54	23.	-18 +29	24.	+27 +36
25.	18 12	26.	12 18	27.	247 139	28.	139 247

2.5 SUBTRACTION OF POLYNOMIALS

We have just learned that after replacing the subtrahend with its additive inverse, the operation of subtraction becomes the operation of addition. This information will enable us to perform the operation of subtraction with polynomials.

Example 1

Find the difference in the exercise below.

$$(-5a^2 b) - (+6a^2 b)$$

Explanation Following the usual procedure, we replace the $+6a^2 b$ with its additive inverse $-6a^2 b$, and then change the operation from subtraction to addition.

Solution
$$(-5a^2 b) - (+6a^2 b)$$
$$= (-5a^2 b) + (-6a^2 b)$$
$$= -11a^2 b$$

Example 2
Find the difference between the two polynomials in the exercise below.

$$3a^2 - 5a + 3$$
$$\underline{-7a^2 + 6a + 8}$$

Explanation After replacing *mentally* each of the terms in the subtrahend with its additive inverse the answer is found in exactly the same manner as when finding the sum of two polynomials.

Solution

$$+3a^2 - 5a + 3$$
$$\underline{-7a^2 + 6a + 8}$$
$$+10a^2 - 11a - 5$$

Explanation continued Notice that the $-7a^2$ is replaced by $+7a^2$ and then added to $+3a^2$. Similarly, $+6a$ is replaced by $-6a$ and that term added to $-5a$. Finally, $+8$ is replaced by -8 and added to $+3$.

Example 3
Find the difference between the polynomials in the exercise below.

$$(6a - 7b - 4c) - (-8a + 2c - 5d)$$

Explanation As in addition of polynomials it is necessary to write the second polynomial below the first one being careful that like terms fall in the same column. Since there is no d term in the first row, $-5d$ in the second row will have to be written at the far right.

Solution

$$6a - 7b - 4c$$
$$\underline{-8a \quad\quad + 2c - 5d}$$
$$14a - 7b - 6c + 5d$$

Explanation continued As there is no term in the subtrahend below the $-7b$, we consider the $-7b$ as being added to 0 thus giving a sum of $-7b$. When we come to $-5d$ in the subtrahend, the first task is to replace it with its additive inverse $+5d$. The $+5d$ is then added to the 0 that is understood to be in the minuend above it, thus giving a sum of $+5d$.

EXERCISES 2.5

A

Find the difference in each of the following exercises.

1. $(+5a) - (+2a)$
2. $(+12b) - (+7b)$
3. $(10c) - (+3c)$
4. $(+12x) - (7x)$

2.5/Subtraction of Polynomials

5. $(+3y) - (+5y)$
6. $(+6x^2) - (+9x^2)$
7. $(-5ab) - (-7ab)$
8. $(-3c^2) - (-12c^2)$
9. $(+7a^3) - (-4a^3)$
10. $(-6b^2) - (-5b^2)$
11. $(+9a^2b) - (10a^2b)$
12. $(-20c^4) - (-15c^4)$
13. $(0) - (+3a^5)$
14. $(0) - (-7b^2)$
15. $(+8b^2c) - (0)$
16. $(-9yz) - (0)$
17. $(3a^2) - (2a^2)$
18. $(5a^3) - (9a^3)$

19. $(+4x^2) - (+x^2)$
20. $(+9y) - (-y)$
21. $(-2a^3) - (-a^3)$
22. $(+x^2y) - (-5x^2y)$
23. $(-x) - (2x)$
24. $(-a) - (-a)$
25. $(a) - (-a)$
26. $(-a) - (a)$
27. $(a) - (a)$
28. $(3ab) - (-ab)$
29. $(-xy) - (-4xy)$
30. $(c^2) - (-7c^2)$
31. $(-ab^2c) - (-ab^2c)$
32. $(ab^2c) - (-ab^2c)$

B

Find the difference between the polynomials in each of the following exercises.

1. $6a - 5b + 4c$
    $4a + 3b + 3c$

2. $-3x + 4y - 7w$
    $-9x - 3y + 2w$

3. $5x^2 - 7x - 4$
    $-3x^2 - 7x + 3$

4. $9a^2 - 5a - 2$
    $10a^2 + 5a - 8$

5. $-7a - 5b + 3c$
    $-7a + 9b - 3c$

6. $4c^2 - 5c - 12$
    $7c^2 + 6c - 13$

7. $4x^2 - 5x - 6$
    $3x^2 - x + 4$

8. $a^2 - 3ab + b^2$
    $2a^2 - 5ab - b^2$

9. $6x + 4y - 10$
    $9x \qquad - 8$

10. $x^2 - xy + y^2$
    $-x^2 - xy$

11. $2a - 3b$
    $4a + 3b - 2c$

12. $\qquad x - 4$
    $3x^2 - 6x - 5$

13. $5a - 2b + 3c$
    $5a$

14. $5a$
    $5a - 2b + 3c$

15. $x^2 \qquad + y^2$
    $x^2 + 2xy + y^2$

C

In each of the exercises below, rewrite the polynomials in column form and find their difference.

1. $(5x - 3y) - (2x - 7y)$
2. $(-3a + 4b) - (5a - 4b)$
3. $(2x^2 - 3x + 4) - (4x^2 + 5x + 5)$
4. $(7a - 2b - 3c) - (-5a + 3b + 3c)$

2 / The Operations of Addition and Subtraction

5. $(9x^2 - 5x - 2) - (11x^2 - 5x - 7)$
6. $(x^2 + 2x + 3) - (3x^2 - 2x + 6)$
7. $(-3x^2 - 7x + 4) - (2x^2 - 5x)$
8. $(8a - 10b + 3c) - (11a - 7c)$
9. $(-2x^2 + 5x) - (4x^2 + 9x - 2)$
10. $(-6a + 2b) - (5a - 3b - 6c)$
11. $(2x^2 - 5y^2) - (4x^2 - 3xy - 2y^2)$
12. $(4a - 3b) - (-6b - 2a)$
13. $(-5x + 7y) - (2y + 3x)$
14. $(-3y^2) - (2y^2 - 6y - 1)$
15. $(a^2 - 2ab + b^2) - (b^2 - ab + a^2)$
16. $(x^2 - y^2) - (x^2 - 2xy + y^2)$

D

In each of the exercises below, determine which polynomial is the minuend, which is the subtrahend, and then find the difference.

1. From $(2a - 3b - c)$ subtract $(5a - 4b)$.
2. From $(+3x^2)$ subtract $(-2x^2 + 5x)$.
3. Subtract $(4a^2 - 2ab + b^2)$ from $(a^2 - 3ab - 2b^2)$.
4. Take $(5x - 4y - 3w)$ from $(5x + 4y)$.
5. Find the difference between $(7a - 2b)$ and $(3a + 4b - 6c)$. (When written in this manner, the first polynomial is the minuend.)
6. Find the difference between $(-2x^2y - 6x)$ and $(-5x^2y - 6x + y^2)$.

2.6 SIMPLIFYING EXPRESSIONS INVOLVING PARENTHESES

As we continue our study of mathematics only rarely will we find situations in which polynomials are arranged vertically and we are asked to perform either the operation of addition or subtraction. We will, however, encounter expressions such as

$$(5x^2 - 3x + 4) + (6x^2 - 7x - 5), \qquad (1)$$

wherein we ourselves must realize without being told that the operation between the two polynomials is addition.

Some time ago, though, we learned that the expression

$$5x^2 - 3x + 4 + 6x^2 - 7x - 5 \qquad (2)$$

also implied the operation of addition. In view of this, expression (2) is simply a shorter way of writing (1). Thus,

$$(5x^2 - 3x + 4) + (6x^2 - 7x - 5) \qquad (1)$$

2.6/Simplifying Expressions Involving Parentheses

can be written as

$$5x^2 - 3x + 4 + 6x^2 - 7x - 5 \qquad (2)$$

and this expression can be shortened by collecting like terms. Hence, (2) becomes,

$$11x^2 - 10x - 1. \qquad (3)$$

The process of eliminating the parentheses in going from step (1) to step (2) is called **simplifying the expression.** It is important to realize that the positive sign before the $6x^2$ in step (2) is *not* the addition sign between the parentheses in step (1) but rather the missing sign of direction that was not written before the $6x^2$ in step (1).

$$(5x^2 - 3x + 4) + (6x^2 - 7x - 5) \qquad (1)$$

$$= 5x^2 - 3x + 4 + 6x^2 - 7x - 5 \qquad (2)$$

Should this expression appear with a subtraction sign between the two polynomials such as

$$(3x - 4y) - (2x - 5y), \qquad (4)$$

we will have to treat this situation with slightly more care than the previous one. Since the operation is subtraction it implies that we must replace each of the terms of the subtrahend with its corresponding additive inverse before addition can take place. Thus, the $+2x$ will have to be replaced with $-2x$ and the $-5y$ will have to be replaced with $+5y$. Once this is done, the work is completed as before.

$$(3x - 4y) - (2x - 5y) \qquad (4)$$

$$= 3x - 4y - 2x + 5y \qquad (5)$$

$$= 1x + 1y \qquad (6)$$

The information above can be summarized as follows.

When simplifying expressions involving parentheses:

a. If the operation is addition, remove the parentheses without altering any of the terms.

b. If the operation is subtraction remove the parentheses but replace each of the terms in the subtrahend with its additive inverse.

Example 1

Simplify and collect terms in the following expression.

$$9x^2 - (-5x^2 + 4x - 3)$$

Explanation Since the operation is subtraction, it is necessary to replace each of the terms in the parentheses with its additive inverse.

Solution
$$9x^2 - (-5x^2 + 4x - 3)$$
$$= 9x^2 + 5x^2 - 4x + 3$$
$$= 14x^2 - 4x + 3$$

Example 2

Simplify and collect like terms in the following expression.

$$(4x^2 - 2) - (3x - 5) + (-2x^2 + 7)$$

Explanation Since the sign before the second polynomial indicates the operation of subtraction, each of the terms in that polynomial will have to be replaced with its additive inverse. The sign before the third polynomial indicates the operation of addition, hence these terms are not altered.

Solution
$$(4x^2 - 2) - (3x - 5) + (-2x^2 + 7)$$
$$= 4x^2 - 2 - 3x + 5 - 2x^2 + 7$$
$$= +2x^2 - 3x + 10$$

As the work in mathematics develops, the expressions do become somewhat more complex to a point where we encounter situations that involve parentheses within brackets, such as

$$5x^2 - [-9x^2 + (5x^2 - 4) - 8]. \qquad (7)$$

Perhaps a simple way in which to consider this situation is to imagine a person enclosed in a room that is itself entirely enclosed within a second room (Figure 2-10). In order for the person to escape from the predicament in which he finds himself, he must first use the key to emerge from the innermost room. He then follows this by using the second key that will get him out of the larger room.

2.6/Simplifying Expressions Involving Parentheses

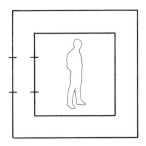

FIGURE 2-10

Similarly, the innermost room in (7) is the parentheses. The key that enables us to emerge from that room is the addition sign before the parentheses telling

$$5x^2 - [-9x^2 + (5x^2 - 4) - 8]$$

us that the terms $5x^2$ and -4 are not to be altered. Hence, (7) becomes

$$5x^2 - [-9x^2 + 5x^2 - 4 - 8] \qquad (8)$$

We next examine the second key, this being the subtraction sign before the brackets.

$$5x^2 - [-9x^2 + 5x^2 - 4 - 8].$$

This operation requires us to replace each of the terms within the brackets with its additive inverse.

$$5x^2 + 9x^2 - 5x^2 + 4 + 8 \qquad (9)$$

As usual, the next step is to collect like terms.

$$9x^2 + 12 \qquad (10)$$

Example 3

Simplify and collect like terms in the following expression.

$$(3x - 5) + [5x - (-3x + 4)]$$

Solution

$$(3x - 5) + [5x - (-3x + 4)]$$
$$= (3x - 5) + [5x + 3x - 4]$$
$$= 3x - 5 + 5x + 3x - 4$$
$$= 11x - 9$$

EXERCISES 2.6

A

Simplify and collect like terms in each of the following exercises.

1. $5x + (2x + 4)$
2. $4a + (3a - 5)$
3. $7b - (8 - 2b)$
4. $8c - (-5c + 3)$
5. $(3x - 4) + 7$
6. $(9x - 2) - 9x$
7. $(3a + 2b) + (5a - 4b)$
8. $(2x^2 - 5x) + (-3x^2 + 4x)$
9. $(5x - 2y) - (x + 2y)$
10. $(x - 3y) - (-x - 3y)$
11. $7a - (3a^2 - 5a - 4)$
12. $-5ab + (-3a - 2ab + 5b)$
13. $0 - (6x - 5y)$
14. $+3y - (4x - 5y)$
15. $-(4x - 5y) + 3y$
16. $-(-a + b) - b$
17. $-(x - y) + (-x + y)$
18. $(x^2 + xy + y^2) + (x^2 - y^2)$
19. $(a^2 + b^2) - (a^2 - ab + b^2)$
20. $-(4x - y)$

B

Simplify and collect like terms in each of the following exercises.

1. $3x + (5x - 4) + (x - 2)$
2. $5a + (2a - 3) - (4a - 7)$
3. $-7c + (3a - 2c) - 4c$
4. $6x - (x + 4) + (2x - 3)$
5. $9b - (-2b - 3c) - (2c - 5b)$
6. $(3a - 5b - 2c) - (4c - 2a + b)$
7. $(6a - b - c) - (a + b) + c$
8. $(x^2 - 3xy + 4y^2) + (x^2 - 5xy) - (-y^2 - x^2)$
9. $(a^2 + 2ab) - (a^2 - 2ab) - (ab - b^2)$
10. $(x^3 - x^2y - xy^2) + (y^3 - x^3) - (x^2y + xy^2)$

C

Simplify and collect like terms in each of the following exercises.

1. $2x + [3x + (4x - 1)]$
2. $[5x - (3x - 2)] + 7x$

Chapter Review

3. $4z - [9z + (7z - 5)]$
4. $-3x^2 - [4 - (2x^2 + 3)]$
5. $(2x - 3) + [5x + (3x + 1)]$
6. $(7x - 2) - [6x - (x - 2)]$
7. $-(a + 3) + [-10a + (-8a + 1)]$
8. $[6a - (-6a + 5)] - (a - 4)$
9. $-[4x + (2x^2 - 3) - 5x^2] + 7x$
10. $[3x - (5x - 9)] - [3x + (5x - 9)]$

CHAPTER REVIEW

1. Use the number line to find the sum in each of the following exercises. By drawing arrows above the number line, show how you arrive at your answer.

   a. $(+6) + (+3)$
   b. $(-2) + (-5)$
   c. $(+4) + (-6)$
   d. $(-7) + (+8)$

2. Use the number line to find the difference in each of the following exercises. By drawing arrows above the number line, show how you arrive at your answer.

   a. $(+10) - (+4)$
   b. $(-8) - (-5)$
   c. $(+2) - (-2)$
   d. $(-7) - (+1)$

3. Find the sum in each of the following exercises.

   a. $(+9) + (-5)$
   b. $(+6) + (-8)$
   c. $(-4) + (-6) + (+10)$
   d. $(-3) + (+8) + (-9)$

4. Find the sum in each of the following exercises.

   a. -5
      -17
   b. -14
      14
   c. 159
      -227

5. Find the sum of the following monomials.

   a. $(+5x^2) + (-4x^2)$
   b. $(-7a) + (-3a)$
   c. $(-2ab^2) + (2ab^2)$
   d. $(-x^2y) + (-x^2y)$

6. Rewrite the following expression without using parentheses and determine the sum of the monomials.

   $$(+4x) + (-3x) + (-2x) + (+x)$$

7. Identify each of the following as to whether it is a polynomial or not a polynomial.

   a. $2y^3 + \dfrac{3}{y^2} + 5y$ b. $\dfrac{1}{3}x^2 - 5x - 7$

8. Find the sum of the polynomials in each of the following exercises.

   a. $3x - 5y + 4w$
      $\underline{-2x - 7y + 3w}$

   b. $2x^2 - x$
      $-2x^2 + 3x - 5$
      $\underline{-3x^2 - 2x + 5}$

   c. $x^2 - xy + 2y^2$
      $xy - x^2 + y^2$
      $\underline{3y^2 + xy - 2x^2}$

9. Arrange the following three polynomials under each other and determine their sum.

   $3a - 2b - c,\ 3b + 2c - a,\ 2c - 2a - 3b$

10. Collect like terms in each of the exercises below.

    a. $5x - 4y - 7y + x$ b. $a^2 - ab + a^2 + 2a^2 + ab$

11. Find the difference in each of the following exercises.

    a. $(+7) - (+3)$
    b. $(+9) - (+12)$
    c. $(-8) - (-2)$
    d. $(-6) - (+6)$
    e. $(5) - (-4)$
    f. $(-7) - (-7)$

12. Find the difference in each of the following exercises.

    a. $(+6a^2) - (-2a^2)$
    b. $(-3xy) - (5xy)$
    c. $(-x^3) - (-2x^3)$
    d. $(7x^5) - (-8x^5)$

13. Find the difference between the polynomials in each of the following exercises.

    a. $3a + 4b - 7c$
       $\underline{2a + 6b + 3c}$

    b. $5x^2 - 2x - 3$
       $\underline{-x^2\quad\ \ - 3}$

    c. $-4a^2 + 3a$
       $\underline{-5a^2 - 7a + 6}$

14. From $4x^3 - 2x$ subtract $7x^3 - 5x^2 + 1$.

15. Simplify and collect like terms in each of the following exercises.

    a. $7x - (3x - 4)$
    b. $(9x - 5) + (-x + 6)$
    c. $(2x - 3) - (x + 4) + (-7x + 1)$
    d. $5x - [x - (2x - 3)]$
    e. $[3x + (4x - 1)] - (9x - 2)$

The Operations of Multiplication and Division

Having completed the operations of addition and subtraction on both real numbers and polynomials, it is only natural that we would now turn our attention to finding ways in which we can perform the operations of multiplication and division. Our first objective in this chapter will be to examine the operation of multiplication. With this as a background we will discover that the principles of division are an outgrowth of the principles of multiplication.

3.1 MULTIPLICATION OF REAL NUMBERS

PART 1

As in the operations of addition and subtraction our immediate move will be to select any two numbers and attach all possible combinations of signs of direction to these numbers. These situations are listed below.

1. (+4) (+3) =   3. (-4) (+3) =
2. (+4) (-3) =   4. (-4) (-3) =

From the fact that 4 times 3 gives exactly the same product as 3 times 4, we will conclude that if the numbers are interchanged, the product is not altered. Although this was not true in subtraction, we did find it to be true in addition.

The statement of the principle that covers this situation is similar to the one used in addition.

Multiplication is commutative.

Since we want the real numbers to behave in just the same manner as the numbers of arithmetic behave, there is no need to examine those situations in which the positions of the 3 and the 4 are interchanged in the four cases above. For instance, whatever the product for (+4) (-3) may turn out to be, the product of (-3) (+4) should be exactly the same.

We are now prepared to examine case (1). Consider the equation,

$$4 \times 3 = 12$$

Since no sign of direction precedes the number 12 we agreed earlier that this will imply that the sign of direction to be used is the positive sign. The same can be said for the numbers 4 and 3. In view of this, the equation above can be written as

Case (1):   (+4) (+3) = +12

What we have just shown for the product of +4 and +3 will work equally well for the product of any two positive numbers. Hence, we can say that the product of any two positive numbers is a positive number.

Based on this information, we now establish a pattern to show what the product of a positive number and a negative number must be.

Case (2):   (+4) (+3) = +12
            (+4) (+2) = +8
            (+4) (+1) = +4
            (+4) (0) = 0
            (+4) (-1) = ☐

Notice that in each of these situations the number in the first column is always the same; it is always +4. However, each succeeding number in the second column is 1 less than the number preceding it. And, finally, each succeeding number in the third column is 4 less than the preceding number. Thus, as each number in the second column decreases by 1, the product in the third column decreases by 4. This seems to imply that the number that must be placed in the empty square is -4 for this number is 4 less than 0. Similarly,

3.1/Multiplication of Real Numbers

4 times -2 should be 4 less than -4 thus making that product a -8. In the same way, 4 times -3 will have to be 4 less than -8, which makes it a -12.

Case (2):
$(+4)(+3) = +12$
$(+4)(+2) = +8$
$(+4)(+1) = +4$
$(+4)(0) = 0$
$(+4)(-1) = -4$
$(+4)(-2) = -8$
$(+4)(-3) = -12$

Hence, we have shown that the product of +4 and -3 turns out to be -12.

To determine the products in case (3) and case (4) we set up similar patterns. Can you discover what they are by examining the situations below? In case (4) we start with the information we learned in case (3) which is that -4 times +3 is -12.

Case (3):
$(+4)(+3) = +12$
$(+3)(+3) = +9$
$(+2)(+3) = +6$
$(+1)(+3) = +3$
$(0)(+3) = 0$
$(-1)(+3) = -3$
$(-2)(+3) = -6$
$(-3)(+3) = -9$
$(-4)(+3) = -12$

Case (4):
$(-4)(+3) = -12$
$(-4)(+2) = -8$
$(-4)(+1) = -4$
$(-4)(0) = 0$
$(-4)(-1) = +4$
$(-4)(-2) = +8$
$(-4)(-3) = +12$

By writing each of the products found for case (1) through case (4), we should be able to discover the principles to be followed when multiplying two real numbers.

1. $(+4)(+3) = +12$
2. $(+4)(-3) = -12$
3. $(-4)(+3) = -12$
4. $(-4)(-3) = +12$

Should we ignore the signs of direction, the very first thing that strikes us is the fact that in each situation the product turns out to be 12. This number can be determined by multiplying the 4 by the 3 as we had in arithmetic.

Let's turn our attention now to determining what the sign of direction of the product will be. Notice that in case (1) and again in case (4) the sign of the product of the two numbers is positive. In both of these situations the signs of the two numbers are exactly the same. In case (1) the signs of both the 4 and the 3 are positive, while in case (4) they are both negative. Hence,

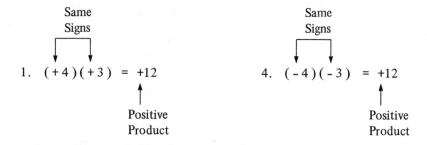

it would seem that if the signs of direction of the *two* numbers are the same, then the sign of the product will be *positive*.

In cases (3) and (4) the signs of direction of the 4 and the 3 are different and in both cases the sign of the product turns out to be negative. This seems

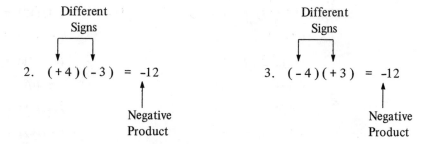

to imply that if the signs of direction of the *two* numbers are *not* the same, the sign of the product will be *negative*.

Putting all the information together, we can state the following principle.

Principle for Multiplication of Real Numbers

When finding the product of *two* signed numbers,

a. If the signs of direction of the two numbers are the same, then the sign of the product is positive.
b. If the signs of direction of the two numbers are *not* the same, then the sign of direction of the product is negative.

Example 1

Determine the product in the exercise below.

$$(-5)(+7)$$

Explanation Since the signs of direction of the two numbers are not the same, the sign of direction of the product will be negative.

Solution $(-5)(+7) = -35$

Example 2

Determine the product in the exercise below.

$$(-2)(-4)(-7)$$

Explanation It is important to realize that the principles of multiplication apply only when finding the product of but *two* numbers. Therefore, in this illustration we begin by finding the product of -2 and -4.

$$(-2)(-4) = +8$$

The +8 is then multiplied by -7 giving a final product of -56.

Solution $(-2)(-4)(-7) = -56$

EXERCISES 3.1 (Part 1)

A

Find the product in each of the following exercises.

1. $(+2)(+5)$
2. $(+7)(+4)$
3. $(+8)(-3)$
4. $(+6)(-9)$
5. $(-2)(-6)$
6. $(-9)(-7)$
7. $(+5)(-8)$
8. $(-7)(+1)$
9. $(-6)(+10)$
10. $(-3)(-12)$
11. $(+4)(-11)$
12. $(3)(-9)$
13. $(5)(-7)$
14. $(-6)(8)$
15. $(-9)(2)$
16. $(3)(10)$
17. $(+6)(1)$
18. $(-1)(7)$
19. $(0)(+4)$
20. $(-9)(1)$
21. $(-12)(0)$
22. $(+\frac{1}{2})(+12)$
23. $(-15)(+\frac{1}{3})$
24. $(-11)(-10)$
25. $(+12)(-9)$
26. $(0)(-15)$
27. $(-1)(-24)$
28. $(+63)(-1)$
29. $(-\frac{1}{2})(+\frac{1}{2})$
30. $(-\frac{3}{4})(-\frac{3}{4})$

B

Find the product in each of the following exercises.

1. (-2) (+3) (-4)
2. (5) (-1) (+3)
3. (-4) (-3) (+6)
4. (0) (-5) (-4)
5. (3) (2) (-5)
6. (-4) (0) (-2)
7. (-1) (-1) (+8)
8. (-6) (2) (+9)
9. (7) (-1) (3)
10. (+8) (-2) (3)
11. (-3) (-1) (+2) (-5)
12. (+7) (+4) (0) (-3)
13. $(+\frac{1}{2})$ (-8) (4) (-2)
14. (-1) (+3) (-4) (-1) (+2)
15. (-6) $(+\frac{1}{3})$ (-2) (3) (-6)

PART 2

There is only one small step in applying what we already know about multiplication to evaluating expressions involving exponents. This is shown in the following examples.

Example 1

Determine the product in the expression below.

$$(-3)^4$$

Explanation The only point of importance here is to recall that the exponent 4 implies that the number -3 is to appear 4 times in the process of multiplication. The work is then completed by multiplying the first number by the second; that answer by the third number; and that answer by the fourth number.

Solution
$$(-3)^4$$
$$= (-3)(-3)(-3)(-3)$$
$$= +81$$

Example 2

Find the value of $3a^2 - 2ab^3$ where $a = -4$ and $b = -1$.

Solution
$$3a^2 - 2ab^3$$
$$= [3a^2] + [-2ab^3]$$
$$= [(3)(-4)^2] + [(-2)(-4)(-1)^3]$$
$$= [(3)(+16)] + [(-2)(-4)(-1)]$$
$$= [+48] + [-8]$$
$$= +40$$

3.1/Multiplication of Real Numbers

Explanation It is best to rewrite the expression $3a^2 - 2ab^3$ in terms of its meaning and that is that $[3a^2]$ is to be added to $[-2ab^3]$. It is also advisable that brackets be used at this time rather than parentheses. The parentheses should be reserved to hold the numbers that replace a and b. Notice that when the a is replaced with –4, the exponent is written *outside* the parentheses *not* inside. The same is true when b is replaced with –1 where the exponent 3 is written *outside* the parentheses. After finding the single number that belongs inside the first brackets and the single number for the second brackets, the two signed numbers are then added.

You may have wondered why so much stress is placed on writing the exponent outside the parentheses. This is so because mathematicians have agreed that,

$$-4^2 \quad \text{means} \quad -4 \cdot 4 \quad \text{which equals} \quad -16$$

while,

$$(-4)^2 \quad \text{means} \quad (-4)(-4) \quad \text{which equals} \quad 16$$

In Example 2, it is the value of a that must be squared and therefore, it is –4 that is squared. This means that –4 must be multiplied by –4 and hence, must be written as $(-4)^2$.

EXERCISES 3.1 (Part 2)

A

Find the value of each of the following expressions.

1. $(-2)^2$
2. $(+5)^2$
3. $(-4)^2$
4. $(-9)^2$
5. $(+11)^2$
6. $(-15)^2$
7. $(-1)^2$
8. $(+2)^3$
9. $(+5)^3$
10. $(-1)^3$
11. $(-4)^3$
12. $(+1)^4$
13. $(+3)^4$
14. $(+17)^2$
15. $(-3)^3$
16. $(-5)^4$
17. $(+2)^5$
18. $(-2)^5$
19. $(+1)^7$
20. $(-1)^9$
21. $(+3)^2 + (+5)^2$
22. $(-2)^2 + (-3)^2$
23. $(-4)^3 + (+2)^4$
24. $(-2)^3 + (-3)^2$
25. $(+2)(+3)^3$
26. $(-1)(-4)^2$
27. $(-3)^2 (+1)$
28. $(-2)^2 (+3)^2$
29. $5(-1)^3 (+2)^2$
30. $(-5)^2 (-2)^3 (-1)^4$

B

Find the value of each of the following expressions under the conditions stated.

1. a^2 where $a = -5$
2. b^3 where $b = +2$
3. x^5 where $x = -2$
4. $3a^2$ where $a = -4$
5. $-2b^2$ where $b = +6$
6. a^2b^2 where $a = +4$ and $b = 2$
7. $-3x^2y$ where $x = -2$ and $y = 5$
8. $+7b^2c^3$ where $b = +1$ and $c = -2$
9. $a^2 + 1b^2$ where $a = +7$ and $b = -7$
10. $2a + 3b$ where $a = -5$ and $b = -2$
11. $a^2 - 2b$ where $a = +3$ and $b = +4$
12. $3x - 5y$ where $x = -4$ and $y = +6$
13. $2x^2 - 4xy$ where $x = +3$ and $y = -1$
14. $-2x^2 + y^2$ where $x = -5$ and $y = -3$
15. $-3x^3 + 5x^2y$ where $x = -2$ and $y = -3$

3.2 MULTIPLICATION OF MONOMIALS

Having determined the principles for finding the product of two signed numbers, we turn our attention to finding the product of two monomials. To do this, we must first recall the meaning of an exponent. Thus,

$$a^2 = a \cdot a \quad \text{and} \quad a^3 = a \cdot a \cdot a. \tag{1}$$

On the other hand, an expression of the form $a \cdot a \cdot a \cdot a \cdot a$ can be written in the much shorter form of a^5. That is,

$$a \cdot a \cdot a \cdot a \cdot a = a^5. \tag{2}$$

With this information as a background, we are prepared to find the product of a^2 and a^3. To do this we need simply replace both the a^2 and a^3 with their meanings as shown in (1).

3.2/Multiplication of Monomials

$$a^2 \cdot a^3 = a \cdot a \cdot a \cdot a \cdot a \tag{3}$$

Then by examining the right side of equation (3) we discover that this can be written in the much shorter form of a^5 as is done in (2) above. Hence,

$$a^2 \cdot a^3 = a \cdot a \cdot a \cdot a \cdot a$$
$$= a^5 \tag{4}$$

By placing the original exercise next to its product

$$a^2 \cdot a^3 = a^5$$

we can readily see that the variable in the exercise is the same as the variable in the product while the exponent in the product is the sum of the two exponents in the exercise.

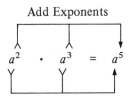

Variables Are the Same

What will happen, though, if the variables are not the same? Let us examine this under the conditions where the quantities to be multiplied are x^2 and y^4. Then,

$$x^2 \cdot y^4 = x \cdot x \cdot y \cdot y \cdot y \cdot y$$
$$= x^2 \cdot y^4 \tag{5}$$

Since the variables in the expression $x \cdot x \cdot y \cdot y \cdot y \cdot y$ are not the same, it is *not* possible to rewrite this expression as either x^6 or y^6. The only simpler way in which this can be rewritten is as $x^2 y^4$, which is identical to the manner in which it is written originally.

Now let us consider finding the product of any two monomials in which the variables are the same. As an example,

$$(-3a^5)(+8a^4). \tag{6}$$

We learned earlier that an expression of the form $5y$ implies that the operation between the 5 and the y is multiplication. In view of this, in (6) above the -3

is to be multiplied by the a^5, while the +8 is to be multiplied by the a^4. Hence, (6) can be rewritten as

$$(-3)(a^5)(+8)(a^4). \qquad (7)$$

Also, from the properties of multiplication, we know that the quantities to be multiplied can be arranged in any order we desire without affecting the product of the quantities. Hence, (7) can be rewritten as

$$(-3)(+8)(a^5)(a^4). \qquad (8)$$

This expression can now be restated in a much shorter form by finding the product of -3 and +8 and also finding the product of a^5 and a^4.

$$(-24)(a^9) \qquad (9)$$

And as a final step, (9) can be shortened to

$$-24a^9 \qquad (10)$$

As soon as this answer is written next to the two monomials whose product we were finding,

$$(-3a^5)(+8a^4) = -24a^9, \qquad (11)$$

we discover that the product can be found as shown below.

$$\text{Add Exponents}$$
$$(-3a^5)(+8a^4) = -24a^9 \qquad (12)$$
$$\text{Multiply Coefficients}$$

We have but one more situation to examine in our investigation of the product of two monomials. Consider the case where $-2a^3b^2$ is to be multiplied by $-5a^4c^2$.

$$(-2a^3b^2)(-5a^4c^2) \qquad (13)$$

3.2/Multiplication of Monomials

Following the same pattern used in the illustration above, (13) is rewritten as

$$(-2)(a^3)(b^2)(-5)(a^4)(c^2)$$
$$= (-2)(-5)(a^3)(a^4)(b^2)(c^2)$$
$$= (+10)(a^7)(b^2)(c^2)$$
$$= +10a^7b^2c^2. \tag{14}$$

Here again, by placing this answer next to the two monomials whose product we were seeking,

$$(-2a^3b^2)(-5a^4c^2) = +10a^7b^2c^2, \tag{15}$$

we find that, although there are more variables than in the earlier example, the coefficient in the product is determined by multiplying the coefficients in the monomials and, when the variables are the same, we simply add their respective exponents.

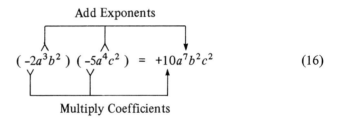

$$(-2a^3b^2)(-5a^4c^2) = +10a^7b^2c^2 \tag{16}$$

We often seem to overlook the exponent in a monomial such as

$$x.$$

As we know, the exponent 3 in the expression x^3 indicates that the variable x must appear three times in the process of multiplication; that is, $x \cdot x \cdot x$. And, in general, the exponent indicates the number of times that the base must appear in the product. It should follow then that x^1 implies that the variable x will appear but once in the product. Therefore,

$$x^1 = x.$$

This, in turn, implies that *if a variable is written and no exponent is indicated, the exponent that the variable is understood to have is the number 1.*

Based on the above information and what we have learned earlier, when a variable appears without a sign of direction, or a coefficient, or an exponent, the following will be understood to be true:

a. Its sign of direction is positive.
b. Its coefficient is 1.
c. Its exponent is 1.

At this stage of our work it would be well to write in any of the above three that may be missing.

Example 1

Determine the product in the exercise below.

$$(5a^3b)(-ab^2c^2)$$

Explanation The positive sign before the 5 is missing. The exponent of 1 for the b in the first monomial is missing. The coefficient of 1 before the a in the second monomial is missing. And, finally, the exponent of 1 for the a in the second monomial is missing. Once these are supplied, the exercise becomes

$$(+5a^3b^1)(-1a^1b^2c^2).$$

Solution

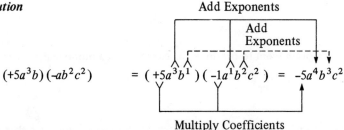

Example 2

Determine the product in the exercise below.

$$(-4a^5b)^2$$

Explanation The exponent of 2 outside the parentheses implies that $-4a^5b$ is to be written twice in the process of multiplication. After doing this the product is found as in Example 1.

3.2/Multiplication of Monomials

Solution

$$(-4a^5b)^2$$
$$= (-4a^5b^1)^2$$
$$= (-4a^5b^1)(-4a^5b^1)$$
$$= +16a^{10}b^2$$

Example 3

Determine the product in the exercise below.

$$(3a^2c)(-4b^3c^5)(-2bc^3)$$

Explanation After supplying the missing sign of direction and the missing exponents, the answer is found by first determining the product of the three coefficients. We then write every variable that appears in any term and determine its exponent by adding the exponents. Notice that the variables in the product are written in alphabetical order, for this is the custom.

Solution

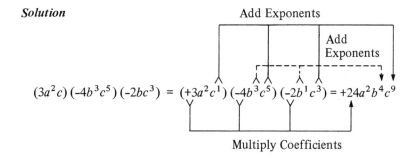

Example 4

Determine the product in the exercise below.

$$(-5a^4b)^2(-2b^3)$$

Explanation The exponent 2 of the first monomial indicates that the monomial is to be written twice. After doing this, the product is found in the same manner as in Example 3.

Solution

$$(-5a^4b)^2(-2b^3)$$
$$= (-5a^4b^1)(-5a^4b^1)(-2b^3)$$
$$= -50a^8b^5$$

EXERCISES 3.2

A

Find the product in each of the following exercises.

1. $(+3a^2)(+4a^5)$
2. $(-7a^3)(-2a^4)$
3. $(+6a^4)(-3a)$
4. $(+5a)(+7a^2)$
5. $(+a^3)(-4a^5)$
6. $(-a^3)(+2a^5)$
7. $(-2a)(-7a)$
8. $(3a^2)(-4a^8)$
9. $(-a)(+a)$
10. $(a)(-a)$
11. $(+3a^2b^3)(-4a^3b^4)$
12. $(-5a^5b)(-2a^2b^2)$
13. $(+6a^3b^2)(-7a^2c^3)$
14. $(+4a^2)(2b^2)$
15. $(-5ab)(9c)$
16. $(6a^2)(+5a^3b^2c)$
17. $(-9x^3y^2)(-2x^4z^5)$
18. $(+6x)(-7xy)$
19. $(x)(5yz)$
20. $(-8)(2xy)$
21. $(+3a^2)(-2a^3)(+4a^7)$
22. $(-a^2)(+3a)(-a^4)$
23. $(-4x^2y)(-2xy^2)(-2xy)$
24. $(-5x^2)(x^3)(4)$
25. $(3x^4)(-x^5)(-2x)$
26. $(7xy^2)(-3x^2y)(-y^3)$
27. $(-x^2yz)(-2xz)(-y^2z^2)$
28. $(-4y^3)(5xz^2)(-2xy^4)$
29. $(x)(-y)(-xy)(x)$
30. $(-4x^2)(-5y^2)(xy^2)(-y^3)$

B

Find the product in each of the following exercises.

1. $(+2a^3)^2$
2. $(-3a^2b^4)^2$
3. $(-ab^2)^2$
4. $(2x^2y^3)^3$
5. $(-4x^2y)^3$
6. $(+x^2yz^3)^3$
7. $(3x^3)^4$
8. $(-2a^2c)^4$
9. $(-x^2)^5$
10. $(a^3b^4)^6$
11. $(-2a^3)(3a^2)^2$
12. $(-4x^2)(-2x^3)^2$
13. $(a^5)^2(2a^3)$
14. $(-3a^3)^2(-2a^4)$
15. $(+5a^2b^3)(+2b^3)^2$
16. $(-4a^2b)^2(-2a^4)$
17. $(5a^4b^5)^2(-3a)$
18. $(3a^2)^2(4a^3)^2$
19. $(-5a)^3(2a^5)$
20. $(-4a^5)(-a^2b^3)^3$

3.3 MULTIPLICATION OF POLYNOMIALS

PART 1

Early in chapter 2 when learning how to find the sum of two monomials we made use of a principle we called the distributive principle of multiplication over addition. At that time we showed that 6 times 4 added to 6 times 5 will give exactly the same result as when 6 is multiplied by 4 added to 5. That is,

$$6 \cdot (4 + 5) = 6 \cdot 4 + 6 \cdot 5. \tag{1}$$

When we add the 24 to the 30 on the right side of the equation we obtain the same answer as when 6 is multiplied by 9. In both cases the answer is 54.

In reality, the expression on the left side of the equality sign represents a monomial being multiplied by a binomial where the 6 is the monomial and the 4 + 5 is the binomial. The expression on the right side of the equality sign tells us just how this product is found. According to this expression, it seems that the monomial 6 is first multiplied by the first term of the binomial, the 4; this is then followed by multiplying the 6 by the second term of the binomial, the 5.

Hence, it appears that *finding the product of a monomial and a binomial involves very little more than multiplying the monomial by each term of the binomial.*

Example 1

Determine the product in the exercise below.

$$-3a^2 b^3 (-5a^4 + 2b^2)$$

Explanation In this situation, as in general, it is necessary to multiply the monomial by *each* term of the binomial.

$$\underbrace{-3a^2 b^3}_{} \underbrace{(-5a^4}_{} + 2b^2) = +15a^6 b^3 - 6a^2 b^5$$

$-3a^2 b^3$ times $-5a^4$

$-3a^2 b^3$ times $+2b^2$

Solution $\quad -3a^2 b^3 (-5a^4 + 2b^2) = +15a^6 b^3 - 6a^2 b^5$

The method for finding the product of a monomial and a binomial works equally well were the polynomial a trinomial or a polynomial of any number of terms. As a general principle we can say,

---
Principle for Multiplying a Monomial by a Polynomial

When finding the product of a monomial and a polynomial, multiply the monomial by each term of the polynomial.

Example 2

Determine the product in the following exercise.

$$-2a^3 (3a^2 - 5a - 6)$$

Explanation The product here is found by multiplying the monomial $-2a^3$ by each term of the trinomial, the $+3a^2$, the $-5a$, and the -6.

Solution
$$-2a^3 (3a^2 - 5a - 6)$$
$$= -6a^5 + 10a^4 + 12a^3$$

Example 3

Determine the product in the following exercise.

$$(5x^2 - 3x)(-2x^3)$$

Explanation From the fact that multiplication is commutative, we can interchange the positions of the monomial and the binomial thus obtaining

$$(-2x^3)(5x^2 - 3x).$$

Solution
$$(5x^2 - 3x)(-2x^3)$$
$$= -2x^3 (5x^2 - 3x^1)$$
$$= -10x^5 + 6x^4$$

EXERCISES 3.3 (Part 1)

Find the product in each of the following exercises.

1. $+3a^2 (5a^3 + 2a^2)$
2. $-4a^3 (6a^5 - 7a^4)$
3. $+2a (-9a^2 + 5a^3)$
4. $-5x^2 (3x + 4)$
5. $6x (-7x - 3)$
6. $3 (x - 2)$
7. $-5 (3a - 1)$
8. $4ab (2ab - 5)$
9. $-2xy (x - y)$
10. $xy (x + 4)$

3.3/Multiplication of Polynomials 85

11. $(3x^2 + 5x^3)(2x^2)$
12. $(-7a^2 + 2a)(+3a^4)$
13. $(5a^2 - 6b^2)(-2ab)$
14. $(-4 - 3x)(-x)$
15. $(a - b)(-ab)$
16. $+5a^2 (3a^2 + 2a + 4)$
17. $-6x^3 (4 - 3x + 7x^2)$
18. $3x^2y^2 (4x^2 - 2x^2y^2 + 5y^2)$
19. $-7ab (5a^2 + 4ab - 3b^2)$
20. $2xy (x^2 - xy + y^2)$

21. $7 (x^2 - 3xy - 4y^2)$
22. $-4x (3a - 4b - c)$
23. $-5 (3 - 7x - 10x^2)$
24. $(a^2 - ab + b^2)(-4ab)$
25. $(3a^2 - 2ab - b^2)(-1)$
26. $-6x^2y (-x^3 + x^2y + 2y^3)$
27. $(x^5 - x^4 + x^3 - x^2)(x)$
28. $(-a)(a^3 - 3a^2b + 3ab^2 - b^3)$
29. $(-1)(a - b - c + d)$
30. $-2x (x + y - z + w)$

PART 2

As we continue our study of algebra we will often run across situations that involve the need for performing the operation of multiplication followed by the operation of addition. This will occur under the conditions shown in the examples below.

Example 1

Simplify and collect like terms in the expression below.

$$-2a (3a + 5) + 9a^2$$

Explanation In this exercise we are asked to find the product of a monomial and a binomial at the outset—that is what is meant by the word "simplify" in this situation. After completing the "simplifying" we are asked to "collect like terms" as we did in addition.

Solution
$$-2a (3a + 5) + 9a^2$$
$$= -6a^2 - 10a + 9a^2$$
$$= +3a^2 - 10a$$

Example 2

Simplify and collect like terms in the expression below.

$$- (4a^2 - 7a) + 3a^2 (-5a + 2)$$

Explanation In the same manner as we inserted the coefficient of 1 in a term such as $-x$ and rewrote it as $-1x$, so, too, is it possible to insert a 1 in the expression $-(4a^2 - 7a)$ and write it as $-1(4a^2 - 7a)$. The example simply expresses the fact that the monomial -1 is to be multiplied by the binomial $4a^2 - 7a$. This is followed by determining the product of the monomial $3a^2$ with the binomial $-5a + 2$. After determining the two products we collect like terms in the same manner as is done in Example 1.

Solution
$$-(4a^2 - 7a) + 3a^2(-5a + 2)$$
$$= -1(4a^2 - 7a) + 3a^2(-5a + 2)$$
$$= -4a^2 + 7a - 15a^3 + 6a^2$$
$$= -15a^3 + 2a^2 + 7a$$

Example 3

Simplify and collect like terms in the expression below.

$$5x^2 - 4[2x^2 - 3(x - 5)]$$

Explanation Recall that in the operation of subtraction we ran into a situation similar to the one above. At that time it was necessary to start with the innermost quantity and "work our way out of each succeeding room." In this case the "innermost" quantity is $(x - 5)$ and the "key" that enables us to "work our way out of this room" is the -3 that must be multiplied by the $(x - 5)$. After completing this multiplication we then turn to the key that will enable us to "work our way out of the room" containing the brackets. We do this by multiplying the -4 by the quantity in the brackets. The process is then completed by collecting like terms.

Solution
$$5x^2 - 4[2x^2 - 3(x - 5)]$$
$$= 5x^2 - 4[2x^2 - 3x + 15]$$
$$= 5x^2 - 8x^2 + 12x - 60$$
$$= -3x^2 + 12x - 60$$

EXERCISES 3.3 (Part 2)

A

Simplify and collect like terms in each of the following exercises.

1. $3x(2x + 3) + 5x$
2. $-4(3x - 5) - 7x^2$
3. $-2(5x - 4) - 9$
4. $4(3 - 2x) + 3x$
5. $5x + 4(2x + 3)$
6. $-3x + 5x(x - 3)$

3.3/Multiplication of Polynomials

7. $9x - 2x(4 - 5x)$
8. $xy - x(x - y)$
9. $-8x^2 - 2x(x + 3y)$
10. $2ab(a - b) - a^2b$
11. $-7xy(x^2 - 2y^2) - 4xy^3$
12. $5x^2y(x - xy + y) + 5x^3y^2$
13. $3(a + 2) + 5(a - 2)$
14. $4(x - y) - 3(x + y)$
15. $-(a - 4) + 2(a + 3)$
16. $2(a + 7) - (a - 5)$
17. $3a(a + b) - 2a(a - b)$
18. $-5x(x - y) + 2y(x + y)$
19. $(x + 3) - 3(x - 4)$
20. $7x(x - 5) + (x - 7)$
21. $-a(2a - 3) - (5a + 4)$
22. $2a(a - b) - b(b - a)$
23. $3(a^2 - ab + b^2) + 4(a^2 - b^2)$
24. $2(x - y) - (x + y) - 5y$

B

Simplify and collect like terms in each of the following exercises.

1. $2[4a + 2(a - b)]$
2. $3[5a - 3(a - 2b)]$
3. $4a + 2[a + 3(a - 1)]$
4. $5x^2 + x[x - 4(x - 2)]$
5. $6x^2 - x[3 - 5(3 - x)]$
6. $2x[5 - 4(x - 1)] + 8x^2$
7. $2[3(x - 4) - 5]$
8. $-x[2(5 - x) + 3]$
9. $3[x + 4(x + 1)] - 2[x - 3(x - 1)]$
10. $x[3x - 7(x - 4)] - [5x - 2(x - 2)]$

PART 3

Having found a method for determining the product of a monomial and a polynomial, our next move would seem to be to determine some way in which we might multiply any polynomial by another polynomial. To do this, let's examine the distributive principle of multiplication over addition where the monomial x is to be multiplied by the binomial $a + 2$.

$$x \cdot (a + 2) \tag{1}$$

From Part 1 of this section we know that (1) can be rewritten in the form,

$$x \cdot (a + 2) = x \cdot a + x \cdot 2 \qquad (2)$$

Since x is a variable that is holding the place for any quantity we so desire, it is possible for us to replace it with the binomial $(a + 5)$. In doing this, equation (2) becomes,

$$x \cdot (a + 2) \;=\; x \cdot a \;+\; x \cdot 2 \qquad (2)$$
$$(a + 5) \cdot (a + 2) \;=\; (a + 5) \cdot a \;+\; (a + 5) \cdot 2 \qquad (3)$$

Step (3) shows that wherever x appeared earlier, the binomial $(a + 5)$ now appears. By examining (3) we find that the left side of the equation shows a binomial $(a + 5)$ being multiplied by a binomial $(a + 2)$. The right side, on the other hand, tells us that this can be done by multiplying the binomial $(a + 5)$ by *each* term of the binomial $(a + 2)$. That is, first multiply $(a + 5)$ by the a and then multiply $(a + 5)$ by the 2. And as a last step the *sum* of the two products is found.

$$(a + 5)(a + 2) \;=\; (a + 5) \cdot a \;+\; (a + 5) \cdot 2 \qquad (4)$$
$$= a^2 + 5a + 2a + 10$$
$$= a^2 + 7a + 10$$

To find the product above we usually prefer to arrange the binomials vertically.

$$\begin{array}{r} a + 5 \\ \underline{a + 2} \end{array} \qquad (5)$$

We now multiply $(a + 5)$ by the a first and then by the 2 as is shown below.

$$\begin{array}{rl} & a \;\; + 5 \\ & \underline{a \;\; + 2} \\ \text{Line 1} \longrightarrow & a^2 + 5a \\ \text{Line 2} \longrightarrow & \underline{ + 2a + 10} \\ \text{Line 3} \longrightarrow & a^2 + 7a + 10 \end{array}$$

Notice that line 1 shows the product of $(a + 5)$ with a; line 2 shows the product of $(a + 5)$ with 2; and finally, line 3 shows the *sum* of these two products.

Example 1

Determine the product in the exercise below.

$$(3x - 4)(2x + 5)$$

Solution

$$
\begin{array}{r}
3x - 4 \\
2x + 5 \\
\hline
\end{array}
$$

Line 1 → $6x^2 - 8x$
Line 2 → $ + 15x - 20$
Line 3 → $6x^2 + 7x - 20$

Explanation Line 1 is determined by multiplying the binomial $(3x - 4)$ by $2x$. Line 2 is determined by multiplying $(3x - 4)$ by $+5$. Line 3 is the sum of line 1 and line 2.

Example 2

Determine the product in the exercise below.

$$(3x^2 - 5x - 4)(2x - 3)$$

Solution

$$
\begin{array}{r}
3x^2 - 5x - 4 \\
2x - 3 \\
\hline
\end{array}
$$

Line 1 → $6x^3 - 10x^2 - 8x$
Line 2 → $ - 9x^2 + 15x + 12$
Line 3 → $6x^3 - 19x^2 + 7x + 12$

Explanation Line 1 is the product of the trinomial $(3x^2 - 5x - 4)$ with $2x$. Line 2 is the product of $(3x^2 - 5x - 4)$ with -3. Line 3 is the sum of line 1 and line 2.

Example 3

Determine the product in the exercise below.

$$(a - b - c)^2$$

Solution

$$
\begin{array}{r}
(a - b - c)^2 \\
a - b - c \\
a - b - c \\
\hline
\end{array}
$$

Line 1 → $a^2 - ab - ac$
Line 2 → $ - ab + b^2 + bc$
Line 3 → $ - ac + bc + c^2$
Line 4 → $a^2 - 2ab - 2ac + b^2 + 2bc + c^2$

Explanation Notice that the terms in line 2 and line 3 are arranged with blank spaces between some of the terms. This came about because of the necessity of placing like terms in the same column. Where there is no term similar to a term such as b^2 in line 2, that term has to be placed by itself at the far right.

EXERCISES 3.3 (Part 3)

A

Find the product in each of the following exercises.

1. $3x + 2$
   $4x + 3$

2. $5x + 7$
   $3x + 1$

3. $3a - 4$
   $2a + 3$

4. $7a + 5$
   $4a - 2$

5. $3y - 5$
   $4y - 6$

6. $-2x^2 + 4$
   $3x^2 + 2$

7. $-5x^3 - 8$
   $-2x^3 - 3$

8. $y^2 + 3$
   $y^2 - 4$

9. $-a^3 + 7$
   $a^3 - 5$

10. $2b^2 - 3$
    $b^2 - 1$

11. $4x + 2y$
    $5x + 4y$

12. $3x + y$
    $x + 2y$

13. $2x - y$
    $3x + y$

14. $5x - 2y$
    $3x - 5y$

15. $2x + 3y$
    $2x - 3y$

16. $a - b$
    $a + b$

17. $3a^2 - 5b$
    $5a^2 - 3b$

18. $x - y^2$
    $x - y^2$

19. $3a^2 + 4$
    $2a - 3$

20. $a + b$
    $c - d$

B

Find the product in each of the following exercises.

1. $(2x + 3)(3x + 5)$
2. $(3x + 4)(2x - 1)$
3. $(2x^2 + 3y)(5x^2 - 2y)$
4. $(3a - 5b^2)(3a + 5b^2)$

3.4/Special Products

5. $(2a + 3)^2$
6. $(4x - 5)^2$
7. $(xy^2 - 4)(2xy^2 + 5)$
8. $(3a^2 - 2bc)(2a^2 - 3bc)$
9. $(a + b)^2$
10. $(a - b)^2$

C

Find the product in each of the following exercises.

1. $2x^2 + 3x + 2$
   $2x + 5$

2. $3a^2 - 4a + 2$
   $3a - 5$

3. $x^2 + xy + y^2$
   $x + y$

4. $4a^2 + 6a + 9$
   $2a - 3$

5. $x^2 + xy + y^2$
   $x - y$

6. $x^2 + 2x + 3$
   $3x^2 + 2x + 1$

7. $5a^2 - 3a - 4$
   $2a^2 + 3a + 2$

8. $2a^2 + ab + 3b^2$
   $a^2 + ab + b^2$

9. $a^2 + 2ab + b^2$
   $a^2 - 2ab + b^2$

10. $x^3 - 3x^2 + 3x - 1$
    $x^2 + 2x - 1$

D

Find the product in each of the following exercises.

1. $(2x^2 + 5x - 4)(x + 3)$
2. $(-3x^2 - 2x + 1)(-x - 2)$
3. $(2a + b + c)^2$
4. $(5a^2 - 2ab - b^2)(2a - b)$
5. $(a + 2)^3$
6. $(2x - 3y)^3$

3.4 SPECIAL PRODUCTS

PART 1

Finding the product of two binomials where the terms in the first binomial are similar to the terms in the second binomial occurs so often that a method has been developed to shorten the computation. For instance, in the two binomials below,

$$(2a + 3)(5a + 4), \qquad (1)$$

the first term in the first binomial and the first term in the second binomial are like terms; they are 2a and 5a respectively. Similarly, the second term, 3, in the first binomial and the second term, 4, in the second binomial are also like terms. When this occurs we prefer to find the product in (1) without rewriting the binomials below each other.

To discover what this method is, let us first find the product of these two binomials in the manner with which we are familiar.

$$
\begin{array}{r}
2a + 3 \\
5a + 4 \\
\hline
\text{Line 1} \longrightarrow \quad 10a^2 + 15a \\
\text{Line 2} \longrightarrow \quad + 8a + 12 \\
\hline
10a^2 + 23a + 12 \longleftarrow \text{Product}
\end{array}
\qquad (2)
$$

Our objective now is to try to discover some way we can write the product immediately without having to write line 1 and line 2 in the process. Finding the first term in the product, the $10a^2$, is rather simple for it is merely the product of $2a$ and $5a$ which are the *first terms* in the two binomials.

$$\text{First Term} \times \text{First Term} = \text{First Term}$$

$$(2a + \underline{}) (5a + \underline{}) = 10a^2 + \underline{} + \underline{}$$

Similarly, finding the last term, the +12, is also relatively simple, for it is the product of the +3 and the +4, which are the *last terms* in the two binomials.

$$\text{Last Term} \times \text{Last Term} = \text{Last Term}$$

$$(\underline{} + 3) (\underline{} + 4) = \underline{} + \underline{} + 12$$

Finding the middle term, though, involves a bit more difficulty.

$$
\begin{array}{r}
2a + 3 \\
5a + 4 \\
\hline
\text{Line 1} \longrightarrow \underline{} + 15a
\end{array}
\qquad (\underline{} + 3) (5a + \underline{}) \qquad (1)
$$

Notice that in line 1 above the $+15a$ is found by multiplying the +3 by the $5a$. When we examine these terms in (1) we notice that they are the terms that one might call the *inner* terms in the two binomials for they appear as the two "inner" terms in the lineup of the four terms in (1).

$$
\begin{array}{r}
2a + 3 \\
5a + 4 \\
\hline
\text{Line 1} \longrightarrow \underline{} + \underline{} \\
\text{Line 2} \longrightarrow \underline{} + 8a \underline{}
\end{array}
\qquad (2a + \underline{}) (\underline{} + 4) \qquad (2)
$$
$$+8a$$

3.4/Special Products

Now notice that the +8a in line 2 is found by multiplying the 2a by the +4. By examining the terms in (2) we see that the 2a and the +4 are the "outer" terms of the two binomials. Finally, the +23a is found by adding the two terms, +15a and +8a.

$$
\begin{array}{r}
2a + 3 \\
5a + 4 \\
\hline
\underline{} + 15a \\
+ 8a + \underline{} \\
\hline
\underline{} + 23a + \underline{}
\end{array}
$$

$(2a + 3)(5a + 4)$ with inners giving +15a and outers giving +8a

Middle Term
Inners: +15a
Outers: +8a
+23a

Example 1

Determine the product in the exercise below.

$$(3a + 7)(a + 2)$$

Explanation

First Term times First Term = First Term

$(3a + \underline{})(a + \underline{}) = 3a^2 + \underline{} + \underline{}$

Last Term times Last Term = Last Term

$(\underline{} + 7)(\underline{} + 2) = \underline{} + \underline{} + 14$

$(3a + 7)(a + 2) = \underline{} + 13a + \underline{}$

with inners +7a and outers +6a

Middle Term
+7a
+6a
+13a

Solution $(3a + 7)(a + 2)$
$= 3a^2 + 13a + 14$

Middle Term
Inners: +7a
Outers: +6a
+13a

As we practice determining the product of two binomials we will discover that the three steps involved in finding the middle term can all be done mentally.

At this stage of our work it would be best if we write down those three steps as is done at the right in the solution above.

Since this method for finding the product of two binomials involves finding the product of the

        First Terms   Outer Terms   Inner Terms   Last Terms,

it is frequently referred to as the **FOIL** method.

Example 2

Use the FOIL method to determine the product below.

$$(5x - 4y)(3x + 2y)$$

Explanation This exercise is in no way different from the one in Example 1 except that greater care must be taken when finding the sign of direction for each of the terms in the product.

Solution

$$(5x - 4y)(3x + 2y)$$
$$= 15x^2 - 2xy - 8y^2$$

Middle Term

Inners: $-12xy$
Outers: $+10xy$
$-2xy$

EXERCISES 3.4 (Part 1)

Use the FOIL method to find the product in each of the following exercises.

1. $(2x + 3)(3x + 2)$
2. $(5x + 4)(2x + 3)$
3. $(4x - 1)(3x + 4)$
4. $(5x - 2)(2x + 5)$
5. $(7x + 3)(2x - 1)$
6. $(3x + 5)(5x - 2)$
7. $(2x - 3)(4x - 5)$
8. $(7x - 2)(5x - 4)$
9. $(x + 2)(x + 3)$
10. $(x + 5)(x + 4)$
11. $(x + 7)(x - 9)$
12. $(x^2 + 10)(x^2 - 4)$
13. $(x - 6)(x - 5)$
14. $(x^3 - 8)(x^3 - 2)$
15. $(3x + 2y)(5x + 3y)$
16. $(2x - 3y)(3x - 5y)$
17. $(5x - 2y^2)(4x - y^2)$
18. $(6xy - 5)(2xy + 1)$

3.4/Special Products

19. $(3ab - 2)(ab - 3)$
20. $(4 + x)(5 + x)$
21. $(1 + x)(9 + x)$
22. $(3 - x)(5 - x)$
23. $(4 - 3x^2)(2 + 5x^2)$
24. $(7 + ab)(3 - ab)$
25. $(6 - 5xy)(4 - xy)$
26. $(3a^2 - 4)(5a^2 - 2)$
27. $(2b^2 + 5)(3b^2 - 5)$
28. $(3 - 2a^2)(2 - 3a^2)$
29. $(4 - a^2)(5 + 2a^2)$
30. $(5x^2 + 2y^2)(3x^2 + y^2)$
31. $(2x^2 - y^2)(x^2 - y^2)$
32. $(x^2 - 5y^2)(x^2 + y^2)$
33. $(4ab + 3c)(5ab + c)$
34. $(2a - 3bc)(3a + 4bc)$
35. $(xy - z)(xy - 2z)$
36. $(3x^2y + 7w)(2x^2y + 5w)$
37. $(2xy^2 - 3w)(xy^2 + 3w)$
38. $(-2x + 3y)(3x - 5y)$
39. $(-4x - 3y)(-2x + 5y)$
40. $(-5x - y)(-2x - 3y)$

PART 2

Certain rather interesting things occur when finding the product of two binomials if those binomials are of a special nature. To illustrate, consider the situation below where the product of these binomials is found by the FOIL method.

$(x + 5)(x - 5)$ Middle Term   (1)
$= x^2 - 25$ Inners: $+5x$
    Outers: $\underline{-5x}$
    0

When determining the middle term we find that it turns out to be 0 and when this number is added to $x^2 - 25$ it leaves $x^2 - 25$ unchanged. Hence, the 0 need not be written as the middle term. Rather than leave the large gap between the x^2 and the -25, we write these terms close to each other.

The question now arises as to what are the conditions under which we will automatically write the last term next to the first term knowing that there will be no middle term. By examining the computation for the middle term at the right above we realize that the middle term will be 0 only if the product of the inner terms is the additive inverse of the product of the outer terms.

We now look at the binomials in (1) to see what led to this odd situation where the product of the inners is the additive inverse of the product of the outers. Notice that the binomials are identical except for the fact that *the second term of the first binomial is the additive inverse of the second term of the second binomial.* Thus, it appears that the identical nature of the two binomials except for this one feature leads to the condition where the product of the two binomials has no middle term.

Example 1

Find the product of the two binomials below.

$$(4x^2y^3 - 7)(4x^2y^3 + 7)$$

Explanation The second term of the first binomial is the additive inverse of the second term of the second binomial, hence the product of these two binomials will have no middle term. We find the first and last terms only and ignore the middle term as it is 0.

Solution
$$(4x^2y^3 - 7)(4x^2y^3 + 7)$$
$$= 16x^4y^6 - 49$$

It is also true that if the binomials are *identical* except for the fact that the *first* term of the first binomial is the additive inverse of the *first* term of the second binomial, then again the product of the two binomials will have no middle term.

There is but one more special situation that is of interest when finding the product of two binomials. There is quite a simple way by which it is possible to *square a binomial*. To determine just what that method is, we will square the binomial $a + b$ and then examine the product.

$$(a+b)^2 = (a+b)(a+b) = a^2 + 2ab + b^2$$

Middle Term

Inners: $+ab$
Outers: $+ab$
$\overline{+2ab}$

Notice that the first term in the product is the square of the first term in the binomial.

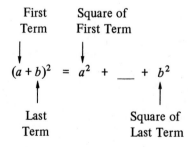

Second, the last term in the product is the square of the last term in the binomial.

Finally, if we were to multiply the first term by the last term and then double that answer, we would have the middle term of the product.

3.4/Special Products

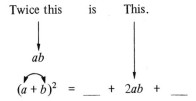

Example 2
Determine the product in the exercise below.

$$(2x - 7y)^2$$

Explanation

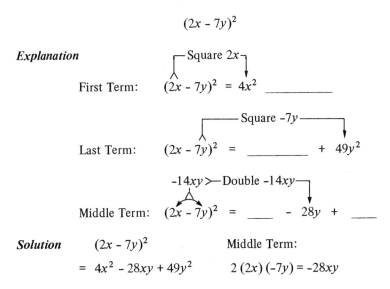

Solution $(2x - 7y)^2$
$= 4x^2 - 28xy + 49y^2$

Middle Term:
$2(2x)(-7y) = -28xy$

Example 3
Simplify and collect like terms in the expression below.

$$(5x - 2)^2 + 2(x - 4)(x + 4)$$

Explanation Notice in step (2) below that the product of $(x - 4)$ and $(x + 4)$ is $x^2 - 16$. This has yet to be multiplied by $+2$, therefore it is written as $+2(x^2 - 16)$. Once this product is found, we can collect like terms as we have in the past.

Solution $(5x - 2)^2 + 2(x - 4)(x + 4)$
$= 25x^2 - 20x + 4 + 2(x^2 - 16)$
$= 25x^2 - 20x + 4 + 2x^2 - 32$
$= 27x^2 - 20x - 28$

EXERCISES 3.4 (Part 2)

A

Find the product in each of the following exercises. Examine each exercise carefully for there will be some in which the middle term is *not* zero.

1. $(2x + 3)(2x - 3)$
2. $(5x - 4)(5x + 4)$
3. $(x - 7)(x + 7)$
4. $(x + 5)(x - 5)$
5. $(2x - 1)(2x + 1)$
6. $(3x^2 - 1)(3x^2 + 1)$
7. $(2x + 1)(2x + 1)$
8. $(5x - 2y)(5x + 2y)$
9. $(x + 3y)(x - 3y)$
10. $(9 - 4x)(9 + 4x)$
11. $(8 + x)(8 - x)$
12. $(12 - x)(12 + x)$
13. $(x^3 + 1)(x^3 - 1)$
14. $(2x + \frac{1}{3})(2x - \frac{1}{3})$
15. $(5x - \frac{2}{7})(5x + \frac{2}{7})$
16. $(7 - 2xy)(7 + 2xy)$
17. $(3x^2 - 2)(3x^2 - 2)$
18. $(\frac{2}{3}x + 5)(\frac{2}{3}x - 5)$
19. $(\frac{1}{2}x - \frac{1}{3}y)(\frac{1}{2}x + \frac{1}{3}y)$
20. $(3x^2 + 2y^2)(3x^2 - 2y^2)$
21. $(4xy - 3w)(4xy + 3w)$
22. $(x^2y^3 - 1)(x^2y^3 + 1)$
23. $(-2x + 5)(-2x - 5)$
24. $(-3x - 4)(-3x + 4)$
25. $(-5x - 3y)(-5x + 3y)$
26. $(2x + 3y)(2x - 5y)$
27. $(4 + 3x)(4 - 3x)$
28. $(3x + 4)(-3x + 4)$
29. $(-5x - 7)(5x + 7)$
30. $(6ab - 5)(-6ab - 5)$

B

Find the product in each of the following exercises.

1. $(2a + 3)^2$
2. $(3a + 4)^2$
3. $(a - 3)^2$
4. $(2a - 5)^2$
5. $(4 - 3a)^2$
6. $(7 - 2a)^2$
7. $(6 - a)^2$
8. $(a + b)^2$
9. $(a - b)^2$
10. $(ab - 1)^2$
11. $(3 + ab)^2$
12. $(6 - ab)^2$
13. $(a^2 - 3)^2$
14. $(5a^2 - 2b)^2$
15. $(4 + ab^2)^2$
16. $(2 - 3a^2b)^2$
17. $(\frac{1}{2}a + 4)^2$
18. $(6a - \frac{1}{2})^2$
19. $(\frac{1}{3}a^3 - 6)^2$
20. $(\frac{1}{3} - 3a^3)^2$

3.5/Division of Real Numbers

C

Simplify and collect like terms in each of the exercises below.

1. $(x + 3)^2 + 5x^2$
2. $(2x + 3)(2x - 3) + 10$
3. $(5x - 1)^2 + 3x - 4$
4. $(x - 2)^2 + 3(x - 5)$
5. $(5x - 3)(5x + 3) - 2(3x - 4)$
6. $7(x - 4) + (2x + 1)^2$
7. $(x + 4)(x + 5) + (3x - 2)^2$
8. $(2x - 1)(2x + 5) + (3x - 4)(3x + 4)$
9. $5x + 2(x - 3)(x + 3)$
10. $4x^2 - 3(2x - 1)(2x + 1)$
11. $3x - 5(3x + 1)(3x - 1)$
12. $9x^2 - 2(x - 7)(x + 7)$
13. $4x^2 - (2x - 5)(2x + 5)$
14. $(5x + 1)^2 + (3x + 1)(3x - 2)$
15. $-(3x - 4)^2 + 3(x + 1)(x + 3)$
16. $-(4x - 1)^2 - (x - 3)(x - 5)$

3.5 DIVISION OF REAL NUMBERS

Earlier, when trying to decide in what manner we should perform the operation of subtraction, we fell back on the scheme of relating the operation of subtraction to the operation of addition. Now that we are trying to develop a few principles that will enable us to perform the operation of division as it pertains to real numbers we find that it will be helpful to relate this operation to the operation of multiplication.

As an illustration, in considering the situation in which 12 is divided by 2, we are in reality asking ourselves, what number when multiplied by 2 will give

Dividend   Divisor   Quotient
↓   ↓   ↓
12   ÷   2   =   ?

a product of 12. From our knowledge of multiplication we know that this number must be 6. Hence, we say that the quotient of 12 and 2 is 6.

The four situations below represent all the possible combinations of signs that can be attached to the numbers 12 and 2. Using the same approach as above we can find the quotient for each of these exercises.

$$(+12) \div (+2) = ? \qquad (1)$$
$$(+12) \div (-2) = ? \qquad (2)$$
$$(-12) \div (+2) = ? \qquad (3)$$
$$(-12) \div (-2) = ? \qquad (4)$$

Thus, in case (1) we know that the number to be multiplied by +2 to give +12 as an answer has to be +6. Hence, +6 is the quotient of +12 and +2.

$$(+12) \div (+2) = +6 \qquad (1)$$

Similarly, in case (2) the number to be multiplied by -2 to give +12 as an answer has to be -6. Hence, -6 is the quotient of +12 and -2.

$$(+12) \div (-2) = -6 \qquad (2)$$

The quotients for all four cases are shown below.

$$(+12) \div (+2) = +6 \qquad (1)$$
$$(+12) \div (-2) = -6 \qquad (2)$$
$$(-12) \div (+2) = -6 \qquad (3)$$
$$(-12) \div (-2) = +6 \qquad (4)$$

If we ignore the sign, it is easy to see that the number 6 in each answer can be found by simply dividing the 12 by the 2. Also, close examination reveals the fact that the *sign* of each answer can be determined in exactly the same manner as it is in multiplication. In cases (1) and (4) where the signs are the *same*—both are positive in (1) while both are negative in (4)—the sign of the quotient is positive. In cases (2) and (3) where the signs are *not the same*—one is positive while the other is negative—the sign of the quotient is negative.

Example 1

Determine the quotient in the exercise below.

$$(-24) \div (+4)$$

Explanation Since the signs of direction of the dividend and divisor are not the same, the sign of direction of the quotient will be negative.

Solution $\qquad\qquad (-24) \div (+4) = -6$

Example 2

Determine the quotient in the exercise below.

$$\frac{-45}{-9}$$

Explanation The only difference between Examples 1 and 2 is that in the first, the dividend and divisor are written horizontally on a single line while in this example, the division is shown in fraction form.

Solution $\qquad \dfrac{-45}{-9} = +5$

Example 3

Determine the quotient in the exercise below.

$$\frac{-8}{+12}$$

Explanation Since the signs of direction are not the same, the sign of the quotient will be negative. Rather than divide 8 by 12, we simply reduce the fraction $\frac{8}{12}$ to lowest terms and leave the quotient in that form.

$$\frac{-8}{+12} = -\frac{2}{3}$$

EXERCISES 3.5

A

Find the quotient in each of the following exercises.

1. $(+6) \div (+3)$
2. $(+10) \div (+2)$
3. $(-18) \div (-9)$
4. $(-12) \div (-4)$
5. $(+24) \div (-6)$
6. $(+5) \div (-5)$
7. $(-28) \div (+7)$
8. $(-36) \div (+4)$
9. $(+42) \div (-7)$
10. $(+33) \div (-11)$
11. $(24) \div (+12)$
12. $(-32) \div (8)$
13. $(-12) \div (12)$
14. $(-42) \div (6)$
15. $(-54) \div (-9)$
16. $(-72) \div (12)$
17. $(-64) \div (+8)$
18. $(60) \div (-6)$
19. $(0) \div (+5)$
20. $(0) \div (-7)$

B

Find the quotient in each of the following exercises.

1. $\dfrac{+10}{+2}$
2. $\dfrac{-12}{+4}$
3. $\dfrac{+16}{-8}$
4. $\dfrac{-20}{+20}$
5. $\dfrac{24}{-6}$
6. $\dfrac{-30}{5}$
7. $\dfrac{+32}{8}$
8. $\dfrac{36}{-9}$
9. $\dfrac{-40}{+10}$
10. $\dfrac{0}{+3}$
11. $\dfrac{+4}{+8}$
12. $\dfrac{-6}{+12}$
13. $\dfrac{+3}{-9}$
14. $\dfrac{-4}{-16}$
15. $\dfrac{0}{-5}$
16. $\dfrac{+6}{+8}$
17. $\dfrac{-6}{+10}$
18. $\dfrac{-9}{-15}$
19. $\dfrac{+5}{-20}$
20. $\dfrac{12}{-15}$
21. $\dfrac{-8}{-10}$
22. $\dfrac{+15}{+10}$
23. $\dfrac{+9}{+1}$
24. $\dfrac{10}{-1}$

3.6 DIVISION OF MONOMIALS

Following the same path we pursued in each of the previous operations, our next move will be to determine a method for finding the quotient of two monomials. To do this, we first consider the situation in which a number is to be divided by itself such as

$$9 \div 9. \qquad (1)$$

From our knowledge of arithmetic, we know that this quotient is 1. That is,

$$9 \div 9 = 1 \qquad (2)$$

Also, from our knowledge of arithmetic, we know that $9 \div 9$ can be written in the form of $9 \times \frac{1}{9}$, and hence we can say that the product of these two numbers is 1.

$$9 \div 9 = 9 \times \frac{1}{9} = 1 \qquad (3)$$

3.6/Division of Monomials

The number $\frac{1}{9}$ is said to be the *multiplicative inverse* of 9 and, in general, the number $1/x$ is said to be the multiplicative inverse of x. The distinguishing feature of any number and its multiplicative inverse is the fact that if a number is multiplied by its multiplicative inverse, the product of the two numbers is 1.

Several examples of this are, if 23 is multiplied by $\frac{1}{23}$, the product is 1; if x is multiplied by $1/x$, the product is 1; if x^2 is multiplied by $1/x^2$, the product is 1; and finally, if $-2x^2y$ is multiplied by $1/-2x^2y$, the product is 1.

Returning again to (3) above, we stated there that $9 \div 9$ is exactly the same as $9 \times \frac{1}{9}$. Actually, this interpretation of division

10 divided by 2 means the same as 10 multiplied by $\frac{1}{2}$

is the meaning that mathematicians give to the operation of division. Notice that the operation has been changed from division to multiplication and that the 2 has been replaced with its multiplicative inverse, the number $\frac{1}{2}$.

In general, this principle is expressed with symbols as

$$a \div b \text{ is the same as } a \cdot \frac{1}{b}$$

With this background, we are now in a position to find the quotient of two monomials where the variables are the same but the exponents are different. Consider

$$x^5 \div x^2. \qquad (4)$$

This can be rewritten as

$$x^5 \cdot \frac{1}{x^2} \qquad (5)$$

However, from our knowledge of multiplication, x^5 is the same as $x^3 \cdot x^2$. Therefore, we can rewrite (5) as

$$x^3 \cdot x^2 \cdot \frac{1}{x^2} \qquad (6)$$

We just learned, though, that the product of any number and its multiplicative inverse is 1. Since $x^2 \cdot (1/x^2)$ is equal to 1, (6) becomes

$$x^3 \cdot 1. \qquad (7)$$

But any number multiplied by 1 gives the number itself; therefore the product of x^3 and 1 is x^3. Hence, (7) becomes

$$x^3. \qquad (8)$$

Thus, the division situation in (4) led to the quotient shown in (8). When these are placed side by side,

$$x^5 \div x^2 = x^3, \qquad (9)$$

we can see that the variable in the quotient is exactly the same as the variable in both the dividend and divisor; it is x in all of them. The exponent 3 in the quotient can be obtained from the exponents in the divisor and dividend by subtracting the 2 from the 5.

Same Variable "x"

Example 1

Determine the quotient in the exercise below.

$$y^7 \div y^3$$

Explanation The exponent of the y in the quotient is found by subtracting 3 from 7.

Solution $\qquad y^7 \div y^3 = y^4$

Example 2

Determine the quotient in the exercise below.

$$(-24x^9) \div (-6x^7)$$

Explanation It might be well to say to yourself that you are looking for *four* different things when determining the quotient of two monomials:

1. The sign of direction,
2. The coefficient,
3. The variable,
4. The exponent.

In this situation, the sign is positive for the two signs of direction are the same. The coefficient is 4 because $24 \div 6$ is 4. The variable is x and the exponent will be 2 since $9 - 7 = 2$.

3.6/Division of Monomials

Solution $(-24x^9) \div (-6x^7) = +4x^2$

Example 3

Determine the quotient in the exercise below.

$$\frac{-30x^5}{+2x^5}$$

Explanation Following the procedure in Example 2, we find the coefficient to be -15. Now, though, since the exponents of the variables are the same we consider this situation one in which x^5 is being divided by x^5. We discovered earlier that this quotient must be 1. Multiplying this 1 by the -15 gives a product of -15. Hence, usually the 1 does not appear in the solution as we have shown below. Rather than this, the final answer is determined immediately.

Solution
$$\frac{-30x^5}{+2x^5}$$

$$= -15 \cdot 1$$

$$= -15$$

Example 4

Determine the quotient in the exercise below.

$$\frac{+32a^3b^2c^7}{-8ab^2}$$

Explanation Since the exponent for the *a* in the divisor is not written, we understand it to be 1. It is best to write it in the original exercise. Subtracting the 1 from the 3 the exponent for the *a* in the quotient will be 2. As in Example 3, when b^2 is divided by b^2 the quotient is 1. Since there is no variable *c* in the divisor, the c^7 in the dividend is not changed by the division.

Solution
$$\frac{+32a^3b^2c^7}{-8ab^2}$$

$$= \frac{+32a^3b^2c^7}{-8a^1b^2}$$

$$= -4a^2 \cdot 1 \cdot c^7$$

$$= -4a^2c^7$$

3/The Operations of Multiplication and Division

EXERCISES 3.6

A

Find the quotient in each of the following exercises.

1. $(x^9) \div (x^5)$
2. $(y^{10}) \div (y^2)$
3. $(x^6) \div (x)$
4. $(x^4) \div (x^4)$
5. $(y) \div (y)$
6. $(x^2y) \div (x^2y)$
7. $(3x^2) \div (3x^2)$
8. $(+12x^5) \div (+2x^3)$
9. $(+16x^7) \div (+8x^6)$
10. $(-10x^{10}) \div (-2x^2)$
11. $(-20x^8) \div (+4x)$
12. $(+24y^6) \div (-6y)$
13. $(-8x^5) \div (-4)$
14. $(+18y) \div (-9)$
15. $(-15y^3) \div (y^2)$
16. $(-30x^5y^3) \div (-5x^2y^2)$
17. $(+32x^4y^7) \div (-4x^3y)$
18. $(-16x^3y^5) \div (+2x^3y^2)$
19. $(-36x^9y^2) \div (-4y)$
20. $(39xy) \div (-3x)$
21. $(+x^2yz^2) \div (x^2z)$
22. $(-48xyz) \div (xy)$
23. $(-12xy) \div (-1)$
24. $(-2x^3) \div (-4x^2)$
25. $(-3x^5) \div (-12x^5)$
26. $(0) \div (-3x^2)$
27. $(-4x^2y) \div (-4x^2y)$
28. $(-6xy^2) \div (6xy^2)$
29. $(-15x^3y^3) \div (-15x^3y^2)$
30. $(+50x^2yz) \div (-10x^2z)$

B

Find the quotient in each of the following exercises.

1. $\dfrac{14x^5}{7x^3}$
2. $\dfrac{-24y^3}{-12y^3}$
3. $\dfrac{+18x^7}{-6x^5}$
4. $\dfrac{15x^2y}{-5x^2}$
5. $\dfrac{-16x^3y^5}{-8y^4}$
6. $\dfrac{-xyz}{+xyz}$
7. $\dfrac{0}{-7y^2}$
8. $\dfrac{36x^3y^2w}{-9yw}$
9. $\dfrac{-28x^2y^3}{7x^2y^3}$
10. $\dfrac{-8x^5}{-16}$
11. $\dfrac{+3x^2y}{-1}$
12. $\dfrac{9x^4y^3}{-12xy}$

3.7/Division of a Polynomial by a Monomial

13. $\dfrac{xyz^2}{-xyz^2}$

14. $\dfrac{-9x^7yz^3}{-9y}$

15. $\dfrac{-6xy^3z}{6y^2z}$

16. $\dfrac{-5wz^2}{20wz^2}$

3.7 DIVISION OF A POLYNOMIAL BY A MONOMIAL

Based on our understanding of division the expression

$$a \div b \quad \text{means} \quad a \cdot \dfrac{1}{b}. \tag{1}$$

Similarly, should we encounter an expression such as

$$x \cdot \dfrac{1}{y} \tag{2}$$

we would know that this means

$$x \div y \tag{3}$$

Thus, an expression such as $x^5 \cdot 1/x^2$ can be interpreted as $x^5 \div x^2$, which, after division, gives a quotient of x^3.

Using these few pieces of information we can say that,

$$(b^4 + b^3 + b^2) \div b = (b^4 + b^3 + b^2) \cdot \dfrac{1}{b} \tag{4}$$

But the expression at the right indicates that a polynomial is to be multiplied by a monomial. When this operation is performed, the right side of (4) becomes,

$$b^4 \cdot \dfrac{1}{b} \quad + \quad b^3 \cdot \dfrac{1}{b} \quad + \quad b^2 \cdot \dfrac{1}{b} \tag{5}$$

However, from what we just discovered, (5) can be expressed as,

$$b^4 \div b \quad + \quad b^3 \div b \quad + \quad b^2 \div b \tag{6}$$

By placing (6) directly beside the expression in (4) from which it originally came we immediately spot the fact that on the left we see a polynomial being

$$(b^4 + b^3 + b^2) \div b = b^4 \div b + b^3 \div b + b^2 \div b \tag{7}$$

divided by b while the quantity equal to it on the right shows that *each term of the polynomial is being divided by b.*

Hence, the following principle seems to apply.

Principle of Dividing a Polynomial by a Monomial

When dividing a polynomial by a monomial, the quotient is found by dividing each term of the polynomial by the monomial.

Example 1

Determine the quotient in the following exercise.

$$\frac{-12x^4 + 24x^3}{-4x^2}$$

Explanation Since each of the terms of the dividend is divided by $-4x^2$, there will be two terms in the quotient. The first of these is found by dividing $-12x^4$ by $-4x^2$, the second by dividing $24x^3$ by $-4x^2$.

Solution $\quad\dfrac{-12x^4 + 24x^3}{-4x^2}$

$= +3x^2 - 6x$

Example 2

Determine the quotient in the following exercise.

$$(+16x^5y^4 - 10x^4y^3 - 2x^2y^2) \div (-2x^2y^2)$$

Solution $\quad (+16x^5y^4 - 10x^4y^3 - 2x^2y^2) \div (-2x^2y^2)$

$= -8x^3y^2 + 5x^2y + 1$

Explanation When the third term of the trinomial, the $-2x^2y^2$, is divided by the monomial, $-2x^2y^2$, the quotient is 1. That number *must* appear in the answer. If it does not, there will be only two terms in the quotient rather than three as there should be. *It is important that for each term in the dividend there be a corresponding term in the quotient.* There are three terms in the dividend and hence, there must be three terms in the quotient.

Based on the fact that the answer to a division exercise is checked by performing the operation of multiplication, can you state why the dividend and the quotient must have the same number of terms?

3.7/Division of a Polynomial by a Monomial

EXERCISES 3.7

A

Find the quotient in each of the following exercises.

1. $(x^9 + x^7) \div (x^5)$
2. $(x^4 - x^3) \div (-x^2)$
3. $(2x^5 + 6x^3) \div (2)$
4. $(15x^5 - 20x^4) \div (-5x^2)$
5. $(7x^2 - 14x^3) \div (7x)$
6. $(-8x^3 + 4x^2) \div (4x^2)$
7. $(3x^3 + 15x^5) \div (-3x^3)$
8. $(x^2 - xy) \div (x)$
9. $(10x^2y^3 + 15x^3y^4) \div (5x^2y^2)$
10. $(x^2y - xy^2) \div (-xy)$
11. $(8x^5 - 12x^4 - 24x^3) \div (-4x^2)$
12. $(-18y^2 + 30y^3 - 12y^4) \div (-6y)$
13. $(5x - 10x^3 - 15x^5) \div (-5x)$
14. $(6x^2y^3 - 2xy^2 + 10x^3y^5) \div (-2xy^2)$
15. $(7x^2y^3z^4 - 9xy^2z^3 - 11yz^2) \div (yz^2)$
16. $(ax + ay - az + aw) \div (-a)$
17. $(-x + y - z) \div (-1)$
18. $(ax + bx) \div (x)$
19. $[a(c+d) + b(c+d)] \div (c+d)$
20. $[3(x+y)^3 + 5(x+y)^4] \div (x+y)^2$

B

Find the quotient in each of the following exercises.

1. $\dfrac{x^3 - x^2}{x}$
2. $\dfrac{x^5 - x^3}{x^3}$
3. $\dfrac{x^8 - x^{10}}{-x^8}$
4. $\dfrac{8x^5 - 4x^3}{4x^2}$
5. $\dfrac{-6x^2y^3 + 15x^3y^5}{-3y^2}$
6. $\dfrac{-2xy + 8x^2y^2}{-2xy}$

7. $\dfrac{25x^2y^3z - 35x^3z^2}{+5x^2z}$

8. $\dfrac{4x^2y^2 - 6xy - 2}{-2}$

9. $\dfrac{ax + ay + a}{-a}$

10. $\dfrac{(a+b)^3 + (a+b)^4}{(a+b)^2}$

3.8 DIVISION OF POLYNOMIALS

We have now reached the final step in our examination of the operation of division. At this stage we must try to find some way to determine the quotient when a polynomial is divided by another polynomial. To do this, it will prove helpful to follow the steps we used in arithmetic when performing what was called "long division." As an example, let us divide 672 by 21.

$$\begin{array}{r} 32 \\ 21\overline{)672} \\ \underline{-63} \\ 42 \\ \underline{-42} \\ 0 \end{array} \qquad (1)$$

In order to determine the trial quotient of 3, we usually divide the 2 in the divisor into 6 in the dividend. However, these numbers are not 3, 2, and 6, for the 3 in 32 is really 30, while the 2 in 21 is really 20 and the 6 in 672 is really 600. In view of this, let's rewrite each of the numbers in their expanded notation form, that is, 21 as 20 + 1, 672 as 600 + 70 + 2, and 32 as 30 + 2.

$$\begin{array}{r} 30 + 2 \\ 20 + 1\overline{)600 + 70 + 2} \\ \underline{600 + 30} \\ 40 + 2 \\ \underline{40 + 2} \\ 0 \end{array} \qquad (2)$$

As in (1) above, the 30 in the quotient is found by dividing the 600 by 20. Referring to (1) we note that the 63 is found by *multiplying* the 3 by 21. In (2) the number 3 is actually shown in its correct form as 30, however the multiplication is the same as before, that is, the 600 + 30 is found by multiplying 30 by 20 + 1. The next step in (1) involves *subtraction* where 63 is *subtracted* from 67. In (2) this shows up as it should whereby 600 + 30 is subtracted from 600 + 70. Again referring to (1), we know that the 2 in 32 is found by dividing the 2 in the divisor, 21, into the 4 in the number 42. By looking at (2) we find that in reality we have divided the 20 in 20 + 1 into the 40 in 40 + 2 which gave us the quotient of +2. This is then followed by multiplying the 2 by 20 + 1 in the same manner as 2 is multiplied by 21 in (1). As a last step we subtract once again, obtaining a remainder of 0.

3.8/Division of Polynomials

In reality, (2) above shows a trinomial 600 + 70 + 2 being divided by a binomial 20 + 1. If we use the letter t as a replacement for the number 10, and rewrite 600 + 70 + 2 as $6 \times 10^2 + 7 \times 10 + 2$, then this, in turn, can be rewritten as $6t^2 + 7t + 2$. In the same way, 20 + 1 can be rewritten as $2 \times 10 + 1$, which finally can be expressed as $2t + 1$. Now let's examine (2) once again, where the trinomial and the binomial are written in their new form.

$$2t + 1 \overline{)\, 6t^2 + 7t + 2}$$

Let us outline the procedure in a step-by-step fashion, following the pattern that is used in (2).

1. Divide the 20 into the 600; this means, divide the $2t$ into $6t^2$ and place the quotient $3t$ over the $7t$ as the 30 had been placed.

$$\begin{array}{r} 3t \\ 2t + 1 \overline{)\, 6t^2 + 7t + 2} \end{array}$$

2. Multiply the 30 by the 20 + 1; this means, multiply the $3t$ by the $2t + 1$ and write the product of $6t^2 + 3t$ under the $6t^2 + 7t$ exactly as the 600 + 30 is written under the 600 + 70.

$$\begin{array}{r} 3t \\ 2t + 1 \overline{)\, 6t^2 + 7t + 2} \\ \underline{6t^2 + 3t } \end{array}$$

3. Subtract the 600 + 30 from 600 + 70; this means, subtract $6t^2 + 3t$ from $6t^2 + 7t$ and write the difference, $4t$, in the same position as the 40 is written.

$$\begin{array}{r} 3t \\ 2t + 1 \overline{)\, 6t^2 + 7t + 2} \\ \underline{6t^2 + 3t } \\ + 4t \end{array}$$

4. Carry down the 2 which is the third term in the trinomial.

$$\begin{array}{r} 3t \\ 2t + 1 \overline{)\, 6t^2 + 7t + 2} \\ \underline{6t^2 + 3t } \\ 4t + 2 \end{array}$$

5. Divide 20 into 40; this means, divide $2t$ into $4t$ and write the quotient $+2$ above the third term, 2, in the dividend.

$$\begin{array}{r} 3t + 2 \\ 2t + 1 \overline{\smash{)}\, 6t^2 + 7t + 2} \\ \underline{6t^2 + 3t} \\ + 4t + 2 \end{array}$$

6. Multiply the $+2$ by $20 + 2$; this means, multiply the $+2$ by $2t + 1$ and place the product of $4t + 2$ under the $4t + 2$ that is already there.

$$\begin{array}{r} 3t + 2 \\ 2t + 1 \overline{\smash{)}\, 6t^2 + 7t + 2} \\ \underline{6t^2 + 3t} \\ + 4t + 2 \\ + 4t + 2 \end{array}$$

7. Complete the process by subtracting $4t + 2$ from $4t + 2$, thus giving a remainder of 0.

$$\begin{array}{r} 3t + 2 \\ 2t + 1 \overline{\smash{)}\, 6t^2 + 7t + 2} \\ \underline{6t^2 + 3t} \\ + 4t + 2 \\ \underline{+ 4t + 2} \\ 0 \end{array}$$

It is important to realize that when the words subtract, multiply, and divide are used above they imply that these operations are to be applied as we have learned them in these past two chapters.

Example 1

Determine the quotient in the following exercise.

$$(15x^2 + 14x + 2) \div (3x + 4)$$

Explanation The very first step is to arrange the dividend and the divisor in the "long division" form. Each of the steps in the process will be shown in the same manner as in the explanation above.

1. $$\begin{array}{r} 5x \\ 3x + 4 \overline{\smash{)}\, 15x^2 + 14x + 2} \end{array}$$

2. $$\begin{array}{r} 5x \\ 3x + 4 \overline{\smash{)}\, 15x^2 + 14x + 2} \\ 15x^2 + 20x \end{array}$$

3.8/Division of Polynomials

$$
\begin{array}{r}
5x \\
3.\quad 3x+4\overline{\smash{)}15x^2+14x+2} \\
15x^2+20x \\
\hline
-6x+2
\end{array}
$$

$$
\begin{array}{r}
5x-2 \\
4.\quad 3x+4\overline{\smash{)}15x^2+14x+2} \\
15x^2+20x \\
\hline
-6x+2
\end{array}
$$

$$
\begin{array}{r}
5x-2 \\
5.\quad 3x+4\overline{\smash{)}15x^2+14x+2} \\
15x^2+20x \\
\hline
-6x+2 \\
-6x-8
\end{array}
$$

$$
\begin{array}{r}
5x-2 \\
6.\quad 3x+4\overline{\smash{)}15x^2+14x+2} \\
15x^2+20x \\
\hline
-6x+2 \\
-6x-8 \\
\hline
+10
\end{array}
$$

Step 1: $15x^2$ is divided by $3x$ to obtain the $5x$.
Step 2: $5x$ is multiplied by $3x + 4$ to obtain the $15x^2 + 20x$.
Step 3: $15x^2 + 20x$ is *subtracted* from $15x^2 + 14x$ to obtain $-6x$.
The $+2$ is brought down at the same time.
Step 4: $-6x$ is divided by $3x$ to obtain the -2.
Step 5: -2 is multiplied by $3x + 4$ to obtain $-6x - 8$.
Step 6: $-6x - 8$ is subtracted from $-6x + 2$ to obtain the remainder of $+10$.
Notice that in this illustration the remainder is *not* 0 as it had been in the previous illustration.

Solution
$$
\begin{array}{r}
5x-2 \\
3x+4\overline{\smash{)}15x^2+14x+2} \\
15x^2+20x \\
\hline
-6x+2 \\
-6x-8 \\
\hline
-10\ \text{(Remainder)}
\end{array}
$$

There are times when the terms in either the dividend or divisor do not follow some fixed pattern. Unless the terms in both the dividend and divisor are so arranged whereby the variable with the highest exponent appears first, the next highest appears second, and so on down the line, it becomes very difficult to perform the operation of "long division." Hence, the very first thing we must do is to make certain that the terms are arranged correctly before doing any division. If they are not so arranged then it is necessary for us to rearrange them in this order.

When terms are arranged whereby the variable with the greatest exponent appears first and so on down the line, we say that the polynomial is arranged in *descending powers* of the variable. Should there be two variables then we concentrate on one and make certain that the terms are arranged in descending powers of that particular variable.

Example 2

Determine the quotient in the following exercise.

$$(x^2 - 7 + 2x^3) \div (2x - 3)$$

Explanation The first step is to arrange the terms in descending powers of x. After writing the $+2x^3$ term and the $+x^2$ term, we notice that the x to the first power term is missing. In view of this we realize that we must leave a space for this term as a first power term in x will probably be introduced in the process of finding the quotient. This empty space will appear between the x^2 term and the -7 term.

Solution

$$\begin{array}{r} x^2 + 2x + 3 \\ 2x-3\overline{)\,2x^3 + x^2 \; -7} \\ \underline{2x^3 - 3x^2} \\ +4x^2 \\ \underline{+4x^2 - 6x} \\ +6x - 7 \\ \underline{+6x - 9} \\ +2 \;\;(\text{Remainder}) \end{array}$$

Example 3

Determine the quotient in the following exercise.

$$(x^3 - 25y^3) \div (x - 3y)$$

Explanation By looking at the dividend, $x^3 - 25y^3$, we ignore the y variable and notice that there is an x to the second power term missing and an x to the first power term missing. Hence, we leave space for these two missing terms.

Solution

$$\begin{array}{r} x^2 + 3xy + 9y^2 \\ x-3y\overline{)\,x^3 \; -25y^3} \\ \underline{x^3 - 3x^2y} \\ +3x^2y \\ \underline{+3x^2y - 9xy^2} \\ +9xy^2 - 25y^3 \\ \underline{+9xy^2 - 27y^3} \\ +2y^3 \;\;(\text{Remainder}) \end{array}$$

3.8/Division of Polynomials

The question arises as to when do we reach the point at which we know we must stop the process of dividing? This occurs when we attempt to divide the first term of the divisor into the remainder and find that it is not possible to do this. For instance, in the example above it is not possible to divide x into the $+2y^3$ (the remainder). Hence, the division must end at that point.

EXERCISES 3.8

A

Determine the quotient in each of the following exercises. The remainder will be 0 in each of these exercises.

1. $(x^2 + 5x + 6) \div (x + 2)$
2. $(x^2 + 14x + 49) \div (x + 7)$
3. $(x^2 - 7x + 12) \div (x - 4)$
4. $(x^2 - 14x + 48) \div (x - 6)$
5. $(x^2 + 2x - 35) \div (x + 7)$
6. $(x^2 - 7x - 8) \div (x - 8)$
7. $(6x^2 + 17x + 12) \div (2x + 3)$
8. $(12x^2 + 13x + 3) \div (3x + 1)$
9. $(3x^2 + 13x + 12) \div (3x + 4)$
10. $(2x^2 - 13x + 15) \div (x - 5)$
11. $(x^2 + 7xy + 10y^2) \div (x + 5y)$
12. $(x^2 + 3xy - 54y^2) \div (x + 9y)$
13. $(6x^2 + 13xy - 28y^2) \div (3x - 4y)$
14. $(20x^2 + 11xy - 3y^2) \div (4x + 3y)$
15. $(30x^2y^2 - 31xy + 5) \div (5xy - 1)$
16. $(14x^2y^2 + 11xy - 15) \div (2xy + 3)$
17. $(9x^2y^2 - 18xy + 5) \div (3xy - 5)$
18. $(9x + x^2 + 18) \div (x + 6)$
19. $(10x^2 - 21 + 29x) \div (5x - 3)$
20. $(-56 + 15x^2 - 11x) \div (8 + 5x)$

B

Determine the quotient in each of the following exercises. The remainder will be 0 in each of these exercises.

1. $(x^2 - 4) \div (x - 2)$
2. $(9x^2 - 16) \div (3x + 4)$
3. $(25x^2 - 4) \div (5x - 2)$
4. $(16x^2 - 49y^2) \div (4x - 7y)$
5. $(64x^2 - 25y^2) \div (8x + 5y)$
6. $(x^4 - 25) \div (x^2 + 5)$
7. $(9x^4 - 49) \div (3x^2 - 7)$
8. $(2x^3 + 7x^2 + 10x + 8) \div (x + 2)$
9. $(x^3 - 7x^2 + 13x - 15) \div (x - 5)$
10. $(2x^3 - 5x^2 - 5x + 12) \div (2x - 3)$
11. $(6x^3 - 13x^2 + 8x - 3) \div (3x^2 - 2x + 1)$
12. $(6x^3 - 7x^2y - 6xy^2 - y^3) \div (2x^2 - 3xy - y^2)$
13. $(1 - 2x^2 + 15x^3 - 4x) \div (x + 5x^2 - 1)$
14. $(x^3 - 2x + 1) \div (x - 1)$
15. $(6x^3 - 11x^2 + 1) \div (3x - 1)$
16. $(x^3 - 8) \div (x - 2)$
17. $(27x^3 + 1) \div (3x + 1)$
18. $(8x^3 - 27y^3) \div (2x - 3y)$
19. $(-18x^2 + x^4 - 12 - 5x + x^3) \div (x - 4)$
20. $(-3x - 7x^3 - 18 + x^4 + 7x^2) \div (3 + x^2 - 2x)$

C

Determine the quotient in each of the following exercises. The remainder will *not* be 0 in any of these exercises.

1. $(x^2 + x + 1) \div (x + 4)$
2. $(21x^2 - 32x + 4) \div (7x + 1)$
3. $(6x^2 - 19xy - 10y^2) \div (2x - 3y)$
4. $(2x^2 - 5xy + 12y^2) \div (x - 4y)$
5. $(4x^2 - 5) \div (2x - 1)$
6. $(x^3 - 1) \div (x + 1)$

7. $(x^4 + 1) \div (x - 1)$
8. $(x^4 - 2x^3 + 2x^2 - 2x + 2) \div (x - 1)$
9. $(x^3 - 5y^3) \div (x - y)$
10. $(2x^3 - 13x^2y + 11xy^2 + y^3) \div (2x - 3y)$

CHAPTER REVIEW

1. Find the product in each of the following exercises.

   a. $(-5)(-4)$
   b. $(-7)(6)$
   c. $(0)(+2)$
   d. $(-3)(+2)(-5)$
   e. $(-6)^2$
   f. $(-5)^3$
   g. $(+2)(-3)^2$
   h. $(-3)(+2)^2$
   i. $(-1)^6$
   j. $(-3)^2(-4)^2$

2. Find the value of each of the following expressions under the conditions stated.

   a. $x^2 + y^2$ where $x = 2$ and $y = -2$
   b. $2x^2 - 4y^2$ where $x = -1$ and $y = 3$

3. Find the product in each of the following exercises.

   a. $(-6a^3)(+5a^4)$
   b. $(2xy^2)(-3xy^3z)$
   c. $(-3x^2)(+2xy^3)(-yz)$
   d. $(-4a^2b^3)^2$
   e. $(-x^3)^4$
   f. $(-2a^2)^3(5a)$

4. Find the product in each of the following exercises.

   a. $-5a(a^2 - 3a)$
   b. $-3ab(a - b)$
   c. $(-2 + 3x)(-2xy)$
   d. $-1(3x - 2y)$
   e. $(x^2 - xy + y^2)(xy)$
   f. $-5x(x - y - z)$

5. Simplify and collect like terms in each of the following exercises.

   a. $2(3a - 5) + 6a$
   b. $-3x + 2(x - y)$
   c. $3x(x - y) + 3y(x + y)$
   d. $-4(a - b) - (a - 2b)$

6. Simplify and collect like terms in each of the following exercises.

   a. $3[5a + 2(a + b)]$
   b. $3[x + 2(x + 1)] + 2[x - 2(x - 1)]$

7. Find the product in each of the following exercises.

   a. $3x^2 + 2x + 1$
      $2x + 3$
      $\overline{}$

   b. $x^2 - xy - 2y^2$
      $x - y$
      $\overline{}$

8. Using the FOIL method find the product in each of the following exercises.
   a. $(5x + 4)(3x + 2)$
   b. $(2x - 3y)(x + 2y)$
   c. $(6 - x^2)(7 + 2x^2)$
   d. $(3xy - z)(5xy - 2z)$

9. Find the product in each of the following.
   a. $(2x - 5)(2x + 5)$
   b. $(7x - 2y)(7x + 2y)$
   c. $(3ab - 4)(-3ab + 4)$
   d. $(3a + 2)^2$
   e. $(5x - 6)^2$
   f. $(4 - xy)^2$

10. Simplify and collect like terms in each of the following.
    a. $(x - 4)^2 + 3x$
    b. $4(x - 2) + (x + 2)(x + 3)$
    c. $(3x - 4)^2 - 2(x - 3)(x - 4)$

11. Find the quotient in each of the following exercises.
    a. $(+14) \div (-7)$
    b. $(-24) \div (-6)$
    c. $(-8) \div (-1)$
    d. $(x^5) \div (x^3)$
    e. $(-16x^5y^7) \div (+2x^4y^3)$
    f. $(+27x^3y^2z) \div (-9x^3y)$
    g. $(-2x^3) \div (-2x^3)$
    h. $(-5xy) \div (-1)$

12. Find the quotient in each of the following exercises.
    a. $(x^8 + x^5) \div (x^2)$
    b. $(6x^3 - 8x^2) \div (-2x)$
    c. $(16x^2y^3 - 24x^3y^2 - 20x^4y) \div (-4x^2y)$
    d. $(-a + ay - ax) \div (-a)$

13. Find the quotient in each of the following exercises.
    a. $(x^2 + 9x + 20) \div (x + 5)$
    b. $(4x^2 + 4xy - 15y^2) \div (2x - 3y)$
    c. $(28 + 3x^2 - 19x) \div (3x - 7)$
    d. $(x^2 - 9) \div (x + 3)$

14. Find the quotient in each of the following exercises.
    a. $(x^3 - 1 - x - 4x^2 + x^4) \div (x^2 + 2x - 1)$
    b. $(x^3 - 3y^3) \div (x - y)$

The Solution of an Open Sentence in One Variable

In Chapter 1 we examined a method for determining the solution set of an open sentence. You may recall that the process involved a matter of guessing replacements for the variable until we struck upon one that made the open sentence true. Quite apparently, if the replacement for the variable that will make the sentence true is a number such as $-8\frac{2}{3}$, as it is in the equation below, the guessing process might continue for some time before striking the right number.

$$9x + 5 = 3(2x - 7) \tag{1}$$

One of the goals of this chapter is to develop a general approach whereby we can determine the elements in the solution set of an open sentence that has only *one variable*. We are going to limit our search even further by insisting that the only exponent for that variable be the *number 1*. In the equation above there is only one variable, x, and the power to which that variable appears is the first power. An open sentence of this nature is called a **linear equation**.

4.1 SOLUTION OF A LINEAR EQUATION BY THE ADDITION PRINCIPLE

Whenever an equality sign appears between two quantities such as

$$x = y, \tag{2}$$

the implication is that the x and y are merely two different names for exactly the same number. Thus, in the equation

$$4 = \text{IV} \tag{3}$$

the symbol 4 and the symbol IV are but different names for the number we know as four. The first of these is the Arabic name, and the second is the Roman name.

Similarly, when we say

$$7 - 4 = 2 + 1 \tag{4}$$

we are simply stating that the number named by $7 - 4$ is the very same number as that named by $2 + 1$. And, quite apparently, it is the number 3.

The quantity to the left of the equality sign in an equation is called the *left member* of the equation while the quantity to the right is the *right member*.

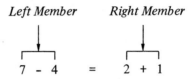

We are interested now in learning what will happen when the same number is added to each member of an equation. Should 12 be added to each member of the equation above, they will become:

Left Member   *Right Member*

$7 - 4 + 12$ $2 + 1 + 12$

$= 15 \longleftrightarrow = 15$

It would appear then that if the same number is added to each member of an equation, the two sums will turn out to be equal. In the example above, each sum is 15. The mathematician states this principle formally in the following manner:

Addition Principle of Equality

If a, b, and c are any three real numbers,
    and if $a = b$,
    then $a + c = b + c$

This principle turns out to be very valuable when determining the solution set of certain mathematical sentences.

4.1/The Addition Principle

Example 1

Determine the solution set of the equation

$$x - 5 = 6.$$

Explanation If we add the additive inverse of −5 to both members of this equation, they become:

Left Member	Right Member
$x - 5 + (+5)$	$6 + (+5)$

However, the addition principle of equality tells us that these two quantities are equal. Hence, we can say

$$x - 5 + (+5) = 6 + (+5).$$

Since $-5 + (+5)$ is 0 and $6 + (+5)$ is 11, this equation can be written as

$$x + 0 = 11.$$

But $x + 0$ is the same as x, therefore,

$$x = 11.$$

Thus we have found the replacement for x that will make the sentence $x - 5 = 6$ a true sentence. That number is 11.

Solution
$$x - 5 = 6$$
$$x - 5 + (+5) = 6 + (+5)$$
$$x + 0 = 11$$
$$x = 11$$

Therefore the solution set is $\{11\}$.

The process of finding the solution set of an equation is called *solving the equation*. *Solving the equation* implies that we are looking for those replacements for the variable that will make the equation true. The numbers in the solution set of an equation are also called the *roots of the equation*. In the example above there is only one root to the equation. It is the number 11. When we test to see if this is so, we find that:

$$11 - 5 \text{ does equal } 6.$$

Example 2

Find the root of the equation

$$x + 7 = 3.$$

Explanation The method used in Example 1 was to *isolate* the variable by adding the additive inverse of −5 to both members of the equation. In this situation since +7 is associated with x in the left member, it will be necessary to add the additive inverse of +7 to both members.

Add −7 to Both Members

$$x + 7 + (-7) = 3 + (-7)$$

The solution is now completed as in Example 1.

Solution

$$x + 7 = 3$$
$$x + 7 + (-7) = 3 + (-7)$$
$$x + 0 = -4$$
$$x = -4 \text{ (Root)}$$

Explanation continued Since 0 added to any number leaves the number unchanged, in the future we will omit the step that shows the variable being added to 0. In the solution above the step

$$x + 0 = -4$$

would be omitted.

Example 3

Find the root of the following equation:

$$14 = y - 2.$$

Explanation Since 14 names the same number as $y - 2$, it is possible to interchange the two names and rewrite the equation as $y - 2 = 14$. In isolating the variable in the next step, we seek out the additive inverse of −2. It is positive 2. Positive 2 is then added to both members of the equation.

Solution

$$14 = y - 2$$
$$y - 2 = 14$$
$$y - 2 + (+2) = 14 + (+2)$$
$$y = 16 \text{ (Root)}$$

Example 4

Find the root of the following equation:

$$-6 + z = -5.$$

Explanation Here, as before, the variable z is isolated by adding the additive inverse of negative 6 to both members of the equation.

Solution
$$-6 + z = -5$$
$$-6 + z + (+6) = -5 + (+6)$$
$$z = +1 \text{ (Root)}$$

Example 5

Find the root of the following equation:

$$7 = 10 + w.$$

Explanation The solution of this equation requires a combination of the methods used in Examples 3 and 4.

Solution
$$7 = 10 + w$$
$$10 + w = 7$$
$$10 + w + (-10) = 7 + (-10)$$
$$w = -3 \text{ (Root)}$$

EXERCISES 4.1

A

What number must be added to both members of each of the following equations in order to determine the root of each of these equations?

1. $x + 5 = 7$
2. $y + 5 = -8$
3. $a - 3 = 4$
4. $b - 10 = 0$
5. $z + 11 = 2$
6. $w - 9 = -3$
7. $5 = x + 12$
8. $7 = v - 10$
9. $6 + b = 3$
10. $-5 + a = 7$
11. $c - 4 = -8$
12. $-9 = b - 6$
13. $-12 = -5 + y$
14. $10 = 15 + d$
15. $y + 2 = 3 + 4$
16. $x - 7 = 10 - 12$

B

Find the root of each of the following equations.

1. $y + 6 = 10$
2. $x + 8 = 17$
3. $a + 9 = 12$
4. $b + 6 = -5$
5. $w + 5 = -2$
6. $c + 8 = 8$
7. $x + 1 = 0$
8. $w - 5 = 7$
9. $a - 10 = 14$
10. $b - 6 = -4$
11. $c - 9 = -3$
12. $y - 7 = -7$
13. $x - 2 = 0$
14. $m - 11 = 11$
15. $v - 10 = -15$
16. $p - 17 = -23$
17. $16 = a + 10$
18. $19 = d + 16$
19. $7 = c + 9$
20. $3 = y + 16$
21. $12 = a + 12$
22. $0 = x + 6$
23. $-2 = b + 9$
24. $-7 = y + 8$
25. $11 = a - 4$
26. $20 = b - 1$
27. $15 = w - 17$
28. $1 = c - 23$
29. $-5 = d - 14$
30. $8 + x = 5$
31. $12 + y = 10$
32. $17 + a = 6$
33. $6 + a = 17$
34. $-2 + x = -5$
35. $-10 + y = -3$
36. $19 = 10 + a$
37. $12 = -5 + z$
38. $8 = 8 + w$
39. $-5 = -5 + z$
40. $-10 = 10 + b$

C

Find the root of each of the following equations.

1. $x + 7 = 15$
2. $14 = x - 3$
3. $5 + x = 11$
4. $-4 + x = 16$
5. $-6 = 12 + x$
6. $x - 8 = 10$
7. $-9 = -6 + x$
8. $-3 + x = -3$
9. $x + 7 = -7$
10. $5 = x + 9$
11. $x - 4 = 6 + 2$
12. $-10 + 2 = x + 6$
13. $1 + x = -14 + 2$
14. $5 + 3 = -2 + x$
15. $+3 + x = 6 - 3$
16. $+2 - 10 = -5 + x$
17. $-4 - 5 = x - 3$
18. $-2 + x = -7 + 2$
19. $x + 3 + 2 = 5 + 4$
20. $5 + x - 2 = 3 - 8$

4.2 SOLUTION OF A LINEAR EQUATION BY THE MULTIPLICATION PRINCIPLE

You may have noticed that in each of the equations we have examined thus far the operation between the variable and the constant was always addition. Thus, in the equation

$$x + 7 = 15$$

the operation between the variable x and the constant $+7$ is addition. Similarly, in the equation

$$14 = x - 3$$

the operation between the variable x and the constant -3 is also addition. To isolate the variable under this condition, we found that it is necessary to *add* the *additive inverse* of a constant to both members of the equation.

We now want to examine equations in which the operation between the variable and the constant is multiplication. This will arise in an equation such as

$$3x = 15.$$

In order to isolate the variable in an equation of this nature it will be necessary to *multiply* both members of the equation by the *multiplicative inverse* of the constant. First, though, we must examine the principle that will enable us to do this.

Consider the equality

$$7 = 3 + 4.$$

Should we multiply both members of this equation by 2 we obtain the following:

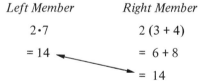

Left Member	*Right Member*
$2 \cdot 7$	$2(3 + 4)$
$= 14$	$= 6 + 8$
	$= 14$

It would appear then that if both members of an equation are multiplied by the same number the two products will turn out to be equal. In the example above, each product is 14. In view of this, we will accept the principle below.

Multiplication Principle of Equality

If $a, b,$ and c are any three real numbers,
and if $\quad a = b,$
then $\quad ca = cb$

4/The Solution of an Open Sentence in One Variable

This principle will enable us to find the root of an equation where the operation between the variable and the constant is multiplication.

Example 1

Find the root of the following equation.

$$6x = 18$$

Explanation Since the operation between the 6 and the x is multiplication, we realize we must multiply both members of the equation by the multiplicative inverse of 6.

Left Member	Right Member
$\frac{1}{6} \cdot 6x$	$\frac{1}{6} \cdot 18$

However, the multiplication principle of equality tells us that these two quantities are equal. Hence we can say

$$\frac{1}{6} \cdot 6x = \frac{1}{6} \cdot 18.$$

Since $\frac{1}{6} \cdot 6$ is 1 and $\frac{1}{6} \cdot 18$ is 3, this equation can be written as

$$1x = 3.$$

But $1x$ is the same as x and therefore,

$$x = 3.$$

Thus we have found that the root of the equation $6x = 18$ is 3.

Solution
$$6x = 18$$
$$\frac{1}{6} \cdot 6x = \frac{1}{6} \cdot 18$$
$$1x = 3$$
$$x = 3 \text{ (Root)}$$

Explanation continued Since $1x$ in the third step will always be replaced by x in the fourth step, there is really no need for us to include the third step in our solution. In the future that step will be deleted.

4.2/The Multiplication Principle

Example 2
Find the root of the following equation:

$$-8x = 32.$$

Solution

$$(-\tfrac{1}{8})(-8x) = (-\tfrac{1}{8})(32)$$

$$x = -4 \text{ (Root)}$$

Explanation Since the multiplicative inverse of -8 is $-\tfrac{1}{8}$, both members of the equation must be multiplied by $-\tfrac{1}{8}$. On the right side of the equation the signs of direction of the two numbers are different, hence the sign of the product is negative.

Example 3
Find the root of the following equation:

$$-x = -7$$

Explanation From our earlier work we know that $-x$ is the same as $-1x$. Hence the equation above can be written as

$$-1x = -7.$$

Solution

$$-x = -7$$

$$-1x = -7$$

$$-\tfrac{1}{1}(-1x) = -\tfrac{1}{1}(-7)$$

$$x = +7$$

Example 4
Find the root of the following equation:

$$-26 = -6x$$

Explanation Since -26 and $-6x$ name the same number, it is possible to interchange the two names and rewrite the equation as $-6x = -26$. The solution then follows the same pattern as the earlier ones.

Solution

$$-26 = -6x$$

$$-6x = -26$$

$$(-\tfrac{1}{6})(-6x) = (-\tfrac{1}{6})(-26)$$

$$x = +4\tfrac{1}{3} \text{ (Root)}$$

Example 5

Find the root of the following equation:

$$\frac{2}{5}x = 12.$$

Explanation The only concern here is in realizing that the multiplicative inverse of $\frac{2}{5}$ is $\frac{5}{2}$ for the product of these two numbers is 1. In general, the multiplicative inverse of a fraction is the *reciprocal* of the fraction.

Solution
$$\frac{2}{5}x = 12$$
$$\frac{5}{2} \cdot \frac{2}{5}x = \frac{5}{2} \cdot 12$$
$$x = 30 \quad \text{(Root)}$$

EXERCISES 4.2

A

By what number must both members of each of the following equations be multiplied in order to determine the root of each of these equations?

1. $7x = 35$
2. $9x = 18$
3. $6x = -24$
4. $5x = -30$
5. $-2x = 10$
6. $-3x = 27$
7. $-8x = -32$
8. $-12x = -24$
9. $4x = 15$
10. $-7x = -16$
11. $-5x = 14$
12. $11x = -15$
13. $16 = 4x$
14. $-20 = 5x$
15. $18 = -2x$
16. $-28 = -7x$
17. $\frac{2}{3}x = 20$
18. $\frac{3}{5}x = 15$
19. $-\frac{4}{5}x = 40$
20. $-16 = \frac{2}{9}x$
21. $-24 = -\frac{3}{8}x$
22. $\frac{3}{7}x = \frac{3}{5}$
23. $-\frac{4}{9}x = -\frac{5}{9}$
24. $-\frac{1}{2} = -\frac{2}{3}x$

4.2/The Multiplication Principle

B

Find the root of each of the following equations.

1. $5x = 20$
2. $4x = 36$
3. $6x = -30$
4. $9x = -45$
5. $-7x = -42$
6. $-10x = -50$
7. $-6x = -84$
8. $-3x = 33$
9. $-8x = 56$
10. $-1x = -4$
11. $-1x = 5$
12. $-x = 2$
13. $-x = -6$
14. $-x = +10$
15. $5x = 26$
16. $7x = -16$
17. $-3x = -10$
18. $-6x = 20$
19. $4x = -95$
20. $-2x = -81$
21. $18 = 2x$
22. $54 = 6x$
23. $-40 = 5x$
24. $-56 = 4x$
25. $32 = -8x$
26. $28 = -7x$
27. $-60 = -10x$
28. $-44 = -11x$
29. $12 = -1x$
30. $-15 = -1x$
31. $+16 = -x$
32. $-17 = -x$
33. $-x = 9$
34. $-8x = 18$
35. $-19 = 6x$
36. $-23 = -10x$
37. $26 = -x$
38. $-37 = -5x$
39. $-9x = +40$
40. $+6x = -8$

C

Find the root of each of the following equations.

1. $\frac{2}{3}x = 6$
2. $\frac{1}{5}x = 10$
3. $-\frac{3}{4}x = 12$
4. $\frac{2}{5}x = -14$
5. $-\frac{1}{6}x = -7$
6. $-\frac{4}{7}x = 28$
7. $-\frac{2}{9}x = -36$
8. $\frac{1}{2}x = -5$
9. $\frac{1}{4}x = \frac{3}{2}$
10. $-\frac{5}{4}x = -\frac{1}{4}$

11. $-21 = \frac{3}{7}x$ 14. $\frac{5}{3} = -\frac{3}{10}x$

12. $35 = -\frac{5}{7}x$ 15. $-\frac{2}{3}x = -\frac{2}{3}$

13. $-\frac{3}{4} = \frac{1}{8}x$ 16. $\frac{5}{9}x = -\frac{3}{5}$

4.3 SOLUTION OF A LINEAR EQUATION USING BOTH THE ADDITION AND MULTIPLICATION PRINCIPLES

PART 1

Thus far we have examined the solution of a linear equation where only the additive principle was involved, as in

$$x + 6 = 10$$

or only the multiplicative principle was involved, as in

$$-3x = 20.$$

Our interest now turns to solving equations where both of these principles are needed to determine the root of the equation.

In general, these equations will take one of the following forms:

$$3x + 7 = 25 \qquad \text{I}$$

$$9x = 7x - 14 \qquad \text{II}$$

$$6x - 5 = 4x + 11 \qquad \text{III}$$

Notice that in equation I a constant term appears on both sides of the equation. In equation II a term with a variable appears on both sides. In equation III, a combination of both of these situations is involved. Once we have mastered the technique for solving linear equations such as these three, we will be able to solve all linear equations for all can be reduced to one of these forms.

Fortunately, the solutions of all of these equations follow the same pattern. The initial step is to apply the addition principle in such a way as to transform whatever equation we are faced with to one such as

$$7x = 43.$$

Here the left member of the equation is a variable term and the right member is a constant term. Once the equation is in this form we recognize it as one that can be solved by using the multiplication principle.

Consider the equation

$$3x + 7 = 25. \tag{1}$$

4.3/The Addition and Multiplication Principles

The first step is to apply the addition principle in order to eliminate the +7 from the left side of the equation. We do this by adding negative 7 to both members of the equation.

$$3x + 7 + (-7) = 25 + (-7). \qquad (2)$$

Now, by collecting like terms on the left side and then collecting like terms on the right side, equation (2) becomes

$$3x = 18. \qquad (3)$$

Equation (3) is of the form whose solution is obtained by using the multiplication principle.

$$\frac{1}{3} \cdot 3x = \frac{1}{3} \cdot 18$$

$$x = 6 \qquad (4)$$

If our work is correct then 6 should be the root of equation (1). That is, when 6 replaces x, the sentence $3x + 7 = 25$ will be a true sentence. Let us determine if this is so.

$$3x + 7 = 25$$

$$3 \cdot 6 + 7 \stackrel{?}{=} 25$$

$$18 + 7 \stackrel{?}{=} 25$$

$$25 = 25$$

Since 25 does equal 25, we made no error in finding the root of equation (1).

Based on the solution above it seems that we can summarize the pattern of solution for a linear equation such as (1) above to a three-step procedure.

---

1. Use the addition principle to isolate the *term* with the variable on the left side of the equation.
2. Collect like terms separately on each side of the equation.
3. Use the multiplication principle to find the value of the variable.

Example 1
Find the root of the following equation:

$$6x - 9 = 39.$$

Solution

$$
\begin{aligned}
6x - 9 &= 39 \\
6x - 9 + (+9) &= 39 + (+9) \quad &\text{Addition Principle} \\
6x &= 48 \quad &\text{Collecting Like Terms} \\
\tfrac{1}{6} \cdot 6x &= \tfrac{1}{6} \cdot 48 \quad &\text{Multiplication Principle} \\
x &= 8 \quad \text{(Root)}
\end{aligned}
$$

Example 2
Find the root of the following equation:

$$17 - \tfrac{2}{3} x = -5.$$

Solution

$$
\begin{aligned}
17 - \tfrac{2}{3} x &= -5 \\
17 - \tfrac{2}{3} x + (-17) &= -5 + (-17) \quad &\text{Addition Principle} \\
- \tfrac{2}{3} x &= -22 \quad &\text{Collecting Like Terms} \\
(-\tfrac{3}{2})(-\tfrac{2}{3} x) &= (-\tfrac{3}{2})(-22) \quad &\text{Multiplication Principle} \\
x &= +33 \quad \text{(Root)}
\end{aligned}
$$

EXERCISES 4.3 (Part 1)

A

Find the root of each of the following equations.

1. $5x + 9 = 29$
2. $4x + 7 = 43$
3. $2x - 5 = 11$
4. $3x - 13 = 2$
5. $6x - 15 = 27$
6. $7 + 3x = 31$
7. $18 + 7x = 81$
8. $-4 + 8x = 36$
9. $-9 + 10x = 21$
10. $5 - 2x = 19$
11. $14 - 9x = 32$
12. $13 - 3x = 46$
13. $2x + 17 = 11$
14. $4x + 25 = 5$

4.3/The Addition and Multiplication Principles

15. $5x + 11 = -4$
16. $12x + 3 = -21$
17. $-3x + 8 = -19$
18. $-7x + 15 = -6$
19. $-x + 2 = -12$
20. $-x - 7 = -2$
21. $3x + 5 = 5$
22. $9x - 16 = -16$
23. $-5x - 14 = -49$
24. $-8x - 23 = 33$
25. $19 - 7x = -2$
26. $14 - 15x = -1$
27. $-7 - 9x = 2$
28. $4x - 18 = 12$
29. $17 + 6x = 57$
30. $16 + 3x = 20$
31. $19 - 5x = 19$
32. $-7x - 10 = -10$
33. $-8x + 6 = -14$
34. $-3x - 8 = -5$
35. $5x - 1 = 2$
36. $2 - 3x = 1$
37. $-4x - 1 = 1$
38. $6x + 7 = 4$
39. $9 - 8x = 11$
40. $-1 - 10x = -5$

B

Find the root of each of the following equations.

1. $\frac{2}{3}x + 1 = 7$
2. $\frac{3}{5}x + 7 = 10$
3. $5 + \frac{1}{2}x = 9$
4. $11 + \frac{3}{4}x = 8$
5. $9 - \frac{2}{5}x = 17$
6. $-7 - \frac{1}{4}x = 2$
7. $-4 + \frac{1}{6}x = -10$
8. $-\frac{5}{6}x + 1 = 11$
9. $-\frac{3}{8}x - 17 = 7$
10. $-\frac{1}{2}x - 19 = -15$
11. $16 - \frac{3}{4}x = 17$
12. $-5 - \frac{3}{2}x = 6$

PART 2

Should an equation be one such as

$$9x = 7x - 14, \tag{1}$$

the approach to its solution is much the same as in Part 1. Here, too, it is necessary to transform this equation to a form such as

$$5x = 23,$$

where the left member is a variable term and the right member is a constant term. This implies that the first step is to isolate *all* terms containing the variable on the left side of the equation.

Since equation (1) has a variable term $+7x$ on the *right* side of the equation, we eliminate $+7x$ from that side by adding its additive inverse $-7x$ to both members of the equation. Thus, equation (1) now becomes

$$9x + (-7x) = 7x - 14 + (-7x). \tag{2}$$

After collecting like terms on each side, we have

$$2x = -14. \tag{3}$$

And, finally, the solution is completed in the usual fashion by multiplying both sides of (3) by the multiplicative inverse of the coefficient of x.

$$\tfrac{1}{2}(2x) = \tfrac{1}{2}(-14)$$

$$x = -7$$

If our work is correct, -7 should **satisfy** equation (1); that is, it should make equation (1) a true sentence. By replacing x with -7 we can determine if this is so.

$$9x = 7x - 14 \tag{1}$$

$$9(-7) \stackrel{?}{=} 7(-7) - 14$$

$$-63 \stackrel{?}{=} -49 - 14$$

$$-63 = -63$$

The fact that -7 does make equation (1) a true sentence implies that no errors were made when finding the root of equation (1).

Example 1

Find the root of the following equation:

$$x = 15 - 4x.$$

Solution

$x = 15 - 4x.$	
$1x + (+4x) = 15 - 4x + (+4x)$	Addition Principle
$5x = 15$	Collecting Like Terms
$\tfrac{1}{5} \cdot 5x = \tfrac{1}{5} \cdot 15$	Multiplication Principle
$x = 3$ (Root)	

4.3/The Addition and Multiplication Principles

Explanation Notice that in step 2 of the solution not only was $+4x$ added to each member of the equation but also 1 was written as the coefficient of x. We did this in order not to forget this number when adding x to $4x$.

In solving the equation above, in solving the equations in Part 1 of this section, and, in fact, in solving any linear equation, a three-step approach is involved:

Step 1: Use the addition principle.
Step 2: Collect like terms.
Step 3: Use the multiplication principle.

Example 2

Find the root of the following equation:

$$-3x = 7x - 6.$$

Solution

$$-3x = 7x - 6$$
$$-3x + (-7x) = 7x - 6 + (-7x) \quad \text{Addition Principle}$$
$$-10x = -6 \quad \text{Collecting Like Terms}$$
$$-\tfrac{1}{10}(-10x) = -\tfrac{1}{10}(-6) \quad \text{Multiplication Principle}$$
$$x = +\tfrac{6}{10} = \tfrac{3}{5} \quad \text{(Root)}$$

Explanation In the very last step the product of $-\tfrac{1}{10}$ and -6 is found to be $+\tfrac{6}{10}$. Why should the sign be positive? The fraction $\tfrac{6}{10}$ is then reduced to lowest terms and renamed as $\tfrac{3}{5}$.

EXERCISES 4.3 (Part 2)

Find the root of each of the following equations.

1. $5x = 6 + 3x$
2. $9x = 20 + 5x$
3. $11x = 18 + 2x$
4. $7x = 15 + 6x$
5. $10x = 7x + 24$
6. $13x = 4x + 36$
7. $8x = 3x - 20$
8. $15x = 2x - 13$
9. $9x = 8x - 7$
10. $6x = x - 15$
11. $4x = x - 36$
12. $3x = 7x - 40$
13. $7x = 9x - 18$
14. $x = 5x + 28$
15. $x = 7x - 42$
16. $-2x = 12 - 5x$

136 4/The Solution of an Open Sentence in One Variable

17. $-8x = 35 - 3x$
18. $-9x = -2x + 56$
19. $-x = 14 - 3x$
20. $7x = -x + 16$
21. $x = -x + 18$
22. $-x = -28 + x$
23. $x = -4x + 5$
24. $2x = 3 - 4x$
25. $-3x = 5x - 4$
26. $2x = 7 - 3x$
27. $-7x = 9 - 5x$
28. $5x = -18 - 3x$
29. $3x + 4x = 20 + 2x$
30. $9x - 7x = 4x - 16$

PART 3

We are now prepared to examine the solution of the most general of linear equations. These are equations in which variable terms appear on both sides of the equation and constant terms also appear on both sides of the equation. An equation of this form is

$$6x - 5 = 4x + 11. \qquad (1)$$

As we might have suspected, the solution of this equation involves the application of the addition principle *twice*. In the first of these we eliminate all *constant* terms from the *left* side of the equation, while in the second we eliminate all *variable* terms from the *right* side of the equation. In reality, we are merely combining what we learned in Part 1 of this section with what we learned in Part 2.

Let's examine this approach using equation (1) above.

$$6x - 5 = 4x + 11 \qquad (1)$$

To eliminate the constant -5 from the left side of this equation, it is necessary to add $+5$ to both members of the equation.

$$6x - 5 + (+5) = 4x + 11 + (+5) \qquad (2)$$

Collecting like terms on each side we get

$$6x = 4x + 16. \qquad (3)$$

To eliminate the $+4x$ from the right side of equation (3), it is necessary to add a $-4x$ to both members of the equation.

$$6x + (-4x) = 4x + 16 + (-4x) \qquad (4)$$

4.3/The Addition and Multiplication Principles

Again collecting like terms, equation (4) becomes

$$2x = 16. \tag{5}$$

The final step is the usual one where both members of the equation are multiplied by the multiplicative inverse of the coefficient of x.

$$\tfrac{1}{2} \cdot 2x = \tfrac{1}{2} \cdot 16$$

$$x = 8 \quad \text{(Root)} \tag{6}$$

Now, again, we are faced with the usual question as to whether the number 8 that we have found is truly the root of equation (1). A test will determine this.

$$6x - 5 = 4x + 11 \tag{1}$$

$$6 \cdot 8 - 5 \stackrel{?}{=} 4 \cdot 8 + 11$$

$$48 - 5 \stackrel{?}{=} 32 + 11$$

$$43 = 43$$

In view of the fact that when 8 is used as a replacement for x in the equation

$$6x - 5 = 4x + 11,$$

the left member is equal to the right member, we can conclude that no errors were made in the solution.

Example 1

Solve and check the following equation:

$$7x + 15 = 11x - 17.$$

Explanation The term "solve and check" implies not only that we are to find the root of the equation but, also, to determine whether the number found actually makes the sentence true.

Solution $\qquad 7x + 15 = 11x - 17$

At this point eliminate $+15$ from the left member.

$$7x + 15 + (-15) = 11x - 17 + (-15) \qquad \text{Addition Principle}$$

$$7x = 11x - 32 \qquad \text{Collecting Like Terms}$$

At this point eliminate $11x$ from the right member.

$$7x + (-11x) = 11x - 32 + (-11x) \quad \text{Addition Principle}$$
$$-4x = -32 \quad \text{Collecting Like Terms}$$
$$-\tfrac{1}{4}(-4x) = -\tfrac{1}{4}(-32) \quad \text{Multiplication Principle}$$
$$x = 8 \quad \text{(Root)}$$

Check for $x = 8$:

$$7x + 15 = 11x - 17$$
$$7 \cdot 8 + 15 \overset{?}{=} 11 \cdot 8 - 17$$
$$56 + 15 \overset{?}{=} 88 - 17$$
$$71 = 71$$

Explanation continued It is very important that we write the question marks above the equality signs as we have in the "check" above. Since we may have made an error in solving the equation, we are not certain that the replacement for x will make the sentence true. We use the question mark to show this uncertainty.

Example 2

Solve and check the following equation:

$$12 - 9x = -11x - 6.$$

Solution

$$12 - 9x = -11x - 6$$
$$12 - 9x + (-12) = -11x - 6 + (-12) \quad \text{Addition Principle}$$
$$-9x = -11x - 18 \quad \text{Collecting Like Terms}$$
$$-9x + (+11x) = -11x - 18 + (+11x) \quad \text{Addition Principle}$$
$$2x = -18 \quad \text{Collecting Like Terms}$$
$$\tfrac{1}{2} \cdot 2x = \tfrac{1}{2} \cdot (-18) \quad \text{Multiplication Principle}$$
$$x = -9 \quad \text{(Root)}$$

Check for $x = -9$:

$$12 - 9x = -11x - 6$$
$$12 - 9(-9) \overset{?}{=} -11(-9) - 6$$
$$12 + 81 \overset{?}{=} 99 - 6$$
$$93 = 93$$

4.3/The Addition and Multiplication Principles

EXERCISES 4.3 (Part 3)

Solve and check each of the following equations.

1. $7x + 8 = 2x + 13$
2. $5x + 5 = 3x + 9$
3. $10x + 2 = 7x + 14$
4. $2x + 3 = 11 - 2x$
5. $x + 7 = 23 - x$
6. $4x - 5 = 2x + 5$
7. $-x + 6 = 3x - 10$
8. $12 - 3x = 2x + 2$
9. $15 + 7x = 3x - 9$
10. $14 - x = 6 + x$
11. $9 - 3x = 9 - 5x$
12. $8 + 7x = -12 - 3x$
13. $6x - 5 = 4x - 5$
14. $4x + 7 = 6x + 11$
15. $7x + 3 = 5 + 9x$
16. $2x + 10 = x + 6$
17. $-5x + 3 = -2x + 12$
18. $-3x + 7 = 3x + 31$
19. $8 - 2x = 3x + 23$
20. $11 - 3x = 7 - 5x$
21. $16 - x = 4 - 3x$
22. $20 + 3x = 12 - x$
23. $2 + 3x = 2 + 3x$
24. $5 + 3x = 2 + 3x$

PART 4

We are now going to examine a number of linear equations that appear to be somewhat more complex than those previously solved. We will find, though, that with very little effort all linear equations can be transformed into equations similar to those whose solutions we know.

Example 1

Find the root of the following equation:

$$5x - 6 - 9x = 17 - 8x + 13.$$

Explanation Notice that once we collect like terms on the left side of the equation and do the same on the right side, the new form of the equation will be identical to those we solved in Part 3 of this section.

Solution

1. $5x - 6 - 9x = 17 - 8x + 13$
2. $-4x - 6 = 30 - 8x$ Collecting Like Terms
3. $-4x = 30 - 8x + (+6)$ Addition Principle
4. $-4x + (+8x) = 30 + (+6)$ Addition Principle
5. $+4x = +36$ Collecting Like Terms
6. $x = 9$ Multiplication Principle

Now, perhaps, we can begin to eliminate some of the work we have been doing when finding the root of a linear equation. In step (3) above we failed to write

$$-6 + (+6)$$

on the left side of the equation as we had in the past. Our purpose in adding a positive 6 to both members of the equation is to eliminate the constant term from the left side of the equation. Hence, knowing that this will happen to the left side of the equation, there seems to be little need for writing $-6 + (+6)$.

It is important, though, that we continue to realize that in going from step (2) to step (3) + 6 had been added to both members of the equation. On the right side this is shown by the fact that the +6 appears there. On the left side it is shown by the fact that the sum of −6 and +6 is 0 and when 0 is added to −4x the answer is −4x.

A similar situation takes place in going from step (3) to step (4). A positive 8x was added to both members of equation (3). In writing equation (4) we can see that +8x was added to the *left* member of the equation, for it actually appears in (4) while it is not in (3). However, +8x was also added to the *right* member of (3) at the same time. How is this shown?

We did but one more thing to lighten our work in the solution of the equation in Example 1. Notice that in going from step (5) to step (6) we eliminated the step

$$\tfrac{1}{4} \cdot 4x = \tfrac{1}{4} \cdot 36$$

For some time now we have been writing the product on the left side of the equation above as x. Now we are suggesting that the right side of the equation be found mentally and that the above step be eliminated.

Example 2

Find the root of the following equation:

$$5(2x - 6) = 4x - 42.$$

Explanation The situation in this example introduces the need for an additional step when solving an equation. This step involves finding the product of the monomial 5 and the binomial $2x - 6$. Once this is completed we follow the same pattern as earlier.

Solution $5(2x - 6) = 4x - 42$

$\qquad\qquad\quad 10x - 30 = 4x - 42 \qquad$ Product of Monomial and Binomial

4.3/The Addition and Multiplication Principles

$$10x = 4x - 42 + (+30) \quad \text{Addition Principle}$$
$$10x + (-4x) = -42 + (+30) \quad \text{Addition Principle}$$
$$6x = -12 \quad \text{Collecting Like Terms}$$
$$x = -2 \quad \text{Multiplication Principle}$$

Example 3

Find the root of the following equation:

$$3x - (5x - 4) = 16.$$

Explanation Before we can attack the solution of this equation we must recall that the sign before the quantity in the parentheses above indicates the operation of subtraction. Since $5x - 4$ is the subtrahend we must find the additive inverse of each of the terms $5x$ and -4 and continue as we have learned in Chapter 2.

Solution
$$3x - (5x - 4) = 16$$
$$3x - 5x + 4 = 16 \quad \text{Subtraction}$$
$$-2x + 4 = 16 \quad \text{Collecting Like Terms}$$
$$-2x = 16 + (-4) \quad \text{Addition Principle}$$
$$-2x = 12 \quad \text{Collecting Like Terms}$$
$$x = -6 \quad \text{Multiplication Principle}$$

Example 4

Find the root of the following equation:

$$-2(3x + 4) + (-3x + 2) = -7x.$$

Explanation The -2 before the $(3x + 4)$ implies the operation of multiplication. The "+" sign before the $(-3x + 2)$ implies the operation of addition.

Solution
$$-2(3x + 4) + (-3x + 2) = -7x$$
$$-6x - 8 - 3x + 2 = -7x$$
$$-9x - 6 = -7x \quad \text{Collecting Like Terms}$$
$$-9x = -7x + (+6) \quad \text{Addition Principle}$$
$$-9x + (+7x) = +6 \quad \text{Addition Principle}$$
$$-2x = +6 \quad \text{Collecting Like Terms}$$
$$x = -3 \quad \text{Multiplication Principle}$$

EXERCISES 4.3 (Part 4)

A

Find the root of each of the following equations.

1. $8x + 6 - 5x = 18$
2. $10x = 15 + 6x + 13$
3. $5x - 7 + 3x = 13 - 2x$
4. $-x + 6 + 4x = 6x - 12$
5. $10 - 6x + 7 = 9x + 2$
6. $7x + 4 - x - 8 = 4x$
7. $3 = 5x - 14 + x - 1$
8. $17 = 19 - x - x - 6$
9. $2x - 5 + 3x = -6 - 4x + 1$
10. $12 - 7x - 5 = 1 - 2x + 1$
11. $-1 + x - 1 = -x + 1 - x$
12. $8x + 9 - 5x = -12 - 3x$

B

Find the root of each of the following equations.

1. $2(x - 4) = 6$
2. $-3(2x - 5) = -21$
3. $3(x + 4) + 10 = 31$
4. $5(2 - x) + 3x = 18$
5. $-2(3x + 5) + 8 = 16$
6. $4x + 2(x - 7) = 10$
7. $x + 3(5 - 2x) = -30$
8. $2x - 3(x + 6) = 9$
9. $12 - 4(2x - 1) = 16$
10. $-(3x - 1) = -23$
11. $-(15 - 2x) = -3x$
12. $-(17 + 5x) = -2$
13. $35 = 5(3x - 2)$
14. $-30 = 6(2x - 5)$
15. $9x = 2(5x + 8)$
16. $7x + 3 = 2(2x + 9)$
17. $6x - 6 = -3(x - 7)$
18. $5x + (9x - 4) = 3x - 26$
19. $4x - (2x + 3) = 9 + 6x$
20. $3(2x - 5) + 6 = 10x + 29$
21. $5x - 4 = 2(7x - 6) + 53$
22. $7 - 2(3x + 1) = 15 - x$
23. $-8 + 6(5 - x) = x + 22$
24. $-(3x - 5) + 12 = 6x - 1$

C

Find the root of each of the following equations.

1. $2(x + 2) + 3(x - 4) = 2$
2. $3(x - 7) + 5(x + 6) = 1$
3. $4(2x + 3) + (x - 5) = -2$
4. $2(x + 6) - (4x - 7) = 3$
5. $3(2x - 5) = 4(2x + 6)$
6. $-5(6 - x) = 2(x - 8)$
7. $(-2x + 3) + 2(2x + 1) = 1$
8. $14 = 3(x - 4) - (5x - 6)$
9. $-4(2x - 7) = -x + 3(x + 6)$
10. $6(-x - 3) - 2x = 14 - 3(5 - 3x)$

4.4 SOLUTION OF AN EQUATION INVOLVING ABSOLUTE VALUES

At the time we investigated the method for determining the sum of two signed numbers we ran across the concept of the *absolute value* of a number. We defined the absolute value of a number as *the distance from the point named by the number to the origin.*

Thus, the term

"The absolute value of +5"

is interpreted as,

"The distance from the point +5 to the origin."

Since this distance is 5 units (Figure 4-1) we say that

"The absolute value of +5 is 5."

FIGURE 4-1

Similarly, the absolute value of -3 is 3 for the distance from the point -3 to the origin is 3 units (Figure 4-2).

FIGURE 4-2

Using symbols, we express the concept that

"The absolute value of -3 is 3."

as follows:

$$|-3| = 3.$$

Thus, whenever vertical lines are drawn on both sides of a number we understand this to imply that the distance from the point named by that number to the origin is being sought.

As an illustration, the expression

$$|+6|$$

is read as,

"The absolute value of +6"

and we understand this to mean,

"The distance from the point +6 to the origin."

In view of the fact that the distance from the point +6 to the origin is 6, we write this information as,

$$|+6| = 6.$$

Now let us examine this in the light of an equation where a variable appears between the absolute bars. For instance, the equation

$$|x| = 5$$

is interpreted with an English sentence as,

The distance from the point named by the variable x to the origin is 5 units.

Hence, were we interested in determining the roots of this equation, we would search for those numbers that will make this equation true. Since the point +5 is 5 units from the origin and so, too, is the point -5, 5 units from the origin, we express this fact by writing

$$x = +5 \quad \text{and} \quad x = -5.$$

Both of these replacements for x will make the equation $|x| = 5$ a true sentence. Thus,

$$|+5| = 5 \quad \text{and} \quad |-5| = 5.$$

Notice that the two numbers +5 and -5 are the additive inverses of each other. This situation had to be true for the distance from any point to the origin is the same as the distance from the additive inverse of that point to the origin (Figure 4-3).

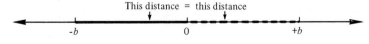

FIGURE 4-3

4.4/Solution of an Equation Involving Absolute Values

In general we can say that the quantity within the absolute bars will name two points:

---
1. The first of these is a point to the right of the origin that is named by a positive number.
2. The second is a point to the left of the origin that is named by a negative number. The second number is the additive inverse of the first number.

Let us apply this information to an equation of the form

$$|2x - 3| = 7. \qquad (1)$$

The quantity $2x - 3$ that falls within the absolute bars names two points. The first of these is the point named by the number $+7$ for this point is 7 units from the origin. The second number is the additive inverse of $+7$, this being -7. Negative 7 is also 7 units from the origin. Hence, equation (1) can be rewritten as *two separate equations*:

$$2x - 3 = +7 \quad \text{and} \quad 2x - 3 = -7. \qquad (2)$$

Each of these equations is a linear equation and the root of each can be determined in the manner we learned earlier.

$$
\begin{array}{lll}
2x - 3 = +7 & \text{and} & 2x - 3 = -7 \qquad (2)\\
2x = +7 + (+3) & & 2x = -7 + (+3) \\
2x = +10 & & 2x = -4 \\
x = 5 & \text{and} & x = -2
\end{array}
$$

In view of this, it would appear that both 5 and -2 are replacements for x that will make equation (1) a true sentence. Let us check to see if this is so.

$$
\begin{array}{ll}
\text{Check for } x = 5: & \text{Check for } x = -2: \\
|2x - 3| = 7 & |2x - 3| = 7 \\
|2 \cdot 5 - 3| \stackrel{?}{=} 7 & |2(-2) - 3| \stackrel{?}{=} 7 \\
|10 - 3| \stackrel{?}{=} 7 & |-4 - 3| \stackrel{?}{=} 7 \\
|7| \stackrel{?}{=} 7 & |-7| \stackrel{?}{=} 7 \\
7 = 7 & 7 = 7
\end{array}
$$

Both 5 and -2 are roots of equation (1) for both make that sentence true.

Example 1

Find the roots of the following equation and check the answers.

$$|3x - 5| = 16.$$

Explanation The very first objective is to rewrite the above equation in terms of the two linear equations that it represents. Once this is done the root of each of the linear equations is found separately.

Solution

$$|3x - 5| = 16$$

$$3x - 5 = +16 \quad \text{and} \quad 3x - 5 = -16$$

$$3x = +16 + (+5) \qquad 3x = -16 + (+5)$$

$$3x = 21 \qquad\qquad 3x = -11$$

$$x = 7 \quad \text{and} \quad x = -3\tfrac{2}{3}$$

Check for $x = 7$:

$$|3x - 5| = 16$$
$$|3 \cdot 7 - 5| \stackrel{?}{=} 16$$
$$|21 - 5| \stackrel{?}{=} 16$$
$$|16| \stackrel{?}{=} 16$$
$$16 = 16$$

Check for $x = -3\tfrac{2}{3}$:

$$|3x - 5| = 16$$
$$|3(-3\tfrac{2}{3}) - 5| \stackrel{?}{=} 16$$
$$|-11 - 5| \stackrel{?}{=} 16$$
$$|-16| \stackrel{?}{=} 16$$
$$16 = 16$$

Example 2

Find the roots of the following equation and check the answers.

$$|5x + 4| + 1 = 10.$$

Explanation You may have noticed that the design of the equation in Example 1 was one where

the absolute value of the quantity equals a number.

This was so arranged for under this condition the number on the right of the equality sign represents the distance from the point to the origin. Thus, in Example 1 we knew that the $3x - 5$ names two points that are 16 units from the origin. In Example 2 we must first determine the distance from the points named by $5x + 4$ to the origin. To determine this it is necessary to isolate $|5x + 4|$ on the left side of the equation. This is done in the usual manner by applying the addition principle.

4.4/Solution of an Equation Involving Absolute Values

Solution $|5x + 4| + 1 = 10$

$|5x + 4| = 10 + (-1)$ Addition Principle

$|5x + 4| = 9$

$5x + 4 = +9$ and $\quad 5x + 4 = -9$

$\quad 5x = +9 + (-4) \qquad 5x = -9 + (-4)$

$\quad 5x = +5 \qquad\qquad 5x = -13$

$\quad x = 1$ and $\quad x = -2\frac{3}{5}$

Check for $x = 1$: $\qquad\qquad$ Check for $x = -2\frac{3}{5}$:

$|5x + 4| + 1 = 10 \qquad\qquad |5x + 4| + 1 = 10$

$|5 \cdot 1 + 4| + 1 \stackrel{?}{=} 10 \qquad |5(-2\frac{3}{5}) + 4| + 1 \stackrel{?}{=} 10$

$|5 + 4| + 1 \stackrel{?}{=} 10 \qquad\quad |-13 + 4| + 1 \stackrel{?}{=} 10$

$|9| + 1 \stackrel{?}{=} 10 \qquad\qquad\quad |-9| + 1 \stackrel{?}{=} 10$

$10 = 10 \qquad\qquad\qquad\quad 10 = 10$

Example 3

Find the roots of the following equation and check your answers.

$10 - |2x - 5| = 3$

Solution (1) $\quad 10 - |2x - 5| = 3$

(2) $\quad -|2x - 5| = 3 + (-10)$ Addition Principle

(3) $\quad -1|2x - 5| = -7$

(4) $\quad |2x - 5| = 7$ Multiplication Principle

$2x - 5 = +7$ and $\quad 2x - 5 = -7$

$2x = +7 + (+5) \qquad 2x = -7 + (+5)$

$2x = +12 \qquad\qquad 2x = -2$

$x = 6$ and $\quad x = -1$

Check for $x = 6$: $\qquad\qquad$ Check for $x = -1$:

$10 - |2x - 5| = 3 \qquad\qquad 10 - |2x - 5| = 3$

$10 - |2 \cdot 6 - 5| \stackrel{?}{=} 3 \qquad\quad 10 - |2(-1) - 5| \stackrel{?}{=} 3$

$10 - |12 - 5| \stackrel{?}{=} 3 \qquad\qquad 10 - |-2 - 5| \stackrel{?}{=} 3$

$10 - |7| \stackrel{?}{=} 3 \qquad\qquad\quad 10 - |-7| \stackrel{?}{=} 3$

$10 - 7 \stackrel{?}{=} 3 \qquad\qquad\qquad 10 - (7) \stackrel{?}{=} 3$

$3 = 3 \qquad\qquad\qquad\quad 10 - 7 \stackrel{?}{=} 3$

$\qquad\qquad\qquad\qquad\qquad\qquad 3 = 3$

Explanation In step 3 we isolated the *negative* of the absolute value of the quantity. Hence, it was necessary to multiply both sides of the equation by the multiplicative inverse of -1 to obtain equation (4).

Before leaving this topic we should point out that mathematicians have agreed that distance will *always* be represented as a positive number. In view of this, were we to encounter a situation such as

$$|3x - 4| = -5,$$

we would know immediately that there are *no* replacements for x that will make this equation true. This is so for the above equation states that the *distance* from the point named $3x - 4$ to the origin is **negative** 5 units. Since distance can *never* be a negative number, the equation is meaningless and therefore has no roots.

An example of this condition arises when trying to solve the equation

$$|7 + 5x - 1| = 4.$$

When the absolute quantity is isolated we obtain

$$|5x - 1| = 4 + (-7)$$
$$|5x - 1| = -3.$$

At this point we realize that it is needless to go any further, for the distance from any point to the origin can never be a *negative* 3. The original equation has no roots.

Other occasions do arise when it is not so apparent that the equation has no roots. Whenever called upon to solve an equation containing absolute values, it is best to check the numbers found. The equation may not have any roots.

EXERCISES 4.4

Determine the roots of each of the following equations. Check your answers.

1. $|x + 2| = 9$
2. $|x - 6| = 5$
3. $|3x| = 12$
4. $|5x| = 10$
5. $|3x| - 9 = 6$
6. $|2x - 3| = 11$
7. $|3x + 9| = 9$
8. $-|2x - 1| = -7$
9. $-|5x - 6| = -24$
10. $|2x| + 3 = 15$
11. $|3x| + 4 = 10$
12. $|7x| - 9 = 5$

13. $9 - |x| = 6$
14. $|x + 10| - 4 = 1$
15. $|2x - 7| = -5$
16. $|x - 6| + 3 = 12$
17. $|3x - 7| + 2 = 17$
18. $6 - |x - 6| = 1$
19. $15 - |x - 4| = 2$
20. $14 - |2x - 3| = 7$
21. $8 + |3x - 1| = 5$
22. $15 = |2x + 1|$
23. $12 = |3 - 3x|$
24. $17 = |2x + 3| + 12$

4.5 SOLUTION OF AN INEQUALITY

Thus far in this chapter we have examined several methods for finding the elements in the solution set of a linear equation. Our attention now turns to determining a general approach for finding the elements in the solution set of a linear *inequality*. Before doing this we will have to develop an addition principle and a multiplication principle for *inequalities* that are similar to those we used for equalities.

Consider the mathematical sentence

$$10 > 7.$$

Recall that the arrowhead, $>$, always points to the smaller number. Hence, we read the above sentence as,

"Ten is greater than seven."

Were we to add 5 to each member of this inequality we would obtain 15 for the left member and 12 for the right member. Since 15 is greater than 12, the arrowhead will continue to point to the right.

$$10 + 5 > 7 + 5$$
or $$15 > 12.$$

What seems to be true here is that:

If the left member of an inequality *is greater* than the right member and should we add the *same* number to both members of the inequality, then the *new* left member *will be greater than* the *new* right member.

The question now arises as to whether this same principle will hold if we add a negative number to each member of the inequality. Let us test this by adding −4 to each member of the inequality.

New Left Member	New Right Member
10 + (-4)	7 + (-4)
= 6	= 3

Since 6 is greater than 3 it appears that this principle is true regardless of whether the number added to both members is positive or negative.

Actually, the principle stated in the preceding paragraph will always be true. The formal statement of this principle is,

---

Addition Principle of Inequality

If a, b, and c are any three real numbers, and if $\quad a > b$, then $\quad a + c > b + c$.

Now let's examine what will happen should we multiply both members of an inequality by the same number. Again we will use the inequality

$$10 > 7.$$

Were we to multiply both members by +2, the left member becomes +20 while the right member will be +14.

$$10 > 7$$
$$(+2)(10) > (+2)(7)$$
$$+20 > +14$$

It appears at first glance that multiplying both sides of an inequality by the same number will leave *the order of the inequality unchanged*; that is, if the left member is greater than the right member, then after multiplication by the same number the *new* left member will be *greater* than the *new* right member.

Unfortunately, the order remains unchanged *only* if the multiplier happens to be a *positive number*. Consider what will happen if we multiply both members of the inequality below by -3.

$$10 > 7$$

New Left Member	New Right Member
(-3)(10)	(-3)(7)
= -30	= -21

4.5/Solution of an Inequality

Since -30 is less than -21, the order of inequality reversed itself; that is, where originally the left member is *greater* than the right member, after multiplication by a *negative* number the left member is *less* than the right member.

$$10 > 7$$
$$(-3)(10) < (-3)(7)$$
$$-30 < -21$$

In view of this, the multiplication principle for inequalities must be separated into two parts. In the first of these we point out that the order of inequality will remain unchanged when the multiplier is a positive number. In the second part we point out that the order of inequality reverses itself when the multiplier is a negative number.

Multiplication Principle of Inequality

If a, b, and c are any three real numbers, and if $a > b$,

then $\quad ca > cb \quad$ if c is positive

or $\quad ca < cb \quad$ if c is negative

Except for the very last step the procedure for finding the solution set of an inequality is identical to that for finding the solution set of an equality. In the last step before applying the multiplication principle we must ask ourselves,

"By what number are we multiplying both members of the inequality?"

If the answer is, "a positive number," then the order of inequality remains unchanged. If the answer is, "a negative number," then we must make certain to reverse the order of inequality.

Example 1

Find the solution set for the following inequality:

$$5x - 6 > 3x + 8.$$

Solution
(1) $\quad 5x - 6 > 3x + 8$
(2) $\quad 5x > 3x + 8 + (+6) \quad$ Addition Principle
(3) $\quad 5x + (-3x) > +8 + (+6) \quad$ Addition Principle
(4) $\quad 2x > +14 \quad$ Collecting Like Terms
(5) $\quad x > +7 \quad$ Multiplication Principle

Explanation The first four steps are identical to those we used when solving an equation. In going from step (4) to step (5), though, we asked ourselves, "By what number must we multiply both members of this inequality?" Since the answer is, "a *positive* $\frac{1}{2}$," the order of inequality *remains unchanged*. As a final point, the inequality (5) informs us that *the set of numbers greater than +7 will satisfy the original inequality* $5x - 6 > 3x + 8$.

Example 2

Find the roots of the following inequality:

$$2x > 5x + 18.$$

Solution

(1) $\quad 2x > 5x + 18$

(2) $\quad 2x + (-5x) > 18 \qquad$ Addition Principle

(3) $\quad \quad \quad -3x > 18 \qquad$ Collecting Like Terms

(4) $\quad \quad \quad \quad x < -6 \qquad$ Multiplication Principle

Explanation Notice that the order of inequality in step (4) is the reverse of that in step (3). This is caused by the fact that both members of the inequality in (3) are multiplied by a **negative** $\frac{1}{3}$. The inequality in (4) implies that all numbers **less than** -6 are roots of the inequality $2x > 5x + 18$.

Example 3

Find the roots of the following inequality:

$$4(2x - 6) - (3x + 5) > 9x - 37.$$

Solution $\quad 4(2x - 6) - (3x + 5) > 9x - 37$

$\quad \quad \quad \quad 8x - 24 - 3x - 5 > 9x - 37$

$\quad \quad \quad \quad \quad \quad 5x - 29 > 9x - 37 \qquad$ Collecting Like Terms

$\quad \quad \quad \quad \quad \quad \quad \quad 5x > 9x - 37 + (+29) \qquad$ Addition Principle

$\quad \quad \quad \quad \quad 5x + (-9x) > -37 + (+29) \qquad$ Addition Principle

$\quad \quad \quad \quad \quad \quad \quad -4x > -8 \qquad$ Collecting Like Terms

$\quad \quad \quad \quad \quad \quad \quad \quad x < 2 \qquad$ Multiplication Principle

Explanation As pointed out earlier, the procedure here is identical to that used in the solution of an equality except for the very last step. In this situation both members of the inequality $-4x > -8$ are multiplied by **negative** $\frac{1}{4}$. Since the multiplier is a negative number, the order of inequality has to be reversed. The inequality $x < 2$ implies that all numbers **less than** 2 will satisfy the original inequality.

4.6/Solution of an Inequality Involving Absolute Values   153

EXERCISES 4.5

A

Find the roots of each of the following inequalities.

1. $5x > 35$
2. $7x > -42$
3. $2x + 6x > 16$
4. $9x < 18$
5. $x + 3x < -24$
6. $6x > -4 - 8$
7. $-2x < 12$
8. $-8x < 32$
9. $-5x > -20$
10. $-x > 14$
11. $5x - 7x < 8$
12. $-3x < -11 - 4$
13. $4x + 5 > 29$
14. $3x - 7 < 11$
15. $19 - 2x > 7$
16. $5x < 3x + 4$
17. $-x > 14 - 3x$
18. $x > 7x - 30$
19. $8x - 11 < 4x - 3$
20. $3x + 5 < 9x - 1$
21. $12 - 2x < x - 9$
22. $5x + 4 - 3x > 8x + 24$
23. $14 - x > 3x + 18 - 6x$
24. $12 - 4x - 5 < 7x - 23 - x$

B

Find the roots of each of the following inequalities.

1. $5x + (x - 2) > 10$
2. $7x - (2x + 9) < 1$
3. $15 - (7 - 2x) < -4$
4. $3(x - 5) > -6$
5. $-2(-1 + x) > 10$
6. $5x < 3(x - 4)$
7. $-3x < 2(2x + 7)$
8. $4x + 3(x + 8) < 3x$
9. $x > 4 - (x - 6)$
10. $3x < 5x - (12 - x)$
11. $2x + (x - 5) > 3x - (x - 4)$
12. $4(2x - 3) > 3(x - 2) - 1$
13. $-(x - 1) < 6 + (x - 3)$
14. $3(x - 5) - 6(2x - 7) < 0$
15. $2(3x + 4) - (2x - 6) > -3x$

4.6 SOLUTION OF AN INEQUALITY INVOLVING ABSOLUTE VALUES

Having examined the solution of a linear equality involving absolute values, it would seem only natural that we would want to investigate the solution of a linear inequality involving absolute values. We will find that there are some rather interesting situations that occur in this investigation.

Consider the equality

$$|x| = 3.$$

The translation of this equation into an English sentence is,

"The points named by x are 3 units from the origin."

And we learned that these points can be either +3 or -3 for both of these points are 3 units from the origin.

Similarly, the inequality

$$|x| > 3$$

is translated into the English sentence as,

"The points named by x are *more than* 3 units from the origin."

Here, again, there are two sets of points that fulfill this condition. Not only are all the points to the right of +3 more than 3 units from the origin but also

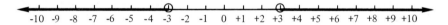

FIGURE 4-4

all those points to the left of -3 are more than 3 units from the origin (Figure 4-4). Hence, the inequality

$$|x| > 3$$

can be expressed in terms of two linear inequalities. These two are written as

$$x > +3 \quad \text{or} \quad x < -3.$$

The word "or" is deliberately written between the two inequalities for the replacements for x in $|x| > 3$ are numbers that are either greater than +3 **or** less than -3.

You may recall that the points indicated by the bold line in Figure 4-4 are called the *graph* of the inequality

$$|x| > 3.$$

The graph of a mathematical sentence is the set of points on the number line that are named by numbers that satisfy the sentence. Circles were drawn around the points +3 and -3 to indicate that these points are *not* to be included in the graph of this inequality. Replacing x with either +3 or -3 will not make this inequality a true sentence.

Example 1

Solve and graph the following inequality:

$$|2x - 5| > 11.$$

Explanation As stated above, we must realize that this inequality can be represented by two linear inequalities. These are

$$2x - 5 > +11 \quad \text{or} \quad 2x - 5 < -11$$

Having established these inequalities, we solve each of them separately. We must keep in mind, though, that the roots of $|2x - 5| > 11$ are the roots of *both* of the above inequalities.

Solution

$$|2x - 5| > 11$$
$$2x - 5 > +11 \quad \text{or} \quad 2x - 5 < -11$$
$$2x > +11 + (+5) \quad\quad 2x < -11 + (+5)$$
$$2x > +16 \quad\quad\quad\quad 2x < -6$$
$$x > +8 \quad \text{or} \quad\quad x < -3$$

Graph

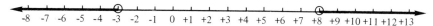

Thus far we have examined the situation where the absolute value of the variable is *greater than* a certain number. What will the situation be, though, if the absolute value of the variable is *less than* a certain number? Consider

$$|x| < 3.$$

After translating this inequality into an English sentence it becomes,

"The points named by x are less than 3 units from the origin."

In view of this, not only must the points be to the left of +3 but they cannot be any further to the left than -3. Were they further to the left than -3, such

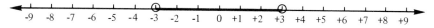

FIGURE 4-5

as -3.5 or -4, then their distance to the origin would be more than 3 units. Hence, not only *must* the points be to the left of +3 but also they *must* be to the right of -3. Figure 4-5 is the graph of

$$|x| < 3.$$

Hence, it would appear that if the absolute value of a variable is *less than* a certain number, then that variable will name points that are *between* two points on the number line.

We can represent the inequality

$$|x| < 3$$

as the two linear inequalities

$$x < +3 \quad \text{and} \quad x > -3.$$

Notice that whereas in the earlier situation the *connective* between the two inequalities was "or," now the *connective* is "and." The connective "and" implies that the roots of

$$|x| < 3$$

must be roots of *both*

$$x < +3 \quad \text{and} \quad x > -3.$$

For instance, the number +2 is a root of $x < +3$ and also a root of $x > -3$. Hence, +2 is a root of $|x| > 3$. Now consider the number -5. Although it is a root of $|x| < +3$ it is not a root of $|x| > -3$; hence it is *not* a root of $|x| < 3$.

Example 2

Solve and graph the following inequality:

$$|2 - x| < 3.$$

Solution

$$|2 - x| < 3$$

$$2 - x < +3 \quad \text{and} \quad 2 - x > -3$$

$$-x < +3 + (-2) \quad\quad -x > -3 + (-2)$$

$$-1x < +1 \quad\quad -1x > -5$$

$$x > -1 \quad \text{and} \quad x < +5$$

4.6/Solution of an Inequality Involving Absolute Values

Graph

```
◄─┼──┼──┼──┼──┼──┼──○──┼──┼──┼──┼──○──┼──┼──┼──┼──┼─►
 -10 -9 -8 -7 -6 -5 -4 -3 -2 -1  0 +1 +2 +3 +4 +5 +6 +7 +8 +9 +10
```

Explanation   In this solution we have shown that *all* the numbers *between* -1 and $+5$ will satisfy the inequality $|2 - x| < 3$.

This summary should help guide our thinking:

---

1. If the absolute value of a quantity is *greater than* a certain number:

   a. The connective will be the word "or."
   b. The graph will be points *to the left* of one point and *to the right* of a second point. (See the graph in Example 1.)

2. If the absolute value of a quantity is *less than* a certain number:

   a. The connective will be the word "and."
   b. The graph will be points *between* two points. (See the graph in Example 2.)

Mathematical sentences where the connective is either "and" or "or" are called *compound sentences*.

Example 3

Solve and graph the following inequality:

$$|2x + 5| - 6 < 1.$$

Solution

$$|2x + 5| - 6 < 1$$

$$|2x + 5| < 1 + (+6)$$

$$|2x + 5| < 7$$

$2x + 5 < +7$	and	$2x + 5 > -7$
$2x < +7 + (-5)$		$2x > -7 + (-5)$
$2x < +2$		$2x > -12$
$x < +1$	and	$x > -6$

Graph

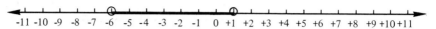

One final word: although we have not investigated them, there are inequalities involving absolute values that do *not* have any roots.

EXERCISES 4.6

Solve and graph each of the following inequalities.

1. $|x| + 2 > 3$
2. $|x| - 1 > 7$
3. $|x| + 3 < 10$
4. $|x| - 5 < 1$
5. $|x + 2| > 3$
6. $|x - 3| > 5$
7. $|2x - 1| > 9$
8. $|2x + 3| < 15$
9. $|3x + 6| < 12$
10. $5 + |2x| < 13$
11. $|x + 1| + 1 > 6$
12. $|2x - 5| - 3 > 8$
13. $|5 - x| < 1$
14. $|7 - 2x| < 3$
15. $|3 - x| > 7$
16. $|9 - 3x| > 3$
17. $|4 - x| + 3 > 8$
18. $|1 - 2x| - 5 < 4$
19. $5 > |2x + 1|$
20. $7 > 2 + |x - 3|$

CHAPTER REVIEW

1. Solve each of the following equations.
   a. $x + 5 = 9$
   b. $x - 6 = 4$
   c. $10 = x + 7$
   d. $-3 = x + 4$

2. Solve each of the following equations.
   a. $7x = 14$
   b. $5x = -30$
   c. $-2x = 16$
   d. $-24 = -3x$
   e. $-15 = 6x$
   f. $\frac{2}{3}x = 4$
   g. $\frac{4}{5}x = -8$
   h. $14 = -\frac{2}{7}x$
   i. $-10 = -\frac{1}{2}x$
   j. $\frac{1}{3}x = -\frac{2}{3}$

3. Find the root of each of the following equations.
   a. $2x + 7 = 16$
   b. $3x - 5 = 10$
   c. $9 - 6x = -21$
   d. $-1 + 8x = 3$
   e. $5x - 9 = 2x$
   f. $24 - 7x = x$
   g. $\frac{1}{2}x - 5 = 1$
   h. $12 - x = x$

Chapter Review

4. Find the root of each of the following equations. Check to determine if your answer is correct.
   a. $5x + 9 = 3x + 15$
   b. $x - 7 = 2x - 10$
   c. $8 - 3x = 12 - x$
   d. $x + 10 - 5x = 4x$

5. Find the root of each of the following equations.
   a. $3(x - 7) = 15$
   b. $-2(2x - 3) = -3x$
   c. $2x - (4x + 5) = 7$
   d. $4 + 3(x - 6) = x$
   e. $3(2x - 3) + 5(5 - x) = 3x - 12$
   f. $-(4 - x) - 2(x - 3) = 4x - 18$

6. Find the roots of each of the following equations. Check to determine if your answers are correct.
   a. $|x + 5| = 7$
   b. $|5x| = 15$
   c. $6 + |2x| = 16$
   d. $|3x - 6| + 1 = 10$
   e. $5 - |x - 4| = 3$
   f. $12 = |3x - 3|$

7. Find the roots of each of the following inequalities.
   a. $2x > 14$
   b. $-3x < 21$
   c. $2x - 3 > 11$
   d. $14 - x < 10$
   e. $15 - x < 3x - 1$
   f. $3x + (x - 4) < 16$
   g. $15 - (7 - 2x) > 20$
   h. $7x - 5(2x + 1) < 2x$
   i. $0 > x - (3x - 8)$
   j. $(x - 5) - 2(3x + 4) > 2$

8. Solve and graph each of the following inequalities.
   a. $|x| + 6 > 9$
   b. $5 + |x| < 7$
   c. $|3x| - 4 > 5$
   d. $|6 - x| > 2$
   e. $|2x - 3| - 5 < 10$
   f. $11 < 6 + |x - 5|$

Problem Solving in Mathematics

In this chapter we are going to take our first step toward applying a few of the concepts we have learned thus far in our study of algebra. There is very little value in knowing that the product of 7 and 6 is 42 if we do not realize that this information can be applied to finding the total cost of 7 apples when each is priced at 6 cents. So, too, there is very little value in being adept at solving equalities if we cannot apply this knowledge when the need arises.

We shall now investigate situations that are often referred to as "narrative problems." These problems are written in "story" form and pertain to numbers. At the close of each narrative a question is posed in which we are usually asked to determine the number or numbers described in the story.

These problems describe a great variety of matters to which numbers can be applied. Our study will touch on only a small fraction of them. Some we will find to be of importance in our daily living and all should prove to be of interest.

5.1 THE MATHEMATICAL PHRASE

Basic to the understanding of problem solving in mathematics is the ability to translate an English phrase into a mathematical phrase. The English phrases that are important to us are those that pertain to the four fundamental operations—addition, subtraction, multiplication, and division. Here are several of these phrases:

Five more than a number, (1)

The sum of a number and two, (2)

Four less than a number, (3)

Six times a number, (4)

Twice a number. (5)

We recognize the fact that words such as "more than" and "sum" imply the operation of addition; "less than" implies subtraction; and, finally, "times" and "twice" pertain to multiplication. Hence, if we permit the variable n to hold the place for the number in each of the above situations, then each can be rewritten with the mathematical phrase below.

$n + 5$ (Five more than a number) (1)

$n + 2$ (The sum of a number and 2) (2)

$n - 4$ (Four less than a number) (3)

$6n$ (Six times a number) (4)

$2n$ (Twice a number) (5)

The only translation above that might raise some question is the translation of (3). If the phrase were, "Two less than seven" we would subtract 2 from 7 to arrive at the number we are seeking. In order to express this subtraction we would write

$$7 - 2.$$

Similarly, the phrase, "Four less than a number" is translated as

$$n - 4$$

Example 1

Translate the phrase "The quotient of five and two."

Explanation The word quotient implies the operation of division and therefore we know that one of these numbers must be divided by the other. By custom, when the phrase is stated in the form above, we know that the first number is to be divided by the second number. This quotient can be written in the form $5 \div 2$ or $\frac{5}{2}$.

Solution $\qquad\qquad\qquad\qquad \frac{5}{2}$

Example 2

Translate the phrase, "Five less than three times a number."

Explanation After allowing the letter n to hold the place for the number, we can represent "three times the number" as $3n$. To express the number that is "five less than" $3n$ we simply write $3n - 5$ for "less than" implies the operation of subtraction.

Solution $\qquad 3n - 5$

Example 3

Consider n to be a whole number. Translate the phrase, "The next whole number after n."

Explanation Should we start with any whole number such as 6 and ask for the next whole number after 6, the answer will be 7. This number is obtained by adding 1 to 6. In general, the next succeeding whole number after any whole number is found by adding 1 to that whole number. Hence, the next succeeding whole number after n is $n + 1$.

Solution $\qquad n + 1$

Example 4

Translate the phrase, "The sum of two consecutive numbers."

Explanation Two consecutive numbers are two integers such as 25 and 26, 49 and 50, 107 and 108, and others where the second number is 1 more than the first. Hence, if we permit n to represent the first, then the second will have to be $n + 1$, for $n + 1$ is 1 more than n. The translation of the phrase, "The sum of two consecutive numbers," involves writing the sum of n and $n + 1$.

Solution $\qquad n + (n + 1)$

EXERCISES 5.1

A

Translate each of the following English phrases into a mathematical phrase.

1. Five more than two.
2. Three more than a number.
3. Nine increased by four.
4. A number increased by six.
5. Ten decreased by seven.
6. A number decreased by five.
7. Two less than nine.

8. Six less than a number.
9. Three added to twelve.
10. A number added to fifteen.
11. Seven subtracted from nine.
12. Ten subtracted from a number.
13. Seven times eight.
14. Four times a number.
15. The double of nine.
16. The double of a number.
17. Twenty divided by five.
18. A number divided by seven.
19. Seven divided by a number.
20. The quotient of twenty-four and eight.
21. The quotient of a number and three.
22. The quotient of three and a number.
23. The difference between seven and four.
24. The difference between four and seven.
25. The difference between a number and twenty.
26. The difference between twenty and a number.
27. Two more than three times a number.
28. Three less than five times a number.
29. A number added to three times the number.
30. A number subtracted from four times the number.

B

If n is allowed to be the placeholder for a number, how would you translate each of the following English phrases.

1. Twice the number n.
2. Five more than the number n.
3. Seven less than the number n.
4. Four added to the number n.
5. The difference between n and five.
6. The difference between seven and n.
7. The quotient of n and four.

8. The number n divided by three.
9. One-half of n.
10. Two more than three times n.
11. Two less than three times n.
12. The next consecutive integer after n.
13. Five times the next consecutive integer after n.
14. Twice the next consecutive integer after n.
15. The difference between twice n and the next consecutive integer after n.
16. The next consecutive integer after twice n.
17. n divided by 3.
18. The quotient of twice n and four.
19. The quotient of three less than twice n and four.
20. Four times the sum of n and twice n.

5.2 NUMBER PROBLEMS

PART 1

Having examined how to translate an English phrase into a mathematical phrase, we now turn our attention to the task of translating an English *sentence* into a mathematical sentence. The English sentences that are of interest to us are usually those in which the verb is one of the following words:

is, will give, gives, will be

From a mathematical standpoint each of these words can be replaced by the word "equals." And in the translation, the word "equals" is replaced by the equality sign (=).

Examples of sentences in which these words appear are:
1. Five times a number *is* thirty-five.
2. Twice a number added to four times the number *will give* fifty-four.
3. Three times a number decreased by six *gives* twenty-seven.
4. Three times a number *will be* eight more than the number.

The translation of each of these sentences requires three steps:
1. Translate the English phrase to the left of the verb.
2. Replace the verb with the equality sign.
3. Translate the English phrase to the right of the verb.

Let us examine each of the sentences separately. In each of the translations we will allow n to represent the number.

1. Five times a number is thirty-five.

$$5 \cdot n = 35$$

2. Twice a number added to three times the number will give fifty-four.

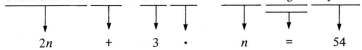

$$2n + 3 \cdot n = 54$$

3. Three times a number decreased by six gives twenty-seven.

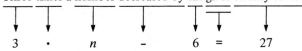

$$3 \cdot n - 6 = 27$$

4. Three times a number will be eight more than the number.

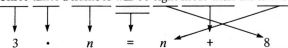

$$3 \cdot n = n + 8$$

EXERCISES 5.2 (Part 1)

Translate each of the following English sentences into mathematical sentences.

1. Seven more than a number is twenty-four.
2. Twice a number is nineteen.
3. A number added to three times the number will give twenty-eight.
4. Five more than three times a number is fifteen.
5. Twice a number will be five more than the number.
6. Four times a number decreased by six gives the number.
7. Six less than the number is eleven.
8. The sum of a number and fourteen is twice the number.
9. Seven less than four times a number is thirty-five.
10. The difference between five times a number and three times the number is twenty.
11. The difference between six times a number and four times the number is ten more than the number.
12. The sum of twice a number and nine is five less than the number.
13. The sum of two consecutive numbers is forty-seven.

5.2/Number Problems

14. If four is added to a number, *the sum is seven.*
15. If twice a number is added to four times the number, *the sum is sixty.*
16. If three times a number is subtracted from five times the number, *the difference is thirty.*
17. If a number is increased by four and then added to the number itself, *the sum will be fifty-six.*
18. If fifty-one is decreased by twice a number, *the difference will be three more than the number.*

PART 2

Once we have made the translation from the English sentence to the mathematical sentence, we have moved a giant step toward our goal. As stated earlier, the narrative usually ends with a sentence requesting that we find the number discussed in the problem. Hence, by solving the mathematical translation, we determine the number called for.

The following examples should help clear up any difficulties.

Example 1

Five more than six times a certain number is forty-seven. What is the number?

Explanation Our initial step is to indicate that we will represent the "certain number" by a variable of our own choice. Frequently the variable is either *n, x,* or *y.* Having done this, we translate the English sentence to the mathematical sentence.

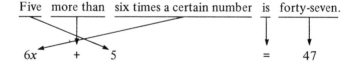

By solving this equation we determine the "certain number."

Solution Let *x* represent the certain number.

Therefore,
$$6x + 5 = 47$$
$$6x = 47 + (-5)$$
$$6x = 42$$
$$x = 7 \quad \text{(the number)}$$

Explanation continued Having found the certain number to be 7, we now check to determine if this number satisfies the conditions of the problem. It is important to realize that we do *not* check our work by replacing *x* with 7 in the equation $6x + 5 = 47$. Since it was *we* who translated the English sentence into $6x + 5 = 47$, there is always the possibility that we may have erred in the translation. In view of this we check our answer by reexamining the *original* English sentence. Hence, we must find the number that is, "Five more than six times 7" and hope that it "is forty-seven." Since it is, we can conclude that 7 is the correct answer.

$$6 \cdot 7 + 5 = 42 + 5 = 47$$

Example 2

The sum of a number and its next consecutive number is ninety-five. What is the number?

Explanation By allowing n to represent the number, then $n + 1$ will represent the next consecutive number. Hence, the translation will be:

"The sum of a number and its next consecutive number is ninety-five."

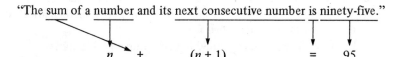

$$n \quad + \quad (n + 1) \quad = \quad 95$$

Solution Let n represent the number.

Therefore,
$$n + (n + 1) = 95$$
$$n + n + 1 = 95$$
$$n + n = 95 + (-1)$$
$$2n = 94$$
$$n = 47 \quad \text{(the number)}$$

Check: Since the sum of 47 and its next consecutive integer 48 is 95, we can conclude that 47 is the correct number.

Example 3

If a number is subtracted from twice its next consecutive number, the difference is eighteen. What is the number?

Explanation Since the "next consecutive number" will be written as $n + 1$, we must be careful that we write twice this number as $2(n + 1)$.

"The difference is eighteen."

$$2(n + 1) \quad - \quad n \quad = \quad 18$$

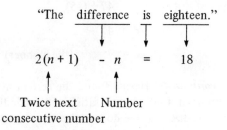

Twice next   Number
consecutive number

5.2/Number Problems

Solution Let n represent the number.

Therefore,
$$2(n + 1) - n = 18$$
$$2n + 2 - n = 18$$
$$2n - n = 18 + (-2)$$
$$n = 16 \quad \text{(the number)}$$

Check: The next consecutive number is 17.
Twice 17 is 34.
The difference is 34 - 16, or 18.
Hence, the answer is correct.

Example 4

If forty-seven is increased by three times a number, the sum is five less than the number. What is the number?

Explanation Translation of the English phrase,

"Forty-seven is increased by three times a number"

is, $\qquad 47 + 3x.$

Since $47 + 3x$ represents the "sum" referred to in the problem, then the translation of the English sentence will be:

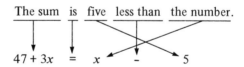

Solution Let x represent the number.

Therefore,
$$47 + 3x = x - 5$$
$$3x = x - 5 + (-47)$$
$$3x + (-x) = -5 + (-47)$$
$$2x = -52$$
$$x = -26 \quad \text{(the number)}$$

Check: 47 is increased by $3(-26) = 47 + (-78) = -31$
5 less than the number $= -26 - 5 = -31$
Since both numbers are the same, the answer is correct.

EXERCISES 5.2 (Part 2)

Find the number requested in each of the following problems.

1. Three times a certain number added to the number is fifty-six. What is the number?
2. Eleven more than twice a number is forty-nine. What is the number?
3. Five times a number is ten more than three times the number. What is the number?
4. Six times a number decreased by ten is four times the number. What is the number?
5. The sum of a number and its next consecutive number is eighty-three. What is the number?
6. Twelve less than a certain number will be four times the number. What is the number?
7. Twice a number is fifty-seven less than five times the number. What is the number?
8. The difference between twice a number and six is twenty. What is the number?
9. The difference between three times a number and nineteen is seven more than the number. What is the number?
10. The difference between seven times a number and three times the number gives fifteen more than the number. What is the number?
11. If nine times a number is added to twice the number, the sum is forty-four. What is the number?
12. If five times a number is increased by thirty-nine, the sum is seventy-four. What is the number?
13. If forty-five is increased by six times a certain number, the sum is the number itself. What is the number?
14. If fifty-six is decreased by a certain number, the difference is seven times the number. What is the number?
15. If a number is increased by sixteen, the sum is twice the next consecutive number. What is the number?
16. When eight is subtracted from three times a certain number, the difference will be the number increased by ten. What is the number?
17. If five more than a certain number is doubled, the result will be thirty-eight. What is the number?
18. If the sum of a certain number and six is multiplied by nine, the result will be twenty-seven. What is the number?

5.2/Number Problems

19. The difference between one-half of a certain number and seventeen is three. What is the number?
20. If three more than one-half of a certain number is multiplied by four, the result is six. What is the number?

PART 3

Thus far we have examined narrative problems in which we were requested to find only one number. There are situations, though, in which we will be asked to determine *two* numbers. In this event, the story will contain *two* pieces of information.

1. One of the pieces of information will relate the size of one of the numbers to the other number.
2. The other piece of information will enable us to determine the equation.

Perhaps the best way of learning what we must do under this circumstance is to take a close look at several problems that involve finding two numbers.

Example 1

The larger of two numbers is five more than the smaller. The sum of the two numbers is forty-seven. What are the two numbers?

Explanation The two pieces of information in this problem are:

1. The larger number is five more than the smaller.
2. The sum of the two numbers is forty-seven.

It is the first piece of information here that relates the larger number to the **smaller**. That is, once we know the smaller number, we immediately know that the larger can be found by adding 5 to that number. Thus, if we represent the smaller by x, then the larger is represented by $x + 5$.

The second piece of information is used to determine the equation. The translation of the English sentence is:

"The sum of the two numbers is forty-seven."

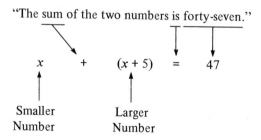

Solution Let x represent the smaller number.
Then $x + 5$ represents the larger number (first piece of information).

172 5/Problem Solving in Mathematics

Therefore, $\quad x + (x + 5) = 47 \quad$ (second piece of information)

$$x + x + 5 = 47$$
$$x + x = 47 + (-5)$$
$$2x = 42$$
$$x = 21 \quad \text{(the smaller number)}$$
$$x + 5 = 26 \quad \text{(the larger number)}$$

Check: The sum of the two numbers 26 and 21 is 47.
Hence, we can conclude the numbers are correct.

In order to make the decision as to which of the two numbers will be represented by x, we ask ourselves this question,

"Which of the two numbers is dependent upon the other?"

In Example 1 it is the larger number that is dependent upon the smaller number. Hence, for lack of a better name, we call the smaller number in this situation the *basic* number. Were the information,

"The smaller number is one-half of the larger number."

then the basic number would be the larger number.

Once we decide which number is the basic number, it is that number we represent by some variable. We then represent the other number in terms of the basic number.

Notice, also, that in the solution above when we found x we had determined only one of the two numbers. Since the other was represented by $x + 5$, we added 5 to 21 to find the larger number.

Example 2

The difference between two numbers is sixteen. The smaller number is fifty less than three times the larger. What are the numbers?

Explanation The two pieces of information are:

1. The difference between numbers is sixteen.
2. The smaller is fifty less than three times the larger.

It is the second piece of information here that relates one number to the other. Since it is the larger number that is the basic number, we represent it by n. The smaller being "fifty less than three times the larger" is represented by $3n - 50$. When writing the equation based on the first piece of information, *we must be careful to subtract the smaller number from the larger number.*

5.2/Number Problems

Solution Let n represent the larger number.
Then $3n - 50$ represents the smaller number (second piece of information).

Therefore, $\quad n - (3n - 50) = 16 \quad$ (first piece of information)
$\qquad\qquad\quad n - 3n + 50 = 16$
$\qquad\qquad\qquad\;\; n - 3n = 16 + (-50)$
$\qquad\qquad\qquad\quad\;\; -2n = -34$
$\qquad\qquad\qquad\qquad\; n = 17 \qquad$ (the larger number)
$\quad 3n - 50 = 3 \cdot 17 - 50 = 51 - 50 = 1 \qquad$ (the smaller number)

Check: The difference between the two numbers 17 and 1 is 16.
Hence, we can conclude the numbers are correct.

Example 3

Of three numbers, the first is nine more than the second number and the third is twice the first. If the sum of the numbers is seventy-five, what are the three numbers?

Explanation Since there are *three* numbers involved in the problem, there will have to be *three* pieces of information. They are:

1. The first number is nine more than the second.
2. The third number is twice the first.
3. The sum of the three numbers is seventy-five.

The first task is to recognize which of the three numbers is the "basic" number. Knowing the second we can represent the first and knowing the first we can represent the third. Once we have represented the three numbers, we use the last piece of information to write the equation.

Solution Let x represent the second number.
Then $x + 9$ represents the first number (first piece of information).
Then $2(x + 9)$ represents the third number (second piece of information).

Therefore, $x + (x + 9) + 2(x + 9) = 75 \qquad$ (third piece of information)
$\qquad\qquad\; x + x + 9 + 2x + 18 = 75$
$\qquad\qquad\qquad\qquad 4x + 27 = 75$
$\qquad\qquad\qquad\qquad\qquad 4x = 75 + (-27)$
$\qquad\qquad\qquad\qquad\qquad 4x = 48$
$\qquad\qquad\qquad\qquad\qquad\; x = 12 \qquad$ (second number)
$\qquad\qquad\qquad\qquad\; x + 9 = 21 \qquad$ (first number)
$\qquad\qquad\qquad\; 2(x + 9) = 2 \cdot 21 = 42 \qquad$ (third number)

Check: The sum of the three numbers 12, 21, and 42 is 75.
Hence, we can conclude that the numbers are correct.

Example 4

John is twice as old as Bill and the sum of their ages is fifty-one. How old is each of them?

Explanation   The only difference between this illustration and the earlier ones is that where previously the term "number" was used, now the term "age" appears. The two pieces of information are:

1. John is twice as old as Bill.
2. The sum of their ages is fifty-one.

From the first piece of information we discover that Bill's age is the "basic" number.

Solution   Let x represent Bill's age (first piece of information).
Then $2x$ represents John's age (second piece of information).

Therefore,
$$x + 2x = 51$$
$$3x = 51$$
$$x = 17 \quad \text{(Bill's age)}$$
$$2x = 34 \quad \text{(John's age)}$$

Check: The sum of their ages 17 and 34 is 51.
Hence, we can conclude that the ages are correct.

EXERCISES 5.2 (Part 3)

Find the numbers requested in each of the following problems.

1. The larger of two numbers is twelve more than the smaller. The sum of the two numbers is twenty-six. What are the two numbers?
2. The smaller of two numbers is fifteen less then the larger. If the sum of the two numbers is thirty-seven, what are the two numbers?
3. The larger of two numbers is seven more than twice the smaller. If the sum of the numbers is sixty-one, what are the numbers?
4. The larger of two numbers is one less than three times the smaller. If the difference of the two numbers is thirteen, what are the numbers?
5. The difference of two numbers is eight. The smaller is twenty less than twice the larger. What are the two numbers?
6. Mary is five years older than Betty. If the sum of their ages is twenty-three, how old is each?
7. Tom's age is two years more than three times his sister's age. The difference of their ages is twelve. How old is each?

8. Hazel is three years older than Cathy. Four times Hazel's age is the same as five times Cathy's age. How old is each?

9. If three years is subtracted from twice Rocco's age, Steve's age is found. The sum of their ages is thirty. What is the age of each?

10. Of three numbers, the second is three more than the first and the third is eight more than the first. If the sum of the numbers is forty-seven, what is each number?

11. The first of three numbers is twice the second while the third is six more than the second. The sum of the three numbers is sixty-two. What are these numbers?

12. The first of three numbers is three more than the third and the second is twice the first. If the sum of the numbers is forty-nine, what are the numbers?

13. Of three numbers, the second is six times the first and the third is one more than three times the first. The sum of twice the third and five times the first is the same as twice the second. What are the numbers?

14. Robin is six years younger than Donna. Gail is six years older than Donna. If Donna's age is doubled and added to Robin's age, the sum will be twice Gail's age. What is the age of each girl?

15. Mona's weight is two-thirds of her brother Clarence's weight. If the difference of their weights is 48 pounds, how much does each weigh?

16. Lynn is twenty-three inches taller than her younger sister, Nora. If Lynn's height is one and one-half times as much as Nora's, how tall is each girl?

17. The sum of three consecutive numbers is twenty-one. What are the numbers? Suggestion: Represent the three numbers as x, $(x + 1)$, and $(x + 1) + 1$.

18. Of three consecutive numbers, the sum of the first and twice the second number plus three times the third number is thirty-eight. What are the three numbers?

5.3 COIN PROBLEMS

Narrative problems related to coins are usually concerned with the "value" of these coins. Specifically, when we ask for the value of a number of coins we are simply inquiring as to how many pennies (cents) can replace these coins. For instance, the value of 6 nickles is 30 cents, since it is possible to replace 6 nickles with 30 cents.

To determine the 30 cents, we multiplied the value of each nickle, which is 5¢, by the number of nickles. Since there were 6 nickles we multiplied 5¢ by 6 for a total of 30¢.

Similarly, to find the value of 3 quarters we would multiply the value of each quarter by the number of quarters. As there are 3 quarters and the value of each quarter is 25¢, then 25¢ times 3 will give us 75¢, which is the value of the three quarters.

In general we can say,

To find the total value of coins that are the same, we multiply the value of one coin by the number of coins.

As an example, the value of seven dimes is found by multiplying 10¢ by 7.

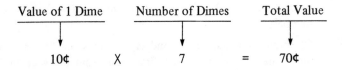

In the same way, to find the value of 6 quarters, we multiply 25¢ by 6.

$$25¢ \cdot 6 = 150¢ \qquad \text{Value of 6 quarters in cents}$$

Hence, to find the value of x quarters, we multiply 25¢ by x.

$$25¢ \cdot x = 25x \qquad \text{Value of } x \text{ quarters in cents}$$

Or to find the value of y dimes, we multiply 10¢ by y.

$$10¢ \cdot y = 10y \qquad \text{Value of } y \text{ dimes in cents}$$

And, finally, to find the value of $(x - 4)$ nickles, we multiply 5¢ by $(x - 4)$.

$$5¢ \cdot (x - 4) = 5(x - 4) \qquad \text{Value of } (x - 4) \text{ nickles in cents}$$

In view of the fact that we are expressing the value of coins in terms of cents, then we must always rewrite every amount of money we encounter in terms of cents. Thus,

$$\$6.72 \quad \text{must be rewritten as} \quad 672¢$$

Similarly,

$$\$4 \quad \text{must be rewritten as} \quad 400¢$$

Now let us take a look at several problems in which we apply this information.

5.3/Coin Problems

Example 1

The number of dimes a person has is three more than the number of nickles. The total of all his money is 75¢. How many nickles and how many dimes does he have?

Explanation The "basic" number here is the number of nickles. We represent the number of nickles by x. Since the number of dimes is three more than the number of nickles, the number of dimes is represented by $x + 3$. The value of the x nickles is $5x$ while the value of the $x + 3$ dimes is $10(x + 3)$. Since the total is 75¢, we need but add the value of the nickles to the value of the dimes to obtain 75¢.

Value of Nickles		Value of Dimes		Total Value
$5x$	$+$	$10(x + 3)$	$=$	75

Solution Let x represent the number of nickles. Then $x + 3$ represents the number of dimes.

Therefore,
$$5x + 10(x + 3) = 75$$
$$5x + 10x + 30 = 75$$
$$5x + 10x = 75 + (-30)$$
$$15x = 45$$
$$x = 3 \quad \text{(number of nickles)}$$
$$x + 3 = 6 \quad \text{(number of dimes)}$$

Check: 3 nickles is 15¢
6 dimes is 60¢
Total value is 75¢

Example 2

A child's coin bank contains only nickles and quarters. There are eleven times as many nickles in the bank as there are quarters. The total amount of money in the bank is $8. How many of each coin are there?

Explanation It is usually easier to translate the piece of information,

"There are eleven times as many nickles as there are quarters."

if we rewrite it as

"The number of nickles is eleven times the number of quarters."

By examining the last sentence we can see that the "basic" number is the number of quarters. If the number of quarters is represented by x, the number of nickles will then be $11x$. Hence, the equation is,

Value of Quarters		Value of Nickles		Total Value
↓		↓		↓
$25x$	$+$	$5(11x)$	$=$	800

Notice that the $8 was changed to 800¢ for the value of the quarters and the value of the nickles are given in terms of cents.

Solution Let x represent the number of quarters.
Then $11x$ represents the number of nickles.

Therefore,
$$25x + 5(11x) = 800$$
$$25x + 55x = 800$$
$$80x = 800$$
$$x = 10 \quad \text{(number of quarters)}$$
$$11x = 110 \quad \text{(number of nickles)}$$

Check: 10 quarters is $2.50
110 nickles is $5.50
Total value is $8.00

Example 3

A person purchased one variety of applies at 32¢ per pound and another variety at 37¢ per pound. The number of pounds at 32¢ per pound is one more than twice the number at 37¢ per pound. How many pounds of each variety were purchased if the total cost was $5.37?

Explanation By representing the number of pounds at 37¢ per pound by x, the number of 32¢ per pound will have to be $(2x + 1)$. In the previous examples we interpreted the total value of w coins where each was valued at 5¢ as

$$5w.$$

In this example we interpret the total value of x pounds where each pound is priced at 37¢ as

$$37x.$$

5.3/Coin Problems

Similarly, the total value of $2x + 1$ pounds where each pound is priced at 32¢ is

$$32(2x + 1).$$

Hence, the equation becomes,

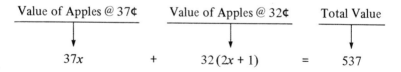

Solution Let x represent the number of pounds of apples at 37¢ per pound. Then $2x + 1$ represents the number of pounds at 32¢ per pound.

Therefore,
$$37x + 32(2x + 1) = 537$$
$$37x + 64x + 32 = 537$$
$$37x + 64x = 537 + (-32)$$
$$101x = 505$$
$$x = 5 \quad \text{(number of pounds @ 37¢)}$$
$$2x + 1 = 11 \quad \text{(number of pounds @ 32¢)}$$

Check: Cost of 5 pounds @ 37¢ is $1.85
Cost of 11 pounds @ 32¢ is $3.52
Total cost is $5.37

Example 4

A collection of twenty coins consists of only nickles and dimes. The total value of the coins is $1.70. How many of each kind are there?

Explanation The only new feature we need examine in this problem is the fact that the total number of coins is 20. If 3 are nickles, then 17 must be dimes. The 17 is arrived at by subtracting the 3 from the total of 20.

$$20 - 3 = 17$$

If 9 are nickles, then 11 are dimes. Here, again, we subtract the 9 from the total of 20.

$$20 - 9 = 11$$

In general, if x is the number of nickles, then to determine the number of dimes we would subtract the x from 20. Thus, in this case

$$20 - x$$

will be the number of dimes. Hence, the equation will be,

Value of the Nickles		Value of the Dimes		Total Value
$5x$	$+$	$10(20-x)$	$=$	170

Solution Let x represent the number of nickles.
Then $20 - x$ represents the number of dimes.

Therefore,
$$5x + 10(20 - x) = 170$$
$$5x + 200 - 10x = 170$$
$$5x - 10x = 170 + (-200)$$
$$-5x = -30$$
$$x = 6 \quad \text{(number of nickles)}$$
$$20 - x = 14 \quad \text{(number of dimes)}$$

Check: 6 nickles is $.30
14 dimes is $1.40
Total value is $1.70

EXERCISES 5.3

Find the numbers requested in each of the following problems.

1. Sarah had ninety cents in her coin purse. There were only nickles and dimes in the purse and the number of nickles was three more than the number of dimes. How many of each coin did she have?

2. A bag contains only pennies and nickles. The number of pennies is four times the number of nickles. If the total value of the money is 63¢, how many pennies and how many nickles are there?

3. A bank teller received $2.70 in dimes and quarters from a depositor. The number of dimes was twice the number of quarters. How many of each of these coins were deposited?

4. A cashier was given $15 in half-dollars and quarters. The number of half-dollars was five more than twice the number of quarters. How many half-dollars and how many quarters did she receive?

5. When counting only the nickles, dimes, and quarters in her cash register, a salesgirl found that she had $5.10. The number of dimes was seven more than the number of quarters and the number of nickles was four times the number of quarters. How many of each coin did she have?

5.4/Percent Problems

6. Susan kept a plate in her kitchen where she collected only quarters and dimes. At the end of one week she found that she had accumulated $6.75. There were five times as many dimes as there were quarters. How many of each coin were there on the plate?

7. In a collection of $2.45 there were four more nickels than dimes. How many of each coin were there?

8. A person spent $8.75 purchasing 13¢ stamps and 9¢ stamps. The number of 13¢ stamps he purchased was twice the number of 9¢ stamps. How many of each did he buy?

9. Mario purchased twice as many 13¢ stamps as he did 11¢ stamps. All told he spent $16.65. How many of each did he buy?

10. A total of $1.74 was spent on the purchase of two different varieties of grapefruit. The number at 12¢ each was two more than the number at 18¢ each. How many of each variety were purchased?

11. A coin bank contains only dimes and quarters and there are thirty coins in all. The total of all the coins is $6.15. How many coins of each variety are there?

12. A person purchased 100 postal cards at a total cost of $12.15. Some were priced at 18¢ each while the rest were priced at 9¢ each. How many of each variety were purchased?

13. Ethel Peters purchased thirty-three oranges for which she paid $3.51. She paid 9¢ each for some of the oranges and 12¢ each for the others. How many of each variety did she purchase?

14. A person spent $6.14 for 18 pounds of grapes. Some of the grapes were purchased at 31¢ per pound while the rest cost 39¢ per pound. How many pounds of each variety were purchased?

15. Joe Santiago prepares a blend of his own coffee for his restaurant. To do this he mixes coffee at $3.25 a pound with coffee that cost him $2.93 a pound. He finds that the cost of the blend is $62.44. If he has 20 pounds in all, how many pounds of each variety were used in the mixture?

5.4 PERCENT PROBLEMS

PART 1

A number that happens to be extremely useful in our daily lives unfortunately also happens to be one that is least understood. It is the percent number. There are many different definitions that we can find for this number. Perhaps the one that will suit our needs best is the one in which we consider a percent number as but another way of writing a fraction. The fraction, though, is a particular one; it is one in which the denominator is 100.

As an example, the number

$$23\%$$

is but another way of writing the fraction

$$\frac{23}{100}.$$

Similarly, the number 42.6% means 42.6/100; 5% means 5/100; .04% means .04/100; and 200% means 200/100.

Since the number 124.8% means 124.8/100 it is possible to express an equality between these numbers,

$$124.8\% = \frac{124.8}{100}.$$

The fraction 124.8/100 implies that the number 124.8 is to be divided by 100. When this is done the quotient is 1.248. Notice that the digits in the quotient 1.248 are exactly the same as in the original number 124.8. However, after 124.8 is divided by 100 the decimal point appears in the quotient two places further to the left than in the original number 124.8.

$$124.8 \div 100 = 1.24{\scriptstyle\wedge}8$$

Actually, whenever any number is divided by 100 the decimal point appears in the quotient two places further to the left than in the original number.

$$36.94 \div 100 = .36{\scriptstyle\wedge}94$$

$$5.6 \div 100 = .05{\scriptstyle\wedge}6$$

Let's take a moment to examine several situations where a percent number is renamed as a fraction number and the fraction number, in turn, is renamed as a decimal.

$$124.8\% = \frac{124.8}{100} = 1.24{\scriptstyle\wedge}8$$

$$36.94\% = \frac{36.94}{100} = .36{\scriptstyle\wedge}94$$

$$5.6\% = \frac{5.6}{100} = .05{\scriptstyle\wedge}6$$

5.4/Percent Problems

Based on these illustrations, it would appear that it is possible to *rename a percent number as a decimal number by merely moving the decimal point two places to the left.* Thus,

$$137.84\% = 1.37{\scriptstyle\curvearrowleft}84$$

$$94.6\% = .94{\scriptstyle\curvearrowleft}6$$

$$7.5\% = .07{\scriptstyle\curvearrowleft}5$$

$$.03\% = .00{\scriptstyle\curvearrowleft}03$$

On the other hand, a decimal number can be renamed as a percent number by reversing the process. That is, *a decimal number can be renamed as a percent number by moving the decimal point two places to the right.*

$$2.647 = 2{\scriptstyle\curvearrowright}64.7\%$$

$$.83 = {\scriptstyle\curvearrowright}83.\%$$

$$.056 = {\scriptstyle\curvearrowright}05.6\%$$

It is only normal to become confused over the direction in which the decimal point is to be moved. A device that may prove helpful in recalling this is to write the first letter of the word "Percent" and the first letter of the word "Decimal" as they appear in alphabetical order. That is, **D** appears before **P** in the alphabet and hence is written first.

D   P

If a number is to be renamed from a *decimal to a percent*, we draw an arrow from **D** to **P**. This reminds us that the decimal point is to be moved to the **right**.

D ⟶ P

If the number is to be renamed from a *percent to a decimal*, we draw an arrow from **P** to **D**. This reminds us that the decimal point is to be moved to the **left**.

D ⟵ P

The great bulk of situations in which we encounter the percent number in our daily lives falls under one of three conditions similar to these:

1. What number is 74% of 396?
2. 43% of what number is 89?
3. 59 is what percent of 43?

In reality, all three are English sentences that can be easily translated into mathematical sentences. Before doing this, we must first determine the mathematical translation for the term "of." This word appears in statements such as,

"Bill is to receive $\frac{2}{3}$ *of* the $9."

To determine how much Bill receives, we *multiply* $\frac{2}{3}$ by 9:

$$\frac{2}{3} \cdot \$9 = \$6.$$

In general, the term "of" will always imply the operation of multiplication. Now let us examine the three situations listed above.

Example 1

What number is 74% of 396?

Explanation By representing the words "what number" by *n*, the translation of this sentence immediately follows:

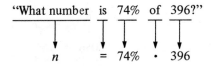

Since we do not know how to perform the operation of multiplication with percent numbers, we must rename the number 74% as a decimal. In doing this, it is necessary to recall that when the decimal point does not appear in a number, it is understood to be at the rightmost end of the number. Thus, the number

52 can be written as 52.

Hence, 74% can be written as 74.%.

Solution Let *n* represent the number.

Therefore,
$$n = 74\% \cdot 396$$
$$n = .74(396)$$
$$n = 293.04$$

Example 2

43% of what number is 89?

Explanation The translation of this sentence is:

"43% of what number is 89?"
43% · *n* = 89.

5.4/Percent Problems

Solution Let n represent the number.

Therefore,
$$43\% \cdot n = 89$$
$$.43n = 89$$
$$n = 206.98$$

Explanation continued In finding the value of n it is necessary to multiply both sides of the equation

$$.43n = 89$$

by the multiplicative inverse of .43. As usual, the product on the left side of the equation turns out to be n. On the right side, though, we get

$$\frac{1}{.43} \times 89.$$

This product is 89/.43, which implies that 89 is to be divided by .43. When this is done the quotient will be 206.98 *to the nearest hundredth.*

Example 3

59 is what percent of 45?

Explanation The translation of this sentence is:

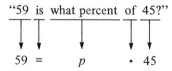

Solution Let p represent the number.

Therefore,
$$59 = p \cdot 45$$
$$59 = 45p$$
$$45p = 59$$
$$p = 1.31 \quad \text{(to the nearest hundredth)}$$
$$p = 131\%$$

Explanation continued Notice that the value of p is found first as the decimal number 1.31. However, the problem called for a percent number as the answer. Hence, the decimal number 1.31 is renamed as the percent number 131%.

EXERCISES 5.4 (Part 1)

Find the number requested in each of the following problems.

1. What number is 56% of 84?
2. What number is 7.8% of 169?
3. How much money is 123% of $420?
4. How much money is 4.5% of $3,400?
5. How much money is $4\frac{1}{2}$% of $2,000?
6. 42 is 25% of what number?
7. 73 is 38% of what number?
8. 7% of what number is 21?
9. $5\frac{1}{2}$% of how much money is $33?
10. 250% of how much money is $94?
11. 6.27% of $134 is how much money?
12. .4% of $2,500 is how much money?
13. 16 is what percent of 80?
14. 13 is what percent of 104?
15. 225 is what percent of 50?
16. What percent of $500 is $75?
17. What percent of $356 is $124?
18. What percent of $20 is $4.50?
19. What percent of $15 is $45?
20. What percent of $12.50 is $64?

PART 2

The problems in Part 1 of this section are somewhat artificial in nature for only rarely do we encounter them in real-life situations. More often the problems will be of the form,

A coat was originally priced at $149.50. During a post-season sale it was marked at $89.50. At what percent of the original price was the coat marked during the sale?

A narrative such as this has to be interpreted in terms of a single sentence similar to one of those in Part 1. Once this sentence has been determined, we translate it into a mathematical sentence which we then solve. For instance, the single important sentence that can replace the above question is,

"The marked priced is what percent of the original price?"

5.4/Percent Problems

After examining the narrative for information, we can translate this sentence as follows:

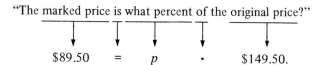

"The marked price is what percent of the original price?"

$89.50 = p · $149.50.

The value of p is found as in Part 1.

Examples 1-5 will help to clarify the method by which we can give meaning to life-situations that involve percent numbers.

Example 1

Employees of a company received an increase in earnings amounting to $1.04 per hour. Their hourly rate had been $5.07. By what percent was their hourly rate increased?

Explanation The rewording of the question and its translation are:

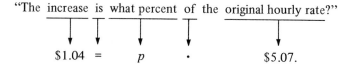

"The increase is what percent of the original hourly rate?"

$1.04 = p · $5.07.

Solution Let p represent the percent of increase.

Therefore, $1.04 = p · $5.07

1.04 = 5.07p

5.07p = 1.04

p = .21   (to nearest hundredth)

p = 21%

Example 2

During a short period of time the price of coffee rose by 85%. The original price per pound of a certain brand of coffee was $1.59. By how much was the price per pound of this coffee raised?

Explanation The rewording of the problem and its translation are:

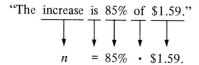

"The increase is 85% of $1.59."

n = 85% · $1.59.

Solution Let n represent the increase in price.

Therefore,
$$n = 85\% \cdot \$1.59$$
$$n = .85 \cdot 1.59$$
$$n = \$1.35 \quad \text{(increase in price)}$$

Example 3

During a sale an article that had been priced at $63.50 sold for $49.50. The discount is what percent of the original price?

Explanation The question needs no rewording.

"The discount is what percent of the original price?"

However, after reading the narrative we notice that the discount does not appear in the statement of the problem. The discount can be determined, though, by subtracting $49.50 from $63.50 giving us $14. With this information, the translation of the above becomes,

$$\$14 = p \cdot \$63.50.$$

Solution Let p represent the percent of discount.

Therefore,
$$\$14 = p \cdot \$63.50$$
$$14 = 63.50p$$
$$63.50p = 14$$
$$p = .22 \quad \text{(to nearest hundredth)}$$
$$p = 22\%$$

Example 4

A merchant increased the price of every article in his store by 15%. A pair of gloves now sells for $5.65. What was the original price of the gloves?

Explanation If an article originally sold for $4, to determine the increase we would multiply 15% by $4. To find the increase in an article that sold for $9 we would multiply 15% by $9. Similarly, to find the increase in an article that originally sold for n dollars we would multiply 15% by n thus obtaining $15\%n$. With this information the narrative and its translation can be written as:

"The original price increased by 15% of the original price is the selling price."

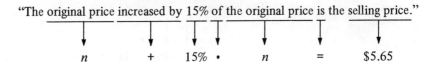

$$n \quad + \quad 15\% \cdot \quad n \quad = \quad \$5.65$$

5.4/Percent Problems

Without the arrows, the equation will be

$$n + 15\% \cdot n = \$5.65$$

Solution Let n represent the original price of the article. Then $15\%n$ represents the increase.

Therefore,
$$n + 15\%n = \$5.65$$
$$1n + .15n = \$5.65$$
$$1.15n = \$5.65$$
$$n = \$4.91 \quad \text{(original price)}$$

Example 5

A one-year-old motorboat is worth 30% less than when it is new. After one year a used motor boat was priced at $9,850. What was the cost of the boat when new?

Explanation The narrative can be reworded as,

"The original price decreased by 30% of the original price is the year-old price."

When translated, this becomes:

$$n - 30\% \cdot n = \$9,850.$$

Solution Let n represent the original price of the article. Then $30\%n$ represents the decrease.

Therefore,
$$n - 30\%n = 9,850$$
$$1n - .30n = 9,850$$
$$.70n = 9,850$$
$$n = \$14,071.43 \quad \text{(original price)}$$

EXERCISES 5.4 (Part 2)

1. During a sale the price of a pair of shoes was decreased by $10. The original price was $36.50. By what percent of the original price were the shoes reduced?

2. The price of a suit is increased by $15. The original price was $125. By what percent of the original price was the suit increased?

3. A pair of ice skates that originally sold for $19.50 now sells for $23.50. The increase is what percent of the original price?

4. A textbook that could be purchased for $8.95 a few years ago now sells for $10.95. The increase is what percent of the original price?

5. A merchant pays $9.50 for a shirt that he sells for $14.95. The gross profit (increase) is what percent of the cost to the merchant?

6. The owner of a department store pays $125 for a sofa that he sells for $279.95. The markup (increase) is what percent of the selling price of the sofa?

7. At a year-end sale a stereo set that had been priced at $259.50 was marked at $189. The discount is what percent of the original price?

8. As a result of wage negotiations the hourly workers at the Bundy Corporation received an increase in earnings of 12%. If the hourly rate had been $5.46, what is the new hourly rate?

9. As a result of a fire the price of every item in a furniture store was reduced by 70%. What is the reduction on a diningroom suite that originally sold for $958?

10. When electronic calculators first came on the market they were selling for approximately $450. Similar calculators can now be purchased at 2% of this price. What is the present price of these calculators?

11. Judith received a 12% increase in her salary. She now earns $184 per week. How much did she earn before she received the raise?

12. Over the past ten years, the population of a town increased by 60%. The present population is 15,000 people. What was the population at the beginning of the ten-year period?

13. During a two-year period the price of gasoline rose by 51%. At the end of the period the price was 59¢ per gallon. What was the price at the beginning of the period?

14. Electric hand-held drills have decreased in price by 60% over a twenty-year period. These drills can now be purchased for $8.95. How much did they cost twenty years ago?

15. A woman's suit that was slightly damaged was reduced 35% in price and sold for $23.95. What was the original price of the suit?

16. The earnings of the Coleman Corporation this year are 18% more than they had been last year. Its earnings last year were $160,000. What are its earnings this year?

17. The earnings of the Eagle Company this year are 23% more than they had been last year. Its earnings this year are $120,000. What were its earnings last year?

CHAPTER REVIEW

A

1. Translate each of the following English phrases into a mathematical phrase.

   a. Five more than a number.
   b. The double of a number.
   c. Twice a number decreased by ten.
   d. The difference between seventeen and a number.

2. Translate each of the following English sentences into a mathematical sentence.

   a. Three more than a number is fourteen.
   b. Twice a number is seventy-three.
   c. Six more than five times a number gives eight.
   d. The difference between five times a number and eleven is four.
   e. If four times a number is subtracted from twenty-five, the difference is 1.

B

Find the number requested in each of the following problems.

1. Twelve more than five times a number is forty-seven. What is the number?
2. If twice a number is added to the number itself, the sum is forty-two. What is the number?
3. If the sum of a certain number and five is doubled, the result will be twenty-four. What is the number?
4. The larger of two numbers is three more than the smaller number. If the smaller is multiplied by four the product is the same as when the larger is multiplied by three. What are the two numbers?
5. Fred's age is one year more than twice Sally's age. The difference of their ages is fourteen. How old is each?
6. Two numbers are consecutive. If the larger is multiplied by two and added to the smaller, the sum will be fifty-three. What are the numbers?
7. The sum of three numbers is fifty-two. The second is seven more than the first and the third is three times the first. What are the three numbers?
8. The second of three numbers is two more than the third. The first is twice the second. When the first is added to twice the second, the sum is five times the third. What are the numbers?
9. A child's bank contains $2.95 in nickles and dimes only. The number of nickles is eleven more than the number of dimes. How many coins of each variety are there?

10. A bag contains eighty-seven coins of which there are only half-dollars and quarters. The total value of these coins is $35.25. How many half-dollars and how many quarters are there?

11. A person spent ten dollars on the purchase of 8¢ stamps and 15¢ stamps. He had ten more 8¢ stamps than 15¢ stamps. How many of each variety did he buy?

12. A blend of 60 pounds of coffee is made from two different varieties of coffee. One variety costs $2.48 per pound while the other costs $2.85 per pound. The total cost of the 60-pound blend is $156.94. How many pounds of each variety are in the blend?

13. How much money is 153% of $260?

14. 56 is 14% of what number?

15. What percent of $2,000 is $180?

16. The original price of a ballpoint pen is 89 cents. A discount store sells the pen at a reduction of 14 cents off the original price. The reduction is what percent of the original price?

17. A heating oil dealer who had been charging his customers 39 cents for a gallon of oil raised the price to 45 cents per gallon. The increase was what percent of the original price?

18. During its spring sale Jonathon's reduced the price of a sofa by 45%. The sale price of the sofa was $260. What was the regular price of this sofa?

Factoring

Earlier in this course we learned how to find the product of two polynomials. In this chapter our interest turns in the direction of reversing this process. For instance, previously we started with two binomials such as

$$(2x + 3) \text{ and } (x - 4),$$

and found their product to be

$$2x^2 - 5x - 12.$$

Now our objective will be to start with a polynomial such as

$$3x^2 - 8x + 5,$$

and seek out the quantities whose product is $3x^2 - 8x + 5$.

We will find that there are some polynomials that can be represented as the product of other polynomials. On the other hand, there are also some in which this cannot be done. Let's start with some relatively simple ideas concerning the product of numbers.

6.1 FACTORS OF A NUMBER

In the study of arithmetic we were occasionally called upon to list the "exact divisors" of a number. For instance, the exact divisors of

$$12 \quad \text{are} \quad 1, 2, 3, 4, 6, 12.$$

These numbers are called the *exact* divisors of 12, for when 12 is divided by any one of these numbers *the remainder is 0*.

There are other numbers, though, that we have not listed that are exact divisors of 12. As an illustration, if 12 is divided by $\frac{1}{2}$, the quotient is 24, and the remainder is 0. Similarly, other fractions such as $\frac{2}{3}, \frac{2}{5}, \frac{2}{7}, \frac{3}{4}, \frac{3}{5}, \frac{3}{7}$, and so forth, are all exact divisors of 12, for when 12 is divided by any one of these numbers the remainder is 0.

If an exact divisor of a number happens to be an *integer*, then that exact divisor is called a *factor of the number*. In the case of the number 12, the numbers 1, 2, 3, 4, 6, and 12 are factors of 12 for each of these exact divisors is an integer. However, the numbers $\frac{2}{3}, \frac{2}{5}, \frac{2}{7}$, and so forth, although they are exact divisors of 12, are not considered factors of 12.

In the same way, the *negative* numbers -1, -2, -3, -4, -6, and -12 are also exact divisors of 12. They, too, are frequently excluded as factors of 12. Unless otherwise indicated, a

---
Factor of a positive whole number will be a positive integer that is an exact divisor of the whole number.

Example 1
List the factors of 18.

Solution Factors of 18: 1, 2, 3, 6, 9, 18

Explanation The numbers 1, 2, 3, 6, 9, and 18 are factors of 18, for they are exact divisors of 18 and they are also integers.

As this topic develops we will find it necessary to express a number in terms of the product of two of its factors. Thus, when asked to determine all the "pairs of factors" of the number 12, we would be seeking those pairs of integers whose product is 12. These pairs are listed below.

$$\begin{array}{ll} \text{Pairs of factors of 12:} & 1 \text{ and } 12 \\ & 2 \text{ and } 6 \\ & 3 \text{ and } 4 \end{array}$$

Notice that we did not list 12 and 1, for 12 and 1 represent the same pair of factors as 1 and 12.

Example 2

List the pairs of factors of 18.

Solution Pairs of factors of 18:   1 and 18
                                        2 and 9
                                        3 and 6

Some rather interesting features come to light when we examine the pairs of factors of numbers such as 9, 16, 25, and 36.

Pairs of factors of 9:	Pairs of factors of 16:
1 and 9	1 and 16
3 and 3	2 and 8
	4 and 4

Pairs of factors of 25:	Pairs of factors of 36:
1 and 25	1 and 36
5 and 5	2 and 18
	3 and 12
	4 and 9
	6 and 6

Each of these numbers contains a pair of factors where the factors are exactly the same number. In the case of 16, this pair of factors is 4 and 4. The number 16 is called a "perfect square," while the number 4 is called "the square root of 16."

Similarly, 25 is a perfect square, while 5 is the square root of 25.

A perfect square is a number that has a pair of equal factors. The square root of a number is one of the factors in the pair of equal factors of the number.

Example 3

Find the square root of 64.

Solution One of the pairs of factors of 64 is 8 and 8. Hence, 8 is the square root of 64.

When we investigate the factors of numbers such as 2, 5, 13, and 29, we find the task quite simple for each of these numbers has exactly two factors. Each of these numbers is called a "prime number."

Factors of 2:   1, 2   Factors of 5:   1, 5
Factors of 13: 1, 13   Factors of 29: 1, 29

A prime number is a number that has exactly two factors.

Notice that the number 1 is not listed as a prime number. Actually 1 is *not* a prime number. In order that a number be prime, it must have exactly two factors — no more, no fewer. The number 1 has but one factor; therefore it is not a prime number.

Let us examine the factors of 12 once again.

$$\text{Factors of 12: } 1, 2, 3, 4, 6, 12$$

Among these factors are the prime numbers 2 and 3. The factors of a number that happen to be prime numbers are called *prime factors of the number*.

Example 4

List the prime factors of 24.

Explanation We first list all the factors of 24 and from that list we select those factors that are prime numbers.

Solution Factors of 24: 1, 2, 3, 4, 6, 8, 12, 24
Prime factors of 24: 2, 3   (Answer)

Our final point at this time is to investigate the meaning of an expression, such as

$$\text{``Factor the number 18.''}$$

A statement of this nature requests that we show how 18 can be written as the product of its *prime factors*. For instance, the prime factors of 18 are 2 and 3. The question now is, how can we write a product that involves only the numbers 2 and 3 such that when these numbers are multiplied the answer is 18? It should be pointed out that each of these numbers may appear several times.

After some thought, we realize that the answer is

$$18 = 2 \cdot 3 \cdot 3.$$

The expression $2 \cdot 3 \cdot 3$ is called the *factored form* of 18.

A somewhat simpler way of "factoring the number 18" is to first rewrite it as the product of *any* two of its factors. Thus,

$$18 = 6 \cdot 3.$$

6.1/Factors of a Number

When we examine the factors 6 and 3 we notice that 6 is *not* a prime number. Hence, we rewrite 6 in terms of the product of a pair of its factors.

$$18 = 6 \cdot 3$$
$$= 2 \cdot 3 \cdot 3$$

At this point we notice that the factors 2, 3, and 3 are all prime numbers and the product of these numbers is 18. Hence, we conclude that we have "factored 18" for we have expressed 18 as the product of its prime factors.

Rather than leave the answer in the form $2 \cdot 3 \cdot 3$, it is usually rewritten in the shortened form of $2 \cdot 3^2$.

$$18 = 6 \cdot 3$$
$$= 2 \cdot 3 \cdot 3$$
$$= 2 \cdot 3^2$$

Example 5

Express 72 in its factored form.

Solution

$$72 = 9 \cdot 8$$
$$= 3 \cdot 3 \cdot 4 \cdot 2$$
$$= 3 \cdot 3 \cdot 2 \cdot 2 \cdot 2$$
$$= 3^2 \cdot 2^3 \quad \text{(factored form of 72)}$$

EXERCISES 6.1

A

List the factors of each of the following numbers.

1. 6
2. 8
3. 15
4. 16
5. 24
6. 30
7. 32
8. 36
9. 48
10. 54
11. 72
12. 84

B

List the pairs of factors of each of the following.

1. 4
2. 5
3. 6
4. 9
5. 16
6. 20
7. 24
8. 28
9. 30
10. 42
11. 45
12. 48
13. 50
14. 56
15. 60

C

Find the square root of each of the following numbers.

1. 4
2. 49
3. 1
4. 64
5. 100
6. 144
7. 121
8. 225
9. 169
10. 289
11. 256
12. 625

D

List the prime factors of each of the following numbers.

1. 3
2. 7
3. 9
4. 14
5. 15
6. 18
7. 27
8. 30
9. 36
10. 40
11. 48
12. 60

E

Express each of the following numbers in its factored form.

1. 6
2. 9
3. 12
4. 20
5. 24
6. 30
7. 36
8. 48
9. 64
10. 84
11. 96
12. 100
13. 108
14. 144
15. 256

6.2 FACTORING FOR THE COMMON FACTOR

PART 1

In the exercises above we were asked to list the factors of 24 and the factors of 36. Let us do this now.

$$24: 1, 2, 3, 4, 6, 8, 12, 24$$
$$36: 1, 2, 3, 4, 6, 9, 12, 18, 36$$

These two numbers have many factors in common. They are:

$$1, 2, 3, 4, 6, \text{ and } 12$$

Factors that two numbers have in common are called **common factors**. The very largest of the factors that two numbers have in common is called the **greatest common factor**. The term "greatest common factor" is often abbreviated by using the letters **G.C.F.**

6.2/Factoring for the Common Factor

Example 1
a. Find all the common factors of 18 and 24.
b. What is the greatest common factor of 18 and 24?

Explanation The solution involves simply listing all the factors of each number. After completing this we select those factors that are common to the two numbers. The largest of these is the G.C.F.

Solution Factors of 18: 1, 2, 3, 6, 9, 18
Factors of 24: 1, 2, 3, 4, 6, 8, 12, 24
a. Common factors of 18 and 24: 2, 3, 6
b. G.C.F. of 18 and 24: 6

Until now we have considered only the factors of constants such as 18, 24, 30, and the like. At this time we will take a look at the factors of variable expressions of the form

$$x^3.$$

The factors of x^3 are those quantities that are the *exact divisors* of x^3. Thus, from our knowledge of division, we know that if x^3 is divided by x^2 the quotient is x and the remainder is 0.

$$x^3 \div x^2 = x$$

Hence, we can conclude that x^2 is a factor of x^3.

Similarly, if x^3 is divided by x, the quotient is x^2 and the remainder is 0. This implies that x is a factor of x^3. Finally, if x^3 is divided by x^3 the quotient is 1 and the remainder is 0. Hence, x^3 is also a factor of x^3. And, of course, 1 is a factor of x^3 for 1 is a factor of every number. Why?

Factors of x^3: 1, x, x^2, x^3

Using this same approach, we can determine the factors of x^4.

Factors of x^4: 1, x, x^2, x^3, x^4

Hence, the monomials x^3 and x^4 have the following factors in common:

1, x, x^2, x^3

In view of this, the greatest common factor of x^3 and x^4 is x^3.

Example 2
What is the G.C.F. of a^2 and a^4?

Solution Factors of a^2: 1, a, a^2
Factors of a^4: 1, a, a^2, a^3, a^4
G.C.F.: a^2

Based on Example 2 and the discussion before it, the greatest common factor of two terms such as

$$y^7 \text{ and } y^{12}$$

can be found immediately by writing the variable to the smaller of the two exponents.

$$\text{G.C.F. of } y^7 \text{ and } y^{12}: y^7$$

Example 3
Find the greatest common factor of x^9 and x^4.

Explanation Since the variables are both x, the G.C.F. will be the variable x raised to the smaller of the two exponents.

Solution G.C.F. of x^9 and x^4: x^4

Example 4
List the factors of x^3y.

Explanation When x^3y is divided by x^2 the quotient is xy and the remainder is 0. Similarly, each of the monomials in the solution below can be shown to be a factor of x^3y.

Solution Factors of x^3y: 1, x, x^2, x^3, xy, x^2y, x^3y, y

Example 5
Find the greatest common factor of x^2y^5 and x^4y^3.

Explanation This solution is a combination of the solutions of Examples 3 and 4. In this situation, we treat each variable separately. Thus, the common factor related to the x variable is x^2. The common factor related to the y variable is y^3. The product of x^2 and y^3 is the greatest common factor.

Solution G.C.F. of x^2y^5 and x^4y^3: x^2y^3

Example 6
Find the greatest common factor of $12a^5$ and $18a^3$.

Explanation In this situation we find the greatest common factor of the coefficients 12 and 18; that factor is 6. We then find the greatest common factor for the variables; that factor is a^3. The product of the two, $6a^3$, is the greatest common factor of these two monomials.

Solution G.C.F. of $12a^5$ and $18a^3$: $6a^3$

6.2/Factoring for the Common Factor

EXERCISES 6.2 (Part 1)

A

List the factors of each of the following.

1. x^2
2. y^3
3. $a^2 b$
4. ab^2
5. $2a$
6. $6a$
7. $3a^2$
8. ab^3
9. $8a^2$

B

Find all the common factors for each of the following pairs of numbers.

1. 4 and 6
2. 8 and 12
3. 12 and 16
4. 15 and 20
5. 18 and 27
6. 18 and 30
7. 24 and 30
8. 24 and 36
9. 40 and 48
10. 36 and 54

C

Find the greatest common factor for each of the following pairs of monomials.

1. x^2 and x^3
2. x and x^2
3. b^4 and b^2
4. b^5 and b
5. $6a^2$ and $9a^3$
6. $9a$ and $4a^2$
7. ab and ac
8. $5a$ and $10b$
9. $x^2 y^3$ and $x^4 y^6$
10. xy^3 and $x^2 y^2$
11. $x^3 y^2$ and $x^2 y^3 z$
12. $10ab^2$ and $12b^3$

PART 2

We are now in a position to express a product relation such as

$$7 \cdot 5 = 35$$

in terms of the new language we have developed. For instance, if 7 is divided into 35, the quotient is 5 and the remainder is 0. Hence, we can conclude that 7 is a factor of 35. Similarly, 5 is a factor of 35, for it, too, is an exact divisor of 35. Thus, if the 35 is divided by either of these two factors, the quotient is the other factor.

Similarly, should we find the product of any two quantities, then each of these quantities is a factor of the product. Furthermore,

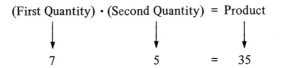

if the product is divided by the first quantity, the quotient will be

$$35 \div 7 = 5$$

the second quantity. And, of course, if the product is divided by the second quantity, the quotient will be the first quantity.

$$35 \div 5 = 7.$$

A similar approach can be used when finding the product of the monomial a with the binomial $b + c$.

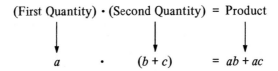

If the product $ab + ac$ is divided by a, the quotient will be $b + c$ and the remainder will be zero. Similarly, when the product $ab + ac$ is divided by $b + c$, the quotient will be a and the remainder will be zero. Hence, we can conclude that both a and $b + c$ are factors of $ab + ac$.

$$ab + ac = a(b + c)$$

The quantity $a(b + c)$ is said to be the *factored form* of $ab + ac$. The a is a *monomial* factor while the $b + c$ is a *binomial* factor. Notice that the monomial factor a is the greatest common factor of the two terms ab and ac:

$$ab + ac.$$

Also, as stated above, the binomial factor is the quotient obtained when a is divided into $ab + ac$.

$$(ab + ac) \div a = b + c$$

6.2/Factoring for the Common Factor

In view of this, to find the factored form of the binomial

$$x + x^3,$$

we would first determine the monomial factor. That factor is the greatest common factor of x and x^3 which is x. Then, to determine the binomial factor, we need simply divide x into $x + x^3$. Thus,

G.C.F. of x and x^3: x
Quotient of $x + x^3$ divided by x: $1 + x^2$
Factored form of $x + x^3$: $x(1 + x^2)$

Example 1

Write the binomial $5a - 10b$ in factored form.

Explanation The first step is to determine the monomial factor. The monomial factor is the G.C.F. of $5a$ and $10b$ which is 5. The binomial factor is found by dividing $5a - 10b$ by 5:

$$(5a - 10b) \div 5 = a - 2b.$$

Hence, $a - 2b$ is the binomial factor.

Solution $\qquad 5a - 10b = 5(a - 2b)$

Example 2

Find the factored form of $x + 2y$.

Explanation The terms x and $2y$ have *no* common factor other than 1. Hence, the only way in which $x + 2y$ can be rewritten as the product of its factors is

$$x + 2y = 1(x + 2y)$$

A polynomial that has but two factors, the number 1 and the polynomial itself, is called a **prime polynomial**. If a polynomial is prime, we say that it is *not factorable*.

Solution $x + 2y$ is a prime polynomial.

Example 3

Find the factored form of $6x + 12y$.

Explanation When we search for common factors of $6x$ and $12y$, we find that they are:

$$2, 3, \text{ and } 6.$$

Were we to use 2 as the monomial factor, we could rewrite

$$6x + 12y \quad \text{as} \quad 2(3x + 6y).$$

The question now arises as to whether $2(3x + 6y)$ is the factored form of $6x + 12y$. We ran across this same situation earlier when writing the number 18 in *factored form*. At that time we learned that the factored form of a number consists of the product of the *prime factors* of the number.

Similarly, the factored form of $6x + 12y$ must be the product of polynomials that are *prime*. The binomial $3x + 6y$ is *not* prime for, among other factors, it has the number 3 as a monomial factor. Hence,

$$2(3x + 6y) \text{ is } not \text{ the factored form of } 6x + 12y.$$

If 3 is used as the monomial factor, again the binomial factor will not be prime. However, if the *greatest common factor* 6 is used as the monomial factor, then the binomial factor will be prime. Hence, the factored form of

$$6x + 12y \text{ is } 6(x + 2y)$$

Solution $\qquad 6x + 12y = 6(x + 2y)$

It is important that the point made in the explanation for Example 3 be emphasized.

The monomial factor must be the *greatest common factor* of the terms of the binomial being factored.

Example 4

Find the factored form of

$$8x^2y^3 - 12x^3y.$$

Explanation The G.C.F. of $8x^2y^3$ and $12x^3y$ is $4x^2y$.

Solution $\qquad 8x^2y^3 - 12x^3y = 4x^2y(2y^2 - 3x)$

Example 5

Find the factored form of

$$15x^3 - 21x^2 + 3x.$$

Explanation Until now the only polynomials we have factored have been binomials. However, everything stated earlier concerning the factoring of binomials holds equally well for any polynomials. The G.C.F. of $15x^3$, $21x^2$, and $3x$ is $3x$. To find the polynomial factor simply divide $3x$ into $15x^3 - 21x^2 + 3x$.

Solution $15x^3 - 21x^2 + 3x = 3x(5x^2 - 7x + 1)$

EXERCISES 6.2 (Part 2)

Factor each of the following polynomials. If any are prime, indicate this by writing the word "prime" to the right of the polynomial.

1. $ax + ay$
2. $3x + 3y$
3. $5a - 15c$
4. $12a + 16b$
5. $ax + a$
6. $xy - x$
7. $x^4 + x^2$
8. $y^3 - y^5$
9. $6x + 3$
10. $15x^2 - 5$
11. $12 - 4x$
12. $3a + 4b$
13. $ab + bc$
14. $x^2y^3 - x^5y^7$
15. $a^2b + ab^2$
16. $x^4 + x^3 + x^2$
17. $b^5 - b^3 + b$
18. $c^2 + c^5 - c^8$
19. $2a^3 - 6a^2 + 10a$
20. $6x^2 - 18x^3 - 12x^4$
21. $20a^2 - 30a - 10$
22. $x^2y^5 + x^3y^4 - x^4y^3$
23. $abc - a^3b^3$
24. $a^2b^3c^4 + a^3b^5c^2$
25. $x^3y^2w^4 + x^2y^4w^3 - xw^2$
26. $5x^3y^2 - 6w^2$
27. $9x^2y^3 + 6xy^4 - 12x^3y^2$
28. $5x^2 - 10x^3 - 25x^5$
29. $40 - 25x^3y^3 - 30x^6y^6$
30. $12x^5y - 4xy - 16x^3y^2w$

PART 3

The type of factoring we did in Part 2 of this section is called "factoring for the common factor" where the common factor is a monomial. The common factor, however, does not necessarily have to be a monomial, as we will discover in the following situation. Consider the polynomial

$$ab + ac. \qquad (1)$$

When this polynomial is written in factored form, it becomes

$$ab + ac = a(b + c). \qquad (2)$$

Now let us replace a in the above equality with the binomial $(x + y)$. Hence, where previously a was a common factor to the quantities ab and ac,

$$\begin{array}{ccccc} ab & + & ac & = & a \cdot (b+c) \\ \downarrow & & \downarrow & & \downarrow \\ (x+y)b & + & (x+y)c & = & (x+y)(b+c) \end{array} \qquad (3)$$

now $(x + y)$ is a common factor to $(x + y)b$ and $(x + y)c$. In addition, where previously the factored form was written as

$$a \cdot (b + c),$$

now the factored form is

$$(x + y) \cdot (b + c).$$

In view of this, it seems that it is possible to factor expressions that have *binomial* common factors in exactly the same manner as we would factor them were they monomial common factors.

Let's examine a few illustrations of this nature.

Example 1

Write the expression below in factored form.

$$(a - b)y + (a - b)w$$

Explanation In this expression, $(a - b)$ is a common factor of both $(a - b)y$ and $(a - b)w$. As earlier, the common factor $(a - b)$ is written as the first factor in the factored form.

First Factor   Second Factor

$$(a - b)y + (a - b)w = (a - b)(\quad)$$

To determine the second factor it is necessary to divide $a - b$ into each term of $(a - b)y + (a - b)w$.

$$a - b \overline{)(a-b)y} \atop \underline{(a-b)y} \atop 0 \qquad \text{and} \qquad a - b \overline{)(a-b)w} \atop \underline{(a-b)w} \atop 0$$

Hence, the second factor turns out to be $y + w$.

Solution $(a - b)y + (a - b)w = (a - b)(y + w)$

Example 2

Write the expression below in factored form.

$$a(x - 2y) - 3(x - 2y)$$

6.2/Factoring for the Common Factor

Explanation When the common factor $(x - 2y)$ is divided into $a(x - 2y)$ the quotient will be a and when it is divided into $-3(x - 2y)$ the quotient will be -3.

Solution
$$a(x - 2y) - 3(x - 2y) = (x - 2y)(a - 3)$$
with the binomial common factor $(x - 2y)$ indicated.

Unfortunately, most polynomials do not come arranged for us in the neat form of

$$b(x + y) + c(x + y). \qquad (4)$$

More often we will find them as

$$bx + by + cx + cy. \qquad (5)$$

When this occurs it will be necessary for us to realize that were we to group the first term with the second and then group the third with the fourth, the polynomial in (5) can be changed into the one in (4). For instance,

$$bx + by + cx + cy \qquad (5)$$

by examining the first two terms only we notice that they have the common factor b. In view of this, these two terms can be rewritten as

$$bx + by = b(x + y).$$

Similarly, by examining the last two terms only we notice that these two terms have the common factor c. Hence, they can be rewritten as

$$cx + cy = c(x + y).$$

Based on this, (5) now appears as

$$bx + by + cx + cy = b(x + y) + c(x + y). \qquad (6)$$

At this point the factoring is completed as in Example 2.

$$\begin{aligned} bx + by + cx + cy &= b(x + y) + c(x + y) \\ &= (x + y)(b + c) \end{aligned}$$

You may have wondered why $b(x+y) + c(x+y)$ is not considered the factored form of $bx + by + cx + cy$. The answer to this lies in the meaning of "the factored form of a polynomial." When a polynomial is expressed in factored form, it must be written as the *product* of other polynomials. The expression $b(x+y) + c(x+y)$ is written as the sum

$$\boxed{\begin{array}{c}\text{This is an addition sign}\\ \downarrow\\ b(x+y) + c(x+y)\end{array}}$$

of the two quantities $b(x+y)$ and $c(x+y)$.

Example 3

Write the expression $2a + 2b + ax + bx$ in factored form.

Solution $2a + 2b + ax + bx = 2(a+b) + x(a+b)$
$= (a+b)(2+x)$

Example 4

Factor the expression $3x + 2a + 6 + ax$.

Explanation   We notice that if we group the first term with the second we find that these terms contain no common factor. The same is true of the last two terms. However, by grouping the first term with the third we discover that these two terms have 3 as a common factor. Similarly, in grouping the second term with the fourth we find that they have a as a factor in common.

Solution $3x + 2a + 6 + ax = 3x + 6 + 2a + ax$
$= 3(x+2) + a(2+x)$
$= 3(x+2) + a(x+2)$
$= (x+2)(3+a)$

Explanation continued   When we wrote the expression $3(x+2) + a(2+x)$ we noticed that the binomials appeared the same except that one was $x+2$ while the other was $2+x$. Since the commutative property of addition permits us to interchange the two numbers, $2+x$ was rewritten as $x+2$.

Example 5

Factor the expression $2x - 2y - ax + ay$.

6.2/Factoring for the Common Factor

Explanation By grouping the first two terms and then grouping the last two terms this expression can be rewritten as

$$2x - 2y - ax + ay = 2(x - y) + a(-x + y).$$

Unfortunately, the binomials are *not* the same. We do notice, though, that the signs of the terms in the second binomial are the opposite of those in the first. To overcome this difficulty, simply use the **negative** value of the common factor rather than the **positive** value. In this situation, rather than considering $+a$ as the common factor, we consider $-a$.

Solution
$$\begin{aligned} 2x - 2y - ax + ay &= 2(x - y) - a(x - y) \\ &= (x - y)(2 - a) \end{aligned}$$

Example 6

Factor $ab + b + a + 1$.

Explanation By grouping the first two terms we see that they have a common factor b. The last two terms have no common factor but the trivial factor 1. However, we use this as a common factor on the possibility that we may recognize something familiar after doing this.

$$b(a + 1) + 1(a + 1)$$

By good fortune, the binomial $(a + 1)$ turns out to be a common factor. This will not always be the case. When this is not so, the original expression is probably prime.

Solution
$$\begin{aligned} ab + b + a + 1 &= b(a + 1) + 1(a + 1) \\ &= (a + 1)(b + 1) \end{aligned}$$

EXERCISES 6.2 (Part 3)

A

Factor each of the following expressions.

1. $(x + y)a + (x + y)b$
2. $(a - b)x + (a - b)y$
3. $a(x + 2) + b(x + 2)$
4. $x(x - 3) + 4(x - 3)$
5. $c(a + 4) - 5(a + 4)$
6. $2a(x - y) - 7(x - y)$
7. $3a(c - 2) + 4b(c - 2)$
8. $x(y + 3) + 1(y + 3)$
9. $b(x - 5) + (x - 5)$
10. $2b(x + 7) - (x + 7)$
11. $1(a + b) + x(a + b)$
12. $(a - b) - y(a - b)$

B

Factor each of the following expressions.

1. $cx + cy + dx + dy$
2. $bx - by + cx - cy$
3. $ax + ay + 5x + 5y$
4. $7x - 7w + bx - bw$
5. $ax + a + bx + b$
6. $3x + a + 3 + ax$
7. $a^2 + b + a + ab$
8. $2ax + 3b + 6a + bx$
9. $ax + a + x + 1$
10. $ax - ay - bx + by$
11. $5x - 5 - ax + a$
12. $x^3 + x^2 + x + 1$
13. $x^3 - 2x^2 + 3x - 6$
14. $5x^3 + 6 - 15x^2 - 2x$

6.3 FACTORING THE DIFFERENCE OF TWO SQUARES

PART 1

In developing the methods for factoring in both this and the next few sections we apply the information we learned earlier concerning the numbers 7, 5, and 35. The numbers 7 and 5 are called factors of their product 35, for each is an exact divisor of 35.

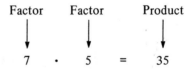

Similarly, when $(x + 6)$ is multiplied by $(x - 6)$ the product is $x^2 - 36$. Hence, both binomials $x + 6$ and $x - 6$ are factors of their product $x^2 - 36$.

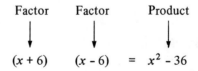

Our interest now is to start with a polynomial such as $x^2 - 36$ and try to seek some way to determine its binomial factors. To do this, let us write the quantity $x^2 - 36$ immediately beside its two binomial factors $(x + 6)$ and $(x - 6)$ and examine them carefully.

$$x^2 - 36 = (x + 6)(x - 6)$$

Notice that the *second* term in each factor is the square root of the *second* term in $x^2 - 36$.

6.3/Factoring the Difference of Two Squares

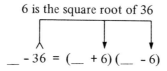

Notice also that the *first* term in each factor is the square root of the first term in $x^2 - 36$.

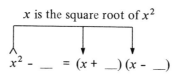

The final point of importance is that the last term in the first factor is the additive inverse of the last term of the second factor.

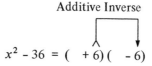

Factoring a binomial such as $x^2 - 36$ becomes a matter of finding the square root of each of the terms of $x^2 - 36$. Quite apparently, then, if either of the terms is *not* a perfect square, it would not be possible to factor the binomial by this method.

In addition, were the operation between the terms addition rather than *subtraction*, we would not be able to factor the binomial. Thus, consider

$$x^2 + 36.$$

We know that its factored form cannot be

$$(x + 6)(x + 6)$$

for the product of these two binomials is $x^2 + 12x + 36$ and not $x^2 + 36$. For the same reason, its factored form cannot be

$$(x - 6)(x - 6).$$

Actually, the binomial $x^2 + 36$ cannot be factored with any numbers with which we are familiar. Hence, it is prime.

Those binomials that can be factored by the method above can be recognized by the fact that *each term is a perfect square* and that *the operation between them is subtraction*.

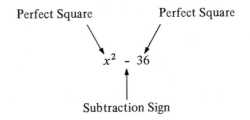

A binomial of this nature is referred to as being

The Difference of Two Perfect Squares

or simply, "The Difference of Two Squares."

Example 1

Factor $4x^2 - 49$.

Explanation Knowing that the factors must be two binomials in which the second term of the first is the additive inverse of the second term of the second we can immediately write

$$(\quad + \quad)(\quad - \quad).$$

The *second* term in each binomial is the square root of 49; this is 7. The first term in each binomial is the square root of $4x^2$; this is $2x$. It would be well to check your answer by finding the product of the two factors to see if it is $4x^2 - 49$.

Solution $\qquad 4x^2 - 49 = (2x + 7)(2x - 7)$

Example 2

Factor $x^6 - 36$.

Explanation The only difficulty here is in finding the square root of x^6. To do this we must recall that the exponent of the product of two quantities such as a^2 and a^3 is determined by adding the exponent 2 to the exponent 3. Thus,

$$a^2 \cdot a^3 = a^5$$

To find the square root of x^6, we must find a pair of factors of x^6 that are exactly the same. This implies that the exponents must be the same and their *sum* must be 6. For this to be true, the exponent of each will have to be one-half of 6. Hence, the square root of x^6 is x^3.

$$x^6 = x^3 \cdot x^3$$

Solution $\qquad x^6 - 36 = (x^3 + 6)(x^3 - 6)$

Example 3

Factor $-25y^2 + x^8$.

Explanation Since $-25y^2$ added to x^8 is the same as x^8 added to $-25y^2$, we rearrange the terms and write the binomial in the familiar form of $x^8 - 25y^2$.

Solution
$$-25y^2 + x^8 = x^8 - 25y^2$$
$$= (x^4 + 5y)(x^4 - 5y)$$

Example 4

Factor $16x^2 + 25$.

Explanation Since $16x^2 + 25$ is the *sum* and *not* the difference of two squares, we cannot factor this binomial. The word "prime" is written immediately to its right.

Solution $\qquad 16x^2 + 25 \qquad$ (Prime)

Example 5

Factor $x^5 - 36$.

Explanation Since x^5 is not a perfect square, we cannot factor this binomial.

Solution $\qquad x^5 - 36 \qquad$ (Prime)

EXERCISES 6.5 (Part 1)

Factor each of the following polynomials. If any are prime, indicate this by writing the word "prime" to the right of the polynomial.

1. $x^2 - 9$
2. $x^2 - 49$
3. $a^2 - 64$
4. $a^2 - 1$
5. $a^2 - 5$
6. $9a^2 - 49$
7. $4x^2 - 25$
8. $16 - x^2$
9. $64 - a^2$
10. $-25 + 81x^2$
11. $4x^2 - 1$
12. $16a^2 - 25b^2$
13. $4x^2 + 25$
14. $a^2b^2 - 1$
15. $9x^2y^2 - 64$
16. $x^2 - .25$
17. $y^2 - .49$
18. $a^2 - .09$
19. $b^2 - .04$
20. $.01x^2 - 25$
21. $-4 + x^2$
22. $-9 + 4y^2$
23. $16x^2y^2 - 25a^2$
24. $x^6 - 9$

25. $y^8 - 4$
26. $16x^{16} - 9$
27. $x^7 - 25$
28. $9 - 100x^4$
29. $-81y^6 + 49x^{10}$
30. $7a^6 - b^4$
31. $36x^4 - 25y^2$
32. $49x^2y^4 - 16$
33. $x^6y^2 - 64w^4$
34. $.09x^2 - .25$
35. $.4x^2 - 49$
36. $144x^4y^2 - 121z^6$
37. $a^2 - b^2$
38. $(x + y)^2 - b^2$
39. $a^2 - (x + y)^2$
40. $9(x - y)^2 - b^2$

PART 2

As might be suspected our next move is to investigate binomials whose factored form involves a combination of factoring for the common factor and factoring the difference of two squares. Actually, no matter what the nature of the factoring may be, we must *always* factor for

---
the greatest common factor **first.**

To illustrate, consider the binomial

$$2x^2 - 50.$$

Since the greatest common factor is 2, the binomial can be rewritten as

$$2(x^2 - 25).$$

The question again arises as to whether $2(x^2 - 25)$ is the factored form of $2x^2 - 50$. The factored form of $2x^2 - 50$ must be the product of polynomials that are *prime*. However, $x^2 - 25$ is not prime for it has the factors $x + 5$ and $x - 5$. Hence,

$$2(x^2 - 25) \ cannot \ \text{be the factored form of} \ 2x^2 - 50.$$

An additional step is required where $x^2 - 25$ is written in factored form. Thus,

$$2x^2 - 50 = 2(x^2 - 25)$$
$$= 2(x + 5)(x - 5)$$
$$\uparrow$$

The arrow above was drawn to emphasize the fact that the monomial factor *must* be included in the final factored form of the polynomial. There is a tendency to overlook the common factor when writing the final step.

6.3/Factoring the Difference of Two Squares

Example 1

Factor $xy^2 - x^3$.

Explanation The G.C.F. of the two terms is x. After factoring for the common factor, we examine the binomial factor to see if it is prime. In this example it turns out to be the difference of two squares $y^2 - x^2$ which is not prime.

Solution $xy^2 - x^3 = x(y^2 - x^2)$
$= x(y + x)(y - x)$

Example 2

Factor $3x^3 + 27x$.

Explanation The G.C.F. is $3x$. After factoring for the common factor, we have

$$3x^3 + 27x = 3x(x^2 + 9).$$

Since $x^2 + 9$ is prime, then $3x(x^2 + 9)$ is the factored form of $3x^3 + 27$.

Solution $3x^3 + 27x = 3x(x^2 + 9)$

Example 3

Factor $x^4 - y^4$.

Explanation In examining $x^4 - y^4$ we notice that the two terms have no common factor. Hence, we factor the binomial as

$$(x^2 + y^2)(x^2 - y^2).$$

Whereas $x^2 + y^2$ is the *sum* of two squares and is prime, $x^2 - y^2$ is the *difference* of two squares and is not prime. Therefore, $x^2 - y^2$ must be replaced by its factored form of $(x + y)(x - y)$.

Solution $x^4 - y^4 = (x^2 + y^2)(x^2 - y^2)$
$= (x^2 + y^2)(x + y)(x - y)$

EXERCISES 6.3 (Part 2)

Factor each of the following polynomials.

1. $ax^2 - ay^2$
2. $bx^2 - b$
3. $3x^2 - 27$
4. $x^3 - x$
5. $7 - 28x^2$
6. $4y^2 - 36$

7. $ab^3 - a^3b$
8. $xy^3 - xy$
9. $2ax^2 - 8a$
10. $3ax^2 - 6a$
11. $a^4 - 1$
12. $1 - b^4$
13. $x^4 - 16$
14. $81a^4 - 1$
15. $16x^4 - 25$
16. $ax^4 - ay^4$
17. $5x^4 + 125$
18. $3b^4 - 48$
19. $16 - 81y^4$
20. $ax^8 - a$

6.4 FACTORING A TRINOMIAL

PART 1

Our interest now lies in examining a trinomial and trying to determine those binomials whose products will yield the trinomial. Saying this somewhat differently, our goal is to find the factored form of a trinomial. As in the past, we will discover that many trinomials are prime. However, if the trinomial can be factored, is there some pattern that can be followed to determine its factors?

When we first examined the method for finding the product of two binomials we determined that product by using the distributive property of multiplication over addition. For instance, in order to multiply $x + 2$ by $x + 3$ we did the computation as follows:

$$(x + 2)(x + 3) = (x + 2)x + (x + 2)3$$
$$= x^2 + 2x + 3x + 6$$

We shall now determine the product of the general binomials $x + a$ and $x + b$ in exactly the same way.

$$(x + a)(x + b) = (x + a)x + (x + a)b$$
$$= x^2 + ax + bx + ab \tag{1}$$

It is now possible for us to group the second and third terms of (1) and factor them for their common factor. Hence,

$$x^2 + ax + bx + ab = x^2 + (a + b)x + ab.$$

6.4/Factoring a Trinomial

Thus, the quantity we have just found is the product of $x + a$ and $x + b$. By placing the product immediately beside the binomials from which it came, we should be able to discover some interesting features about the two.

$$x^2 + (a + b)x + ab = (x + a)(x + b)$$

The very first thing that should strike us is the fact that the first term in each binomial is the square root of the first term in the trinomial.

x is the square root of x^2

$$x^2 + \underline{} + \underline{} = (x + \underline{})(x + \underline{})$$

Second, we notice that the last terms in the two binomials are a pair of factors of the last term of the trinomial.

a and b are a pair of factors of ab

$$\underline{} + \underline{} + ab = (\underline{} + a)(\underline{} + b)$$

And finally we can see that the sum of the pair of factors above turns out to be the coefficient of the middle term.

$$\underline{} + (a + b)x + \underline{} = (\underline{} + a)(\underline{} + b)$$

Thus, factoring a trinomial where the *coefficient of the first term is 1* simply becomes a matter of following the three steps above. The examples below should help clarify this.

Example 1

Factor $x^2 + 7x + 12$.

Explanation (Step 1)—The first term in each binomial is the square root of the first term in the trinomial:

$$x^2 + 7x + 12 = (x \quad)(x \quad).$$

(Step 2)—The last terms in the two binomials are a pair of factors of the last term of the trinomial. The pairs of factors of 12 are:

1 and 12,   2 and 6,   3 and 4.

In order to determine which pair of factors is correct, we must examine Step 3. (Step 3)—The sum of the pair of factors must be the coefficient of the middle term. That coefficient is 7.

$$1 + 12 = 13 \quad \text{Incorrect Pair}$$
$$2 + 6 = 8 \quad \text{Incorrect Pair}$$
$$3 + 4 = 7 \quad \text{Correct Pair}$$

Hence, the factored form of $x^2 + 7x + 12$ is

$$(x + 3)(x + 4).$$

Solution $\qquad x^2 + 7x + 12 = (x + 3)(x + 4)$

Example 2

Factor $y^2 - 11y + 18$.

Explanation (Step 1)—The first term in each binomial is the square root of the first term in the trinomial:

$$y^2 - 11y + 18 = (y\quad)(y\quad)$$

(Step 2)—The last terms in the two binomials are a pair of factors of the last term of the trinomial. From the fact that the sign of 18 is positive we know that both factors will either have to be positive or both factors will have to be negative. The product of two positive numbers is positive and the product of two negative numbers is positive. In Step 3, however we will be looking for a sum that is *negative* 11. Therefore, for the sum to be negative, both factors will have to be negative. The pairs of factors of 18 are:

$$-1 \text{ and } -18, \quad -2 \text{ and } -9, \quad -3 \text{ and } -6.$$

(Step 3)—The sum of the pair of factors must be the coefficient of the middle term. That coefficient is -11.

$$(-2) + (-9) = -11 \quad \text{Correct Pair}$$

Solution $\qquad y^2 - 11y + 18 = (y - 2)(y - 9)$

Example 3

Factor $b^2 + 14b + 24$.

Explanation (Step 1)—The square root of b^2 is b. (Step 2)—The factors of +24 must both have the same sign. Since their sum must be a *positive* 24, the

6.4/Factoring a Trinomial

two signs will have to be positive. The pairs of factors of +24 are:

$$+1 \text{ and } +24, \quad +2 \text{ and } +12, \quad +3 \text{ and } +8, \quad +4 \text{ and } +6.$$

(Step 3)—The sum of the pair of factors must be +14.

$$(+2) + (+12) = +14 \qquad \text{Correct Pair}$$

Solution $\qquad b^2 + 14b + 24 = (b + 2)(b + 12)$

Example 4

Factor $c^2 + 4c - 21$.

Explanation (Step 1)—The square root of c^2 is c.
(Step 2)—One of the factors of -21 must be positive while the other must be negative in order for the product to be a *negative* 21. In Step 3, however, we will be looking for a sum that is *positive* 4. For the sum to be positive, the larger factor will have to be positive. The pairs of factors of 21 are:

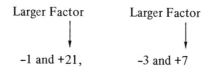

(Step 3)—The sum of the pair of factors must be +4.

$$(-3) + (+7) = +4 \qquad \text{Correct Pair}$$

Solution $\qquad c^2 + 4c - 21 = (c - 3)(c + 7)$

Example 5

Factor $z^2 - 9z - 10$.

Explanation (Step 1)—The square root of z^2 is z.
(Step 2)—The factors of -10 must have different signs. Since their sum must be a *negative* 9, then the sign of the larger factor will have to be negative. The factors of -10 are:

$$+1 \text{ and } -10, \quad +2 \text{ and } -5.$$

(Step 3)—The sum of the pair of factors must be -9.

$$(+1) + (-10) = -9 \qquad \text{Correct Pair}$$

Solution $\qquad z^2 - 9z - 10 = (z + 1)(z - 10)$

Example 6

What can be said of the signs in the factors of $x^2 - 14x - 15$?

Explanation Since the sign of the third term is negative, the signs will

$$x^2 - 14x \overset{\downarrow}{-} 15$$

be different. Since the sign of the middle term is negative, the sign

$$x^2 \overset{\downarrow}{-} 14x - 15$$

of the larger factor will be negative.

Solution $x^2 - 14x - 15 = (\ +\ )(\ -\underset{\uparrow}{\ }\ )$
$$\text{Larger Number}$$

Example 7

What can be said of the signs in the factors of $x^2 - 7x + 6$?

Explanation Since the sign of the third term is positive, the signs

$$x^2 - 7x \overset{\downarrow}{+} 6$$

will be the same. Since the sign of the middle term is negative, the

$$x^2 \overset{\downarrow}{-} 7x + 6$$

two signs will be negative.

Solution $x^2 - 7x + 6 = (\ -\ )(\ -\ )$

EXERCISES 6.4 (Part 1)

A

In each of the exercises below, indicate only what the signs will be within the two factors. If the signs are different, point out the position of the larger number. As examples, exercises 1 and 2 have been completed for you.

1. $x^2 - 7x - 8 = (\ +\ )(\ -\underset{\uparrow}{\ }\ )$
    $$\text{Larger Number}$$

2. $x^2 + 3x - 18 = (\ -\ )(\ +\underset{\uparrow}{\ }\ )$
    $$\text{Larger Number}$$

6.4/Factoring a Trinomial

3. $x^2 + 6x + 5$
4. $x^2 - 5x + 4$
5. $x^2 - 3x - 4$
6. $x^2 + 9x - 10$
7. $a^2 - a - 72$
8. $a^2 - 18a + 45$
9. $b^2 - 6b - 27$
10. $b^2 + 12b + 20$
11. $y^2 - 11y - 12$
12. $x^2 + 15xy + 36y^2$
13. $a^2 - 14ab + 45b^2$
14. $x^2y^2 + 12xy - 28$

B

Factor each of the following trinomials. None of the trinomials is prime.

1. $x^2 + 6x + 5$
2. $x^2 + 8x + 7$
3. $x^2 - 4x + 3$
4. $x^2 - 12x + 11$
5. $x^2 + 6x - 7$
6. $x^2 - 12x - 13$
7. $x^2 + 6x + 8$
8. $x^2 - 7x + 12$
9. $x^2 - 8x + 12$
10. $x^2 + 3x - 10$
11. $x^2 - 6x - 16$
12. $x^2 - 14x + 24$
13. $x^2 - 11x + 24$
14. $x^2 - x - 2$
15. $x^2 + x - 12$
16. $x^2 + 2x + 1$
17. $x^2 - 7x - 30$
18. $x^2 + 6x - 27$
19. $x^2 - 3xy + 2y^2$
20. $x^2 + 7xy - 8y^2$
21. $a^2 - 9ab + 14b^2$
22. $a^2 + 6ab - 16b^2$
23. $b^2c^2 + 7bc - 8$
24. $a^2c^2 - 13ac + 36$
25. $w^2z^2 - 5wz - 36$
26. $b^4 + 15b^2 + 36$
27. $c^4 - 13c^2 + 40$
28. $a^2 - a - 42$
29. $x^2 + x - 30$
30. $y^4 - 17y^2 + 30$
31. $x^4 - 3x^2y - 54y^2$
32. $b^2 - 14bc^2 + 48c^4$
33. $c^2 - 13cd^2 - 48d^4$
34. $x^2 + 17xy + 60y^2$
35. $a^2 - 32ab + 60b^2$
36. $y^4 - 6y^2w - 72w^2$

PART 2

Let us now turn our attention to factoring trinomials where the coefficient of the first term is an integer other than 1. Everything we learned about factoring a trinomial in Part 1 will still be true except for one point. If the sign of the last term is negative, the signs in the two binomials are still different, however the sign of the middle term gives us *no* information.

When trying to factor an expression such as

$$2x^2 - 7x - 15,$$

all the initial steps are exactly as before.

$$2x^2 - 7x - 15 = (\ x +\ \ )(\ x -\ \ )$$

Based on the fact that the sign of the third term is negative the signs in the two binomials will be different. We know also that the square root of x^2 will appear as the variable of the first term in the first binomial and also as the variable of the first term in the second binomial.

At this point, not only must we search for pairs of factors of the 15 as we had before but also for pairs of factors of the coefficient 2.

Pairs of Factors of 2:	Pairs of Factors of 15:
1 and 2	1 and 15, 3 and 5

We are now faced with a matter of "trial and error." We must test each pair of factors of 15 with each pair of factors of 2 until we strike the correct combination. The "right combination" is the one in which the product of the "inners" added to the product of the "outers" gives the middle term.

The testing process is shown below.

$$2x^2 - 7x - 15 \stackrel{?}{=} (2x + 1)(1x - 15) \qquad \text{Check:}$$

$$\begin{array}{rl} +1x & \text{Inners} \\ -30x & \text{Outers} \\ \hline -29x & \end{array}$$

Factors are incorrect.

$$2x^2 - 7x - 15 \stackrel{?}{=} (2x + 15)(1x - 1) \qquad \text{Check:}$$

$$\begin{array}{rl} +15x & \text{Inners} \\ -\ 2x & \text{Outers} \\ \hline +13x & \end{array}$$

Factors are incorrect.

$$2x^2 - 7x - 15 \stackrel{?}{=} (2x + 3)(1x - 5) \qquad \text{Check:}$$

$$\begin{array}{rl} +3x & \text{Inners} \\ -10x & \text{Outers} \\ \hline -7x & \end{array}$$

Factors are correct.

Quite apparently, if both the coefficient of the first term and the coefficient of the last term have many pairs of factors, the process can be long and tedious.

6.4/Factoring a Trinomial

Example 1

Factor $3x^2 - 2x - 8$.

Explanation Pairs of factors of 3: Pairs of factors of 8:

1 and 3 1 and 8, 2 and 4

If either of the two numbers happens to be prime, immediately write the pair of factors of that number in the binomials. Thus, in this situation, since 3 is prime, we can complete the following without any hestitation.

$$3x^2 - 2x - 8 = (3x - \underline{\ \ })(1x + \underline{\ \ })$$

From this point we must test the combinations of 1 and 8, 8 and 1, 2 and 4, and finally 4 and 2. Notice that we must test not only 1 and 8 but also 8 and 1. The first combination will result in different binomials than the second combination. This is shown below:

This binomial is not the same as *this one.*

$(3x - 1)(1x + 8)$ $(3x - 8)(1x + 1)$

This binomial is not the same as *this one.*

Unfortunately, neither of these combinations is the correct factored form of $3x^2 - 2x - 8$.

When testing the pair of factors 4 and 2 we find that the "check" at the right shows a *positive* $2x$ as the middle term.

$3x^2 - 2x - 8 \stackrel{?}{=} (3x - 4)(1x + 2)$ Check:

$-4x$ Inners
$+6x$ Outers
$+2x$

Factors are incorrect.

However, the middle term of the trinomial is a *negative* $2x$. By interchanging the positive and negative signs in the binomial factors we will obtain the correct middle term.

Solution $3x^2 - 2x - 8 = (3x + 4)(1x - 2)$ Check:

$+4x$ Inners
$-6x$ Outers
$-2x$

Example 2

Factor $4x^2 + 12x + 5$.

Explanation Since 5 is prime we immediately write the following as our first step in finding the factored form of $4x^2 + 12x + 5$:

$$4x^2 + 12x + 5 = (\_x + 5)(\_x + 1).$$

We now test the pairs of factors of 4:

4 and 1, 1 and 4, 2 and 2

as the possible coefficients in the first terms. If none yields the middle term, the trinomial $4x^2 + 12x + 5$ is prime.

Solution $4x^2 + 12x + 5 = (2x + 5)(2x + 1)$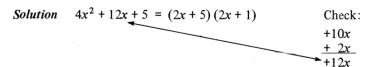

Check:
+10x
+ 2x
+12x

EXERCISES 6.4 (Part 2)

A

Factor each of the following trinomials. None of the trinomials is prime.

1. $3x^2 + 5x + 2$
2. $5x^2 + 11x + 2$
3. $6x^2 + 5x + 1$
4. $8x^2 + 6x + 1$
5. $8x^2 - 2x - 1$
6. $12x^2 + 4x - 1$
7. $7x^2 - 9x + 2$
8. $11x^2 - 9x - 2$
9. $3x^2 + 4x - 7$
10. $7a^2 + 5a - 2$
11. $5a^2 - 6a - 11$
12. $8a^2 - 11a + 3$
13. $6a^2 - a - 5$
14. $7a^2 + a - 8$
15. $5a^2 - 13a + 8$
16. $3a^2 + 11a + 6$
17. $4a^2 - 8a + 3$
18. $2a^2 - 9a + 10$
19. $6a^2 + 17a + 5$
20. $3a^2 - 14a + 8$
21. $3x^2 + 2xy - 8y^2$
22. $4x^2 - 8xy - 5y^2$
23. $3y^2 + 4yz - 4z^2$
24. $5y^2 - 13yz - 6z^2$
25. $3y^2 + 10yz - 8z^2$
26. $6x^2 + 13x + 6$
27. $6x^2 - 25x + 4$
28. $9x^2 + 13x + 4$
29. $9x^2 - 5x - 4$
30. $12x^2 - 20x + 3$
31. $12x^2 - 5x - 3$
32. $8x^2 - x - 9$
33. $9x^2 + 14x - 8$
34. $15x^2 + 4x - 4$
35. $15x^2 - 7x - 4$
36. $15x^2 - 23x + 6$
37. $6x^2 - 19x + 15$
38. $4x^2 - 16x - 9$
39. $8x^2 + 2xy - 15y^2$
40. $15a^2 - 8ab - 12b^2$

6.4/Factoring a Trinomial

B

Factor each of the following trinomials. If any are prime, indicate this by writing the word "prime" to the right of the trinomial.

1. $6x^2 + 5xy - 6y^2$
2. $7x^2 + 10xy - 2y^2$
3. $5x^2 - 8xy - 3y^2$
4. $3a^2 + ab - 10b^2$
5. $6a^2 - 13ab + 5b^2$
6. $8a^2 + 10ab - 3b^2$
7. $5a^2 + 6ab + 8b^2$
8. $10x^2 - 23xy - 5y^2$
9. $6x^2 + 20xy - 7y^2$
10. $7a^2 - 10ab - 8b^2$

PART 3

When factoring certain trinomials, we run across a very special situation. This is the case in the trinomial

$$a^2 + 2ab + b^2.$$

After factoring this trinomial in the usual manner, we discover that its two binomial factors are exactly the same.

$$a^2 + 2ab + b^2 = (a + b)(a + b)$$

In view of the fact that $(a + b)$ is one of a pair of equal factors of the trinomial $a^2 + 2ab + b^2$, then $(a + b)$ is the square root of $a^2 + 2ab + b^2$. Similarly, since $a^2 + 2ab + b^2$ has a pair of equal factors, $(a + b)$ and $(a + b)$, then $a^2 + 2ab + b^2$ is a perfect square. A trinomial such as this is usually called a **perfect square trinomial**.

Rather than write the two binomial factors twice as we have above, we express the product using the exponent 2.

$$a^2 + 2ab + b^2 = (a + b)^2$$

By examining the binomial factor we recognize the fact that a is the square root of a^2. Similarly, the b in the binomial factor is the square root of the b^2 in the trinomial. Hence, if a trinomial is to be a perfect square, we know that both its first term and last term must be perfect squares,

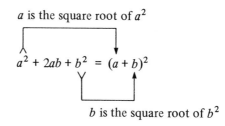

It is not enough, though, that both the first and last terms of the trinomial be perfect squares. In addition to this, the middle term of the trinomial must be twice the product of the terms of the binomial.

$$2ab \text{ must be twice } a \text{ times } b.$$

$$\underline{} + 2ab + \underline{} = (a + b)^2$$

For instance, although the first and last terms of the trinomial below are perfect squares the trinomial itself is *not* a perfect square trinomial.

$$25x^2 + 34x + 9 \stackrel{?}{=} (5x + 3)^2$$

The middle term is *not* twice the product of $5x$ and $+3$:

$$34x \text{ does not equal } 2(5x)(+3).$$

In view of this $(5x + 3)^2$ is not the factored form of $25x^2 + 34x + 9$.

Example 1

Is $49x^2 + 42x + 9$ a perfect square trinomial?

Solution If it is a perfect square trinomial, its factored form can be found by determining the square root of the first term and the square root of the last term.

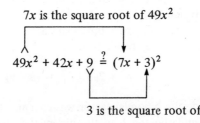

Now we check for the middle term. Is $42x$ equal to twice the product of $7x$ and $+3$?

$$42x \text{ must be twice } 7x \text{ times } 3$$

$$\underline{} + 42x + \underline{} \stackrel{?}{=} (7x + 3)^2$$

Since the answer to this question is yes, then $49x^2 + 42x + 9$ is a perfect square trinomial.

Example 2

Factor $4x^2 - 12x + 9$.

Explanation Since the first and last terms are perfect squares, we test to see if the trinomial is a perfect square trinomial. If it is, the factored form will be

$$(2x - 3)^2.$$

Test for middle term: $2(2x)(-3) = 2(-6x) = -12x$; hence, the trinomial is a perfect square. Notice that the sign of the 3 in the binomial is negative. Were this not so, then twice the product of $(2x)$ and $(+3)$ would not give us the middle term of $-12x$.

Solution $\qquad 4x^2 - 12x + 9 = (2x - 3)^2$

Example 3

Factor $x^2 - 2xy + y^2$.

Explanation If the trinomial is a perfect square, then its factored form will be $(x - y)^2$.

Test for middle term: $2(x)(-y) = 2(-xy) = -2xy$; hence, the trinomial is a perfect square.

Solution $\qquad x^2 - 2xy + y^2 = (x - y)^2$

EXERCISES 6.4 (Part 3)

A

Which of the following trinomials are perfect squares? You may have to rearrange the terms before making your decision.

1. $4x^2 + 20x + 25$
2. $9x^2 - 30x + 25$
3. $2x^2 + 12x + 9$
4. $9a^2 + 24ab + 12b^2$
5. $2x + x^2 + 1$
6. $6y + 9 + y^2$
7. $4x^2 - 29x + 25$
8. $16x^2 + 25 + 40x$
9. $4a^2 + 12a - 9$
10. $36y^2 + 84y + 49$
11. $100x^2 - 1 - 20x$
12. $48xy + 64y^2 + 9x^2$

B

Factor each of the following perfect square trinomials.

1. $x^2 + 8x + 16$
2. $25x^2 + 10x + 1$
3. $a^2 - 2a + 1$
4. $x^2 + 2x + 1$
5. $16x^2 + 56x + 49$
6. $25y^2 - 20y + 4$

7. $16x^2 - 40x + 25$
8. $81a^2 + 36a + 4$
9. $a^2 + 4ab + 4b^2$
10. $9b^2 - 6ab + a^2$
11. $a^2 - 6ab + 9b^2$
12. $49x^2 + 70xy + 25y^2$
13. $9y^2 + 48yz + 64z^2$
14. $49x^2 + 4 + 28x$
15. $25x^2 + 36y^2 - 60xy$
16. $28ab + 49b^2 + 4a^2$

PART 4

You may recall that the final step in our investigation of the methods for factoring a binomial involved a combination of first factoring for the common factor followed by factoring for the difference of two squares. When factoring a trinomial, we follow a similar path. The initial step is to examine the trinomial for any common factors. Thus, in the trinomial

$$3ax^2 + 15ax + 12a \tag{1}$$

we find that the G.C.F. is $3a$. After factoring (1) for $3a$ it becomes

$$3a(x^2 + 5x + 4). \tag{2}$$

At this point we examine the trinomial $x^2 + 5x + 4$ to determine if it is factorable. Since it is, we write (2) in the factored form

$$3a(x + 4)(x + 1). \tag{3}$$

Example 1

Factor $12a^4 - 14a^3 - 6a^2$.

Explanation The G.C.F. is $2a^2$. By factoring the trinomial for the common factor $2a^2$ it becomes

$$2a^2(6a^2 - 7a - 3).$$

After applying the "trial and error" method of factoring to the trinomial $6a^2 - 7a - 3$ we find that it is factorable.

Solution $\quad 12a^4 - 14a^3 - 6a^2 = 2a^2(6a^2 - 7a - 3)$
$\qquad\qquad\qquad\qquad\quad = 2a^2(3a + 1)(2a - 3)$

Example 2

Factor $x^4 - 7x^2 - 18$.

Explanation After examining the trinomial we find that the terms have no common factor. We can, however, factor the trinomial into two binomial factors.

$$x^4 - 7x^2 - 18 = (x^2 - 9)(x^2 + 2)$$

6.4/Factoring a Trinomial

The binomial factor $x^2 - 9$ is not prime for it is the difference of two squares; this implies that $(x^2 - 9)(x^2 + 2)$ is *not* the factored form of $x^4 - 7x^2 - 18$. The process is completed by factoring $x^2 - 9$.

Solution $\quad x^4 - 7x^2 - 18 = (x^2 - 9)(x^2 + 2)$

$$= (x + 3)(x - 3)(x^2 + 2)$$

Example 3

Factor $4ax^4 + 4ax^2 - 80a$.

Explanation Here again, as *always*, we must look for the common factor first. It is $4a$. The procedure now follows a combination of the solutions in Examples 1 and 2.

Solution $\quad 4ax^4 + 4ax^2 - 80a = 4a(x^4 + x^2 - 20)$

$$= 4a(x^2 + 5)(x^2 - 4)$$

$$= 4a(x^2 + 5)(x + 2)(x - 2)$$

EXERCISES 6.4 (Part 4)

Factor each of the following polynomials. *Look for common factor first.*

1. $ax^2 + 5ax + 6a$
2. $bx^2 - 2bx - 8b$
3. $cx^2 - 8cx + 15c$
4. $2x^2 + 10x - 28$
5. $5x^2 + 15x - 50$
6. $2a^2 - 4a - 48$
7. $4a^2 + 40a + 84$
8. $x^3 - 9x^2 + 20x$
9. $20 - 14x + 2x^2$
10. $36 + 30x + 6x^2$
11. $15a^2 - 8a^3 + a^4$
12. $7x^2 + 14x + 7$
13. $2a^2 - 20a + 50$
14. $x^2y - 5xy - 6y$
15. $a^2b - ab - 30b$
16. $x^2y - 6xy^2 + 5y^3$
17. $a^3 + 6a^2b - 27ab^2$
18. $a^2x^2 + 3a^2xy - 40a^2y^2$
19. $x^4 - 3x^2 - 4$
20. $x^4 - 11x^2 + 18$
21. $6x^2 + 16x + 10$
22. $2a^2x - 15ax + 7x$
23. $15a^2 - 35a + 10$
24. $3a^3 + 14a^2 + 8a$
25. $x^3y - 2x^2y^2 - 15xy^3$
26. $6a^2 - 12a - 18$
27. $a^3 - 2a^2 + a$
28. $18a^2 + 12a + 2$
29. $3a^4 - 16a^3 + 5a^2$
30. $20a^2 - 40a - 25$
31. $9x^2 - 18xy + 9y^2$
32. $16x^2 - 48x + 36$
33. $4x^3 - 2x^2 - 6x$
34. $4ax^4 + 11ax^2 - 3a$
35. $x^4 - 13x^2 + 36$
36. $x^5 + 4x - 5x^3$

6.5 SUMMARY OF FACTORING

At this point it might be well to summarize what we know about factoring and to examine a set of exercises where the polynomials to be factored involve a mix of every type we have encountered.

Step 1: Always look for common factor **first**.
Step 2: In the event there is *no* common factor:
    a. If the polynomial is a **binomial**.
        1. It must be the *difference of two squares* to be factorable by our methods.
    b. If the polynomial is a **trinomial**,
        1. The *binomial factors* will have to be found by the "trial and error" method.
        2. Check to determine if it is a perfect square trinomial.
    c. If the polynomial has **four terms**, try to group the terms in pairs to see if there is a binomial common factor.
Step 3: If there is a common factor,
    a. Factor the polynomial for this common factor.
    b. Examine the polynomial factor by following the procedure in step 2.
Step 4: Make certain that each of the factors in your answer is **prime**.

EXERCISES 6.5

Factor each of the following polynomials. If any are prime, indicate this by writing the word "prime" to the right of the polynomial.

1. $2a + 2b$
2. $x^2 - 4$
3. $5x^2 - 10x$
4. $x^2 + 7x + 6$
5. $(x + 2)a + (x + 2)b$
6. $2a^2b + 6ab^2$
7. $1 - 5a + 4a^2$
8. $a^2 - 6a + 9$
9. $9x^2 - 25$
10. $y(a - 4) - 2(a - 4)$
11. $3a^3b - 11a^2b^4$
12. $8x^2 + 6x + 1$
13. $3a^2 + 7a + 2$
14. $9x^2 - 8x - 1$
15. $(x - y)a - (x - y)b$
16. $4a^2 + 9$
17. $x^3 - x^2$
18. $1 - 7x - 18x^2$
19. $12a^2 + a - 1$
20. $9a^3b^2 - 12a^2b^5 - 3ab$
21. $3a^2 + 20a - 7$
22. $3a + 2b - 5c$
23. $x(a + b) + (a + b)$
24. $4x^2 - 20xy + 25y^2$
25. $11x^2 - 5xy - 6y^2$
26. $5a^2 + 20$
27. $2x^2 - 18x + 40$
28. $ax^2 - ay^2$
29. $6x^2 - xy - 7y^2$
30. $a^2x - 3ax - 28x$

31. $a(x - 7) - (x - 7)$
32. $x^4 - 16$
33. $3a^2 - 25$
34. $5 - 25a - 30a^2$
35. $ax + ay + 3x + 3y$
36. $a^2 b^2 - 1$
37. $6x^2 - 17xy + 7y^2$
38. $9a^2 b - 24ab + 16b$
39. $16a^2 + 4a - 6$
40. $x^4 + x^2 - 20$
41. $ax + bx - ay - by$
42. $x^3 - 6x^2 - 40x$
43. $24x^2 - 28x - 20$
44. $ax + 3y - 3x - ay$
45. $2ax^2 - 24ax + 70a$
46. $49x^4 - 81y^4$
47. $3x^4 - 28x^3 + 9x^2$
48. $12a^2 + 28ab - 5b^2$
49. $x^4 - 10x^2 + 9$
50. $16x^5 y - 17x^3 y^3 + xy^5$

6.6 SOLUTION OF THE QUADRATIC EQUATION BY FACTORING

Our background in factoring gives us the opportunity to solve equations other than the linear equation. You may recall that the highest power of the variable that appears in a linear equation is the first power.

$$5x^1 - 3 = 2x^1 + 12 \qquad \text{Linear Equation}$$

In this section we plan to develop a method for finding the roots of a **quadratic equation**. A quadratic equation is an equation where the highest power of the variable is the *second* power. The general form of the quadratic equation is

$$ax^2 + bx + c = 0 \qquad \text{where} \quad a \neq 0.$$

Notice that specific attention is called to the fact that the coefficient of x^2 must *not* equal zero. Were it equal to zero then there would be no "x^2" term and hence the equation would be a linear equation rather than a quadratic equation.

No matter how many terms the quadratic equation may have at the outset it can always be rewritten in a manner similar to one of the three below.

1. $5x^2 - x - 4 = 0$
2. $5x^2 - x = 0$
3. $5x^2 - 4 = 0$

Notice that in each of these equations the highest exponent that appears for x is 2. In the third equation there is no term containing x to the first power. It is not necessary that this term be present in a quadratic equation.

Each of the three equations above is written in what is called **standard form**. A quadratic equation is written in standard form when the left member is arranged in descending powers and the *right member is zero*.

We need a little more information before we can attempt to solve a quadratic equation. In the study of arithmetic we learned that any number multiplied by zero is zero and that zero multiplied by any number is also zero.

$$4 \cdot 0 = 0 \quad \text{and} \quad 0 \cdot 4 = 0$$

More important to us, however, is the fact that the reverse of this statement is also true. That is, if the product of two numbers is zero, at least one *must* be zero and possibly both may be zero.

$$\text{Thus, if} \quad x \cdot y = 0, \quad \text{then}$$
$$\text{either} \quad x = 0$$
$$\text{or} \quad y = 0$$
$$\text{or perhaps both} \quad x = 0 \text{ and } y = 0$$

Expressing this somewhat more generally we can say that if the product of two quantities is zero, then either one of the quantities is zero or both quantities are zero. For instance, in the equation

$$(x - 2)(x - 5) = 0,$$

the product of $(x - 2)$ and $(x - 5)$ is zero. Hence, we can conclude that either

$$x - 2 = 0 \quad \text{or} \quad x - 5 = 0.$$

If this is true, then by solving each of these linear equations we obtain

$$x = 2 \quad \text{or} \quad x = 5.$$

The question now arises as to which of the two numbers, 2 or 5, is the root of the equation

$$(x - 2)(x - 5) = 0?$$

Actually both are, for when we *solve* an *equation* we are looking for *all* numbers that can replace the variable and make the equation true.

Example 1

Solve and check the equation $(x - 7)(x + 2) = 0$.

Explanation Since the product of $x - 7$ and $x + 2$ equals 0, we know that one or both of these factors must be 0. To determine all the roots, we set both factors equal to 0.

6.6/Solution of the Quadratic Equation by Factoring

Solution

$$(x - 7)(x + 2) = 0$$

$x - 7 = 0$	$x + 2 = 0$
$x = 7$	$x = -2$
Check for $x = 7$:	Check for $x = -2$
$(x - 7)(x + 2) = 0$	$(x - 7)(x + 2) = 0$
$(7 - 7)(7 + 2) \stackrel{?}{=} 0$	$(-2 - 7)(-2 + 2) \stackrel{?}{=} 0$
$0 \cdot 9 \stackrel{?}{=} 0$	$(-9)(0) \stackrel{?}{=} 0$
$0 = 0$	$0 = 0$

Example 2

Solve and check the equation $x^2 + 8x - 33 = 0$.

Explanation Our method of attack is based on the fact that we can draw a conclusion only if *the product of two factors is zero*. This implies that $x^2 + 8x - 33 = 0$ must be expressed in factored form.

Solution

$$x^2 + 8x - 33 = 0$$
$$(x + 11)(x - 3) = 0$$

$x + 11 = 0$	$x - 3 = 0$
$x = -11$	$x = 3$
Check for $x = -11$	Check for $x = 3$
$x^2 + 8x - 33 = 0$	$x^2 + 8x - 33 = 0$
$(-11)^2 + (8)(-11) - 33 \stackrel{?}{=} 0$	$(3)^2 + (8)(3) - 33 \stackrel{?}{=} 0$
$121 - 88 - 33 \stackrel{?}{=} 0$	$9 + 24 - 33 \stackrel{?}{=} 0$
$0 = 0$	$0 = 0$

Example 3

Solve the equation $2x^2 + 3x = 20$

Explanation Were we to factor the left member of the equation as it now stands

$$x(2x + 3) = 20,$$

there would be no conclusion of any real value that can be drawn about x and $(2x + 3)$. There are infinitely many pairs of numbers whose product is 20. Some of these are

1 and 20,   2 and 10,   4 and 5,   -1 and -20, $\frac{1}{2}$ and 40, $-\frac{1}{10}$ and -200.

It is imperative that the right member of the equation be 0 in order to conclude that each factor is 0. To make the right member of the equation 0 it is necessary to add −20 to both members of the equation.

Solution

$$2x^2 + 3x = 20$$
$$2x^2 + 3x - 20 = 0$$
$$(2x - 5)(x + 4) = 0$$

$$2x - 5 = 0 \qquad\qquad x + 4 = 0$$
$$2x = 5 \qquad\qquad x = -4$$
$$x = \tfrac{5}{2}$$

Example 4

Solve the equation $4x^2 = 25$.

Explanation As we discovered above, our first move is to write the equation $9x^2 = 25$ in standard form.

Solution

$$9x^2 = 25$$
$$9x^2 - 25 = 0$$
$$(3x - 5)(3x + 5) = 0$$

$$3x - 5 = 0 \qquad\qquad 3x + 5 = 0$$
$$3x = 5 \qquad\qquad 3x = -5$$
$$x = \tfrac{5}{3} \qquad\qquad x = -\tfrac{5}{3}$$

Example 5

Solve the equation $8x^2 = 12x$.

Explanation When the equation is written in standard form we find after factoring the left member that one of the factors happens to be a monomial. This merely implies that the monomial factor will have to be set equal to 0.

Solution

$$8x^2 = 12x$$
$$8x^2 - 12x = 0$$
$$4x(2x - 3) = 0$$

$$4x = 0 \qquad\qquad 2x - 3 = 0$$
$$x = 0 \qquad\qquad 2x = 3$$
$$\qquad\qquad x = \tfrac{3}{2}$$

6.6/Solution of the Quadratic Equation by Factoring

It would not be quite right to leave you with the impression that we can solve all quadratic equations. Our procedure has been to write the equation in standard form and then *factor* the left member. Quite apparently, if the left member *cannot be factored*, then we cannot find the roots of the equation. Quadratic equations where the left member is not factorable will be investigated at a later date.

EXERCISES 6.6

A

Solve and check each of the following equations.

1. $(x - 5)(x - 4) = 0$
2. $(x + 3)(x - 6) = 0$
3. $(x - 7)(x + 7) = 0$
4. $(x - 5)(x - 5) = 0$
5. $(x - 3)^2 = 0$
6. $x(x - 4) = 0$
7. $5x(x + 2) = 0$
8. $(2x + 5)(x + 3) = 0$
9. $(x - 2)(x + 3)(x - 4) = 0$
10. $6x(x - 3)(x + 5) = 0$

B

Solve each of the following equations.

1. $x^2 - 7x + 10 = 0$
2. $x^2 + 9x + 18 = 0$
3. $x^2 - 7x - 8 = 0$
4. $x^2 - 25 = 0$
5. $4x^2 - 49 = 0$
6. $x^2 + 2x - 35 = 0$
7. $x^2 - 5x = 0$
8. $3x^2 - x - 2 = 0$
9. $5x^2 + 12x + 7 = 0$
10. $6x^2 - 18x = 0$
11. $9x^2 - 25 = 0$
12. $12x^2 - 19x + 4 = 0$
13. $x - x^2 = 0$
14. $4x - 6x^2 = 0$
15. $x^2 + 6x + 9 = 0$
16. $15x^2 + 2x - 1 = 0$
17. $4x^2 - 4x + 1 = 0$
18. $6x^2 + 5x - 6 = 0$
19. $x^3 - 4x = 0$
20. $2x^3 - 50x = 0$

C

Solve each of the following equations.

1. $x^2 + 3x = 40$
2. $x^2 - 10x = -21$
3. $x^2 - 24 = 10x$
4. $x^2 - 45 = -4x$
5. $x^2 = 4x + 77$
6. $x^2 = 36$
7. $x^2 = 9x$
8. $3x^2 = 6x$
9. $x^2 = 32 - 4x$
10. $x^2 + 16 = 8x$

11. $2x^2 + 15 = 11x$
12. $4x^2 = 49$
13. $4x^2 = 12x$
14. $x^2 = 7x - 6$
15. $5x = 24 - x^2$
16. $-6 = x - 2x^2$
17. $14 = 13x - 3x^2$
18. $x^3 = 16x$
19. $x(x + 3) - 2(x + 3) = 0$
20. $x^2(x + 5) - 4(x + 5) = 0$

6.7 PROBLEM SOLVING INVOLVING QUADRATIC EQUATIONS

PART 1

Our ability to solve certain quadratic equations enables us to investigate a number of narrative problems that we were unable to examine earlier. The approach to problem solving now is the same as before. At this time though, we will find it necessary to solve a quadratic equation rather than a linear equation.

Example 1

One number is 6 more than another. The product of the two numbers is 40. What are the two numbers?

Solution Let x represent the smaller number.
Then $x + 6$ will represent the larger number.

Therefore,
$$x(x + 6) = 40$$
$$x^2 + 6x = 40$$
$$x^2 + 6x - 40 = 0$$
$$(x + 10)(x - 4) = 0$$

$x + 10 = 0$		$x - 4 = 0$	
$x = -10$	(Smaller)	$x = 4$	(Smaller)
$x + 6 = -10 + 6 = -4$	(Larger)	$x + 6 = 4 + 6 = 10$	(Larger)
Check:		Check:	
$(-10)(-4) = 40$		$(4)(10) = 40$	

Explanation Notice that there are *two* numbers that can be the *smaller* number. They are -10 and 4. Either of these numbers satisfies the conditions of the narrative. For each of these "smaller" numbers there is a corresponding larger number.

Example 2

Two numbers are consecutive even integers. If 2 is added to the smaller and 4 is added to the larger, their product will be 96. What are the two numbers?

6.7/Problem Solving Involving Quadratic Equations

Explanation Pairs of numbers such as 6 and 8, or 24 and 26, or 90 and 92 are consecutive *even* integers. By adding 2 to the smaller of the two numbers we can find the next consecutive even number. Hence, if x is the smaller of two consecutive even numbers, $x + 2$ will be the next larger.

Solution Let x represent the smaller number. Then $x + 2$ will represent the larger number.

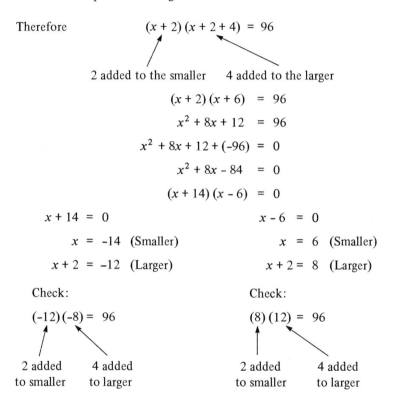

Therefore $(x + 2)(x + 2 + 4) = 96$

    2 added to the smaller   4 added to the larger

$$(x + 2)(x + 6) = 96$$
$$x^2 + 8x + 12 = 96$$
$$x^2 + 8x + 12 + (-96) = 0$$
$$x^2 + 8x - 84 = 0$$
$$(x + 14)(x - 6) = 0$$

$x + 14 = 0$	$x - 6 = 0$
$x = -14$ (Smaller)	$x = 6$ (Smaller)
$x + 2 = -12$ (Larger)	$x + 2 = 8$ (Larger)
Check:	Check:
$(-12)(-8) = 96$	$(8)(12) = 96$
2 added 4 added	2 added 4 added
to smaller to larger	to smaller to larger

Explanation Notice that it was necessary to find the product of $(x + 2)$ and $(x + 6)$ *before* adding -96 to both members of the equation $(x + 2)(x + 6) = 96$.

EXERCISES 6.7 (Part 1)

Find the numbers requested in each of the following problems.

1. If three is added to the square of a number the sum is thirty-nine. What is the number?

2. If eight is added to the square of a number, the sum is six times the original number. What is the number?

3. If a number is added to its square, the sum is fifty-six. What is the number?
4. If twice a number is subtracted from the square of the number, the difference is thirty-five. What is the number?
5. One number is three more than another. Their product is forty. What are the two numbers?
6. One number is six less than another. If their product is fifty-five, what are the numbers?
7. The product of two numbers is thirty-two. If the larger is twice the smaller, what are the numbers?
8. Two numbers are consecutive numbers. Their product is seventy-two. What are the numbers?
9. Two numbers are consecutive numbers. If four is added to the smaller and two is added to the larger, their product will be thirty. What are the numbers?
10. Two numbers are consecutive even numbers. Their product is forty-eight. What are the numbers?
11. Two numbers are consecutive odd numbers. If their product is sixty-three, what are the numbers?
12. A larger number is two more than a smaller number. The square of the smaller added to the larger is fourteen. What are the numbers?
13. Two numbers are consecutive even numbers. The square of the smaller added to twice the larger is twenty-eight. What are the numbers?
14. Two numbers are consecutive odd numbers. The sum of their squares is thirty-four. What are the numbers?

PART 2

Narrative problems related to the area of a rectangle frequently involve the solution of a quadratic equation. A rectangle as you may recall is a four-sided figure such as Figure 6-1. The sides of this figure are lines while the angles are right angles.

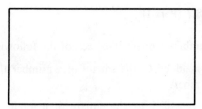

FIGURE 6-1

6.7/Problem Solving Involving Quadratic Equations

Finding the area of the rectangle is done by counting the number of square units in the region enclosed by the four lines. A count will show that there are 18 units in Figure 6-2.

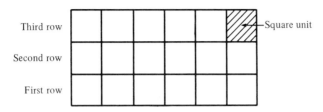

FIGURE 6-2

Rather than resorting to counting, we notice that there are 6 units in each row and that there are 3 rows. Hence, by multiplying 6 by 3 we can find the number of square units very rapidly.

Similarly, if the rectangle were 7" long and 4" wide, as in Figure 6-3, we would know immediately that there are 7 square inches in the first row and

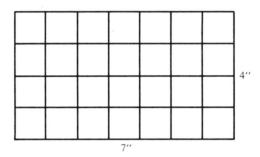

FIGURE 6-3

4 rows in the figure. Hence, the number of square inches in the rectangle is 28. Here, again, this number is found by multiplying 7 by 4.

It would seem then that the area of a rectangle can be found by multiplying the number representing the length of the rectangle by the number representing its width.

$$\text{Area} = \text{``Length''} \times \text{``Width''}$$

Example 1

The length of a rectangle is five inches more than the width. If the area of the rectangle is eighty-four square inches, what are the dimensions of the rectangle?

Explanation Should we use x to represent the width of the rectangle, then its length which is 5 inches more will be $x + 5$. The product of x and $x + 5$ will give the area of the rectangle.

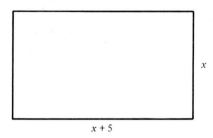

Solution Let x represent the width of the rectangle. Then $x + 5$ represents the length of the rectangle.

Therefore,
$$x(x + 5) = 84$$
$$x^2 + 5x = 84$$
$$x^2 + 5x - 84 = 0$$
$$(x + 12)(x - 7) = 0$$

$x + 12 = 0 \qquad\qquad x - 7 = 0$

$x = -12$ (Discard) $\qquad\qquad x = 7''$ (Width)

$\qquad\qquad\qquad\qquad x + 5 = 12''$ (Length)

Check:

Area $= 12 \cdot 7 = 84$ square inches

Explanation continued Notice that although -12 is a root of equation $x(x + 5) = 84$, it cannot be the width of the rectangle. The width of a rectangle can never be a *negative* number of inches.

The above example points up the fact that we must carefully examine the answers we find in the solution of narrative problems for the numbers may be roots of the quadratic equation but yet not fit the conditions of the problems. For instance, if the narrative dealt with people and one of the answers we find is $5\frac{1}{2}$, this answer must be discarded as being meaningless. Also, an answer such as -4 will have to be discarded were the problem concerned with a certain number of typewriters.

The *perimeter* of any closed figure whose sides are lines is understood to be the sum of the measures of the sides. The perimeter of Figure 6-4 is

$$4'' + 3'' + 5'' + 6'' + 3'' \text{ or } 21''.$$

6.7/Problem Solving Involving Quadratic Equations

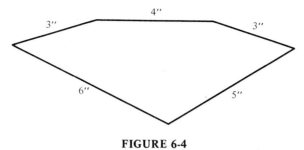

FIGURE 6-4

In the case of a rectangle where the opposite sides are equal, as in Figure 6-5 we can represent the perimeter as

$$a + b + a + b.$$

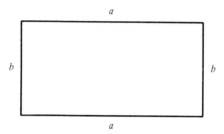

FIGURE 6-5

Collecting like terms, the perimeter of a rectangle becomes

$$2a + 2b.$$

Hence, if the perimeter of a rectangle is 60 inches we can say

$$2a + 2b = 60.$$

To simplify this equation somewhat we multiply both members by $\frac{1}{2}$.

$$\frac{1}{2}(2a + 2b) = \frac{1}{2}(60)$$
$$a + b = 30$$

Since $a + b$ is the sum of the length and width of the rectangle and 30" is half the perimeter, we can say that,

---

Half the perimeter of a rectangle is equal to the sum of the length and width.

In view of this, if the perimeter is 36", then the sum of the length and width of the rectangle is 18". Further, if the width of this rectangle is 6", then the length must be 18" - 6" or 12". Or if the width is 3", then the length must be 18" - 3" or 15". In general, if the width of a rectangle is subtracted from one-half the perimeter, the difference will be the length. We will use this information in the next illustration.

Example 2

The perimeter of a rectangle is 40 inches. The area of this rectangle is 75 square inches. What are the length and width of the rectangle?

Explanation We will represent the width by x. This means that by finding one-half of the perimeter of 40" and subtracting x from that number, the difference will represent the length.

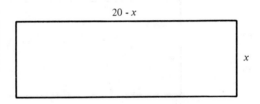

Solution Let x represent the width of the rectangle. Then $20 - x$ will represent the length of the rectangle.

Therefore,

$$x(20 - x) = 75$$
$$20x - x^2 = 75$$
$$-x^2 + 20x + (-75) = 0$$
$$-1(-x^2 + 20x - 75) = -1(0)$$
$$x^2 - 20x + 75 = 0$$
$$(x - 15)(x - 5) = 0$$

$x - 15 = 0$	$x - 5 = 0$
$x = 15"$ (Width)	$x = 5"$ (Width)
$20 - x = 5"$ (Length)	$20 - x = 15"$ (Length)
Check:	Check:
Area = $5 \cdot 15$ = 75 square inches	Area = $15 \cdot 5$ = 75 square inches

6.7/Problem Solving Involving Quadratic Equations

Explanation continued Notice that the length and width simply interchanged positions in the two rectangles. Notice, also, that in order to make the left

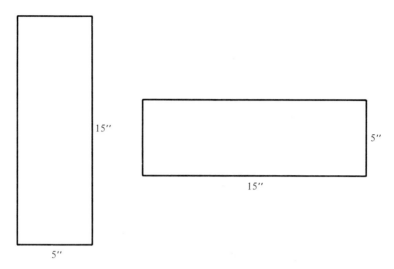

member of the equation $-x^2 + 20x - 75 = 0$ somewhat easier to factor, we multiplied both members of that equation by -1.

Example 3

The length of a square is increased 3 inches while its width is decreased by 3 inches. The area of the new rectangle formed by this change is 72 inches. How long was a side of the original square?

Explanation It is important that we know that a square is a rectangle whose length and width are equal. Hence, if x is the measure of a side of the original square, then $x + 3$ will be the measure of the length of the new rectangle and $x - 3$ will be its width.

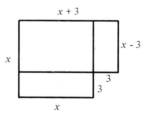

Solution Let x represent the measure of the side of the square. Then $x + 3$ will represent the length of the rectangle. Also $x - 3$ will represent the width of the rectangle.

Therefore,
$$(x + 3)(x - 3) = 72$$
$$x^2 - 9 = 72$$
$$x^2 - 81 = 0$$
$$(x + 9)(x - 9) = 0$$

$x + 9 = 0$ \qquad\qquad $x - 9 = 0$
$x = -9$ (Discard) \qquad $x = 9''$ (Side of Square)

Check:
Area $= (9 + 3)(9 - 3) = 12 \cdot 6 = 72$ square inches

Example 4

Jim Bowers has a piece of sheet metal in which the length is twice the width. He plans to cut $3''$ squares out of each of the corners and fold up the strips to form a box. The area of the bottom of the box will be 56 square inches. What are the dimensions of the original piece of sheet metal?

[Diagram showing a rectangle of length $2x$ and width x, with $3''$ squares at each corner. Inner bottom dimensions labeled $2x - 6$ and $x - 6$.]

Solution (See diagram) Let x represent the width of the rectangle. Then $2x$ represents the length of the rectangle.

Therefore, $\quad\underbrace{(2x - 6)}_{\text{Length of bottom}} \underbrace{(x - 6)}_{\text{Width of bottom}} = 56$

$$2x^2 - 18x + 36 = 56$$
$$2x^2 - 18x + 36 + (-56) = 0$$
$$2x^2 - 18x - 20 = 0$$
$$\tfrac{1}{2}(2x^2 - 18x - 20) = \tfrac{1}{2}(0)$$
$$x^2 - 9x - 10 = 0$$
$$(x - 10)(x + 1) = 0$$

$x - 10 = 0$ \qquad\qquad $x + 1 = 0$
$x = 10$ (Width) \qquad $x = -1$ (Discard)
$2x = 20$ (Length)

Check:
Area of bottom $= (10 - 6)(20 - 6) = 4 \cdot 14 = 56$ square inches

6.7/Problem Solving Involving Quadratic Equations

Explanation Notice that in order to make the left member of the equation $2x^2 - 18x - 20 = 0$ somewhat easier to factor, we multiplied both members of that equation by $\frac{1}{2}$.

EXERCISES 6.7 (Part 2)

Find the numbers requested in each of the following problems.

1. The length of a rectangle is 3 inches more than the width. The area of the rectangle is 54 square inches. What are the length and width?
2. The width of a rectangle is 6 feet less than it is long. If the area of the rectangle is 27 square feet, what are the dimensions of the rectangle?
3. The length of a rectangle is twice its width. The area of the rectangle is 32 square yards. What are the dimensions of the rectangle?
4. The length and width of a rectangle happen to be two consecutive even numbers. If the area is 80 square feet, what are the dimensions of the rectangle?
5. The perimeter of a rectangle is 30 inches. The area of the rectangle is 56 square inches. What are the dimensions of the rectangle?
6. Nick Calleo purchased 40 yards of fencing to enclose a rectangular plot of land he planned to use for a vegetable garden. If the area of the plot is 96 square yards, what are the dimensions of the land?
7. Diane Barsto purchased 40 feet of edging to sew around the raw edge of a carpet she purchased. The rug will cover 99 square feet of flooring in the room. What are the dimensions of the carpet?
8. What are the dimensions of the floor of a square room if its area is 144 square feet?
9. The length of a square was increased by 4 inches while its width remained the same. If the area of the new rectangle is 45 square inches, what was the length of a side of the original square?
10. The length of a square is increased by 6 inches and its width is increased by 4 inches. The area of the new rectangle is 63 square inches. What was the length of a side of the original square?
11. A 1-yard-wide cement walk completely surrounds a rectangular pool. The length of the pool is 3 yards more than the width. The total area of land covered by the pool and walk is 70 square yards. What are the dimensions of the pool? (See figure on page 246.)
12. In order to shut out some of the light from a box that has a glass top, strips of masking tape of the same width are placed completely around the edge of the glass. The area of the glass that is still exposed is 120 square inches. The dimensions of the glass are 16 inches by 14 inches. What is the width of the masking tape? (See figure on page 246.)

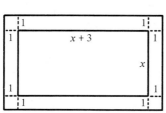
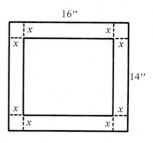

PROBLEM 11 **PROBLEM 12**

13. A grass walk 4 feet wide completely surrounds a rectangular flower garden. The length of the flower garden is twice its width. If the area of the flower garden and walk are 330 square feet, what are the dimensions of the flower garden?

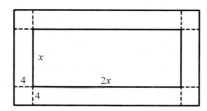
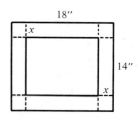

PROBLEM 13 **PROBLEM 14**

14. A rectangular oil painting whose dimensions are 18 inches by 14 inches is masked with a border of cloth tape of uniform width. Still visible are 192 square inches of the painting. What is the width of the tape?

CHAPTER REVIEW

A

1. List the factors of each of the following.

   a. 9   b. 20   c. 28   d. ab

2. List the pairs of factors of each of the following numbers.

   a. 8   b. 10   c. 36

3. List the prime factors of each of the following numbers.

   a. 5   b. 8   c. 42

4. Factor each of the following numbers.

   a. 10   b. 16   c. 30

5. Factor each of the following polynomials for the common factor *only*. If any are prime, indicate this by writing the word "prime" to the right of the polynomial.

   a. $ab + ac$
   b. $ax - a$
   c. $x^5 - x^2$
   d. $x^2y + xy^2$
   e. $8a^2 - 24a + 4$
   f. $7a^2 - 14a^3 - 35a^7$
   g. $b(x - y) + a(x - y)$
   h. $x(a + b) - (a + b)$

6. Factor each of the following polynomials for the difference of two squares only. If any are prime, indicate this by writing the word "prime" to the right of the polynomial.

   a. $x^2 - 36$
   b. $49a^2 - 1$
   c. $16x^2 - 81$
   d. $4a^2 + 9$
   e. $x^6 - 25$
   f. $x^2 - y^2$
   g. $-9 + 49x^2$
   h. $x^2 - 16$
   i. $5a^2 - 36$
   j. $49a^2 - 64b^2$
   k. $25x^2y^4 - z^6$
   l. $(a + b)^2 - c^2$

7. Factor each of the following trinomials. None is prime and none is factorable for a common factor.

   a. $x^2 + 3x + 2$
   b. $x^2 - 8x + 7$
   c. $x^2 - 4x - 5$
   d. $x^2 + 10x - 11$
   e. $12 - 7x + x^2$
   f. $18 - 7x - x^2$
   g. $x^2 - 5xy - 24y^2$
   h. $2x^2 - 11x + 5$
   i. $4x^2 - 12x - 7$
   j. $9x^2 - 12x + 4$
   k. $6x^2 + x - 15$
   l. $9x^2 + xy - 8y^2$
   m. $6x^2 - 11xy - 10y^2$
   n. $8x^2 + 14xy - 15y^2$

8. Factor each of the following polynomials. None is prime.

   a. $5a^2 - 10ab$
   b. $x^2 - 1$
   c. $x^2 - 8x + 15$
   d. $3a^2 - 12$
   e. $ax + 3a + bx + 3b$
   f. $7 - 11a + 4a^2$
   g. $9x^2 + 45$
   h. $a^4 - 81$
   i. $9x^2 - 24xy + 16y^2$
   j. $y^3 - y$
   k. $6x^2 - 23x + 21$
   l. $ax - bx - a + b$
   m. $6x^2 - 21x + 9$
   n. $4x^2y^4 - 25z^4$
   o. $x^4 - 2x^2 - 8$
   p. $4x^4 - 13x^2 + 9$

9. Solve and check each of the following equations.

   a. $(x - 4)(x + 3) = 0$
   b. $3x(x + 7) = 0$
   c. $(x - 7)^2 = 0$
   d. $(x - 1)(x - 3)(x + 5) = 0$

10. Solve each of the following equations.

    a. $x^2 - 10x + 9 = 0$
    b. $x^2 - 49 = 0$
    c. $x^2 - 8x = 0$
    d. $x^2 = 36$
    e. $9x^2 = 6x$
    f. $2x^2 + x = 3$
    g. $3x^2 = 10x - 8$
    h. $11x = 6 + 5x^2$
    i. $x(x - 2) + 3(x - 2) = 0$
    j. $x^3 = 25x$

B

Solve each of the following problems.

1. If 5 is added to the square of a number, the sum is 41. What is the number?
2. If a number is subtracted from its square, the difference is 56. What is the number?
3. The product of two consecutive odd numbers is 63. What are the numbers?
4. One number is 7 more than another. If their product is 44, what are the numbers?
5. Two numbers are consecutive even numbers. If the sum of their squares is 52, what are the numbers?
6. The length of a rectangle is 5 feet more than the width. If the area of the rectangle is 84 square feet, what are the dimensions of the rectangle?
7. The perimeter of a rectangle is 18 yards. If the area of the rectangle is 20 square yards, what are the dimensions of the rectangle?
8. What are the dimensions of a square rug if twice its area is 162 square feet?
9. The length of a square is increased by 3 feet and the width is increased by 2 feet thus forming a new rectangle. When the area of the square is added to the area of the new rectangle, their sum is 58 square feet. What was the length of a side of the square?
10. A 10-foot by 12-foot rug is placed on a floor so that the wooden border surrounding the rug has a uniform width. If the area of the entire floor is 168 square feet, what is the width of the border?

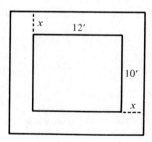

The Operations with Fractions

When we first began the study of arithmetic in elementary school we were concerned only with whole numbers. After we had learned the operations of addition, subtraction, multiplication, and division as they pertained to whole numbers we moved on to the new number called the fraction. Here, again, we found it necessary to learn how to perform these same fundamental operations as they related to fractions.

We find ourselves in the same position at the present time. At the outset of this course we devoted our time to learning addition, subtraction, multiplication, and division of polynomials. Now we plan to investigate the manner in which these operations can be performed when polynomials are found in fractional expressions. For instance, how will we perform the four fundamental operations on expressions such as

$$\frac{3x^2 - 5x - 6}{x^2 - 9}?$$

A fraction such as the one above, in which the numerator is a polynomial and the denominator is also a polynomial, is called a *rational expression*. We will find that each of the fundamental operations on these expressions is dependent on two concepts we must accept.

The first of these involves finding the product of two fractions. Thus, we will agree that the method by which the product of two rational expressions can be found is the same as the method used to find the product of two fractional numbers. The product of $\frac{2}{3}$ and $\frac{5}{7}$ is determined by finding the product of the numerators and writing that answer over the product of the denominators.

$$\frac{2}{3} \times \frac{5}{7} = \frac{2 \times 5}{3 \times 7} \tag{1}$$

We state this principle in general terms as

Multiplication Principle of Fractions

If a, b, c, and d are any real numbers where b does not equal 0 and d does not equal 0, then

1. $\quad \dfrac{a}{b} \cdot \dfrac{c}{d} = \dfrac{a \cdot c}{b \cdot c}$

2. $\quad \dfrac{a \cdot c}{b \cdot d} = \dfrac{a}{b} \cdot \dfrac{c}{d}$

The second concept we must accept is that the product of any number and 1 is the number itself. Thus,

$$5 \cdot 1 = 5, \quad 7 \cdot 1 = 7, \quad \text{and} \quad 125 \cdot 1 = 125.$$

Principle of 1

If a is any real number, then

$$a \cdot 1 = a.$$

However, the number 1 can be named in many different ways since any fraction where the numerator and denominator are the same names the number 1. Each of the fractions below names the number 1.

$$\frac{3}{3}, \quad \frac{5a}{5a}, \quad \frac{-6x^2y}{-6x^2y}, \quad \frac{3x-1}{3x-1}, \quad \frac{(5a-2)(3a-4)}{(5a-2)(3a-4)}.$$

Notice that the "multiplication principle of fractions" points out very specifically that neither b nor d can be equal to 0.

Consider the situation where

$$12 \div 4 = 3.$$

Were we interested in checking to determine whether the quotient 3 is correct, we would multiply the quotient 3 by the divisor 4 to see whether the product is 12.

$$3 \cdot 4 = 12$$

If the divisor happens to be 0, we run into difficulty when trying to check whatever quotient we obtain. For instance,

$$12 \div 0 \stackrel{?}{=} x.$$

Let us say, as a guess, that the quotient x is 6. To check this answer we must multiply 0 by 6. However, 0 times 6 is 0 and *not* 12. In fact, no matter what replacement we make for x, when that number is multiplied by 0, the product is 0 and *not* 12.

Since there is no number that is the quotient when 12 is divided by 0, we say that this fraction is *not defined*. Of course, what we have shown here where 12 is the numerator and 0 the denominator applies to any fraction where 0 is the denominator.

In all of our work henceforth, we will assume that the denominators of the fractions are not zero.

With the "multiplication principle of fractions" and the "principle of 1" as a background, we are ready to begin the investigation of rational expressions.

7.1 REDUCING FRACTIONS TO LOWEST TERMS

PART 1

At the time we learned the process of "reducing a fraction" to lowest terms it involved dividing the numerator and denominator of the fraction by the same number. Thus,

$$\frac{48}{60} = \frac{48 \div 12}{60 \div 12} = \frac{4}{5}$$

In general a fraction is said to be reduced to lowest terms when the numerator and denominator of that fraction have *no common factor other than 1*.

In the situation above, the fraction $\frac{48}{60}$ is *not* reduced to lowest terms, for the numbers 48 and 60 have several common factors other than 1. These are 2, 3, 4, 6, and 12. On the other hand, the fraction $\frac{4}{5}$ is reduced to lowest terms for the numbers 4 and 5 have no common factors other than 1. Two numbers that have no common factor other than 1 are called **relatively prime numbers**.

Hence, *a fraction is reduced to lowest terms when the numerator and denominator are relatively prime.* To reduce a fraction to lowest terms, we must determine the **greatest common factor** of the numerator and the denominator. It is this factor by which the numerator and denominator are divided.

In terms of the multiplication principle of fractions and the principle of 1 what we are actually doing when we reduce a fraction to lowest terms is shown in the following example.

$$\frac{24}{42} = \frac{4 \cdot 6}{4 \cdot 7} = \frac{4}{7} \cdot \frac{6}{6} = \frac{4}{7} \cdot 1 = \frac{4}{7}$$

$$\uparrow \qquad \uparrow \qquad \uparrow \qquad \uparrow$$
$$(1) \qquad (2) \qquad (3) \qquad (4)$$

a. In step (1) 24 and 42 are each written as the product of a pair of factors. One of the factors in each pair is their G.C.F., 6.
b. The multiplication principle of fractions was used to determine step (2).
c. The name "$\frac{6}{6}$" in step (2) was replaced by the name "1" in step (3).
d. The principle of 1 was used to determine step (4).

As a final check, we examine the fraction $\frac{4}{7}$ to make certain that the numerator and denominator are relatively prime.

The process followed above is the very same one that is used when reducing any rational expression to lowest terms.

Example 1

Reduce the following fraction to lowest terms.

$$\frac{12a^2}{16a^5}$$

Explanation Perhaps the best method by which to approach this exercise is to find the G.C.F. of $12a^2$ and $16a^5$ and partly rewrite the fraction as

$$\frac{ \cdot 4a^2}{ \cdot 4a^2} \, .$$

The remaining factors in the numerator and denominator are found by dividing $4a^2$ into each of the terms, $12a^2$ and $16a^5$.

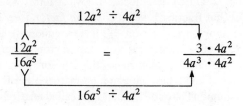

Solution

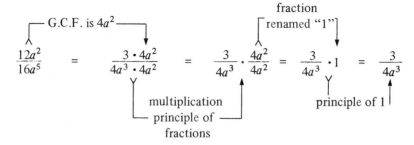

Example 2

Reduce the following fraction to lowest terms.

$$\frac{15x^3yz^2}{10x^5y^3}$$

Solution $\quad \dfrac{15x^3yz^2}{10x^5y^3} = \dfrac{3z^2 \cdot 5x^3y}{2x^2y^2 \cdot 5x^3y} = \dfrac{3z^2}{2x^2y^2}$

Example 3

Reduce the following fraction to lowest terms.

$$\frac{a(x+y)}{b(x+y)}$$

Explanation In this particular situation the G.C.F. happens to be the binomial factor $(x + y)$. We treat this factor the same way in which we treat a monomial factor.

Solution $\quad \underset{(1)}{\dfrac{a(x+y)}{b(x+y)}} = \underset{(2)}{\dfrac{a \cdot (x+y)}{b \cdot (x+y)}} = \underset{(3)}{\dfrac{a}{b} \cdot \dfrac{(x+y)}{(x+y)}} = \underset{(4)}{\dfrac{a}{b}}$

Explanation continued Step (3) may be omitted.

EXERCISES 7.1 (Part 1)

Reduce each of the following fractions to lowest terms.

1. $\dfrac{3a}{6b}$

2. $\dfrac{ab}{ac}$

3. $\dfrac{8a^2}{10a^3}$

4. $\dfrac{x^3y^2}{x^2y^3}$

5. $\dfrac{27xy}{18yz}$ 11. $\dfrac{3(x+y)}{4(x+y)}$

6. $\dfrac{21}{14ab^2}$ 12. $\dfrac{xy(a-b)}{xw(a-b)}$

7. $\dfrac{15x^2z}{15x^2y}$ 13. $\dfrac{16x^2(x-4)}{8x^3(x-4)}$

8. $\dfrac{5x}{x}$ 14. $\dfrac{ab(a-5)}{b}$

9. $\dfrac{a}{ab}$ 15. $\dfrac{(x+5)(x-6)}{(x-1)(x-6)}$

10. $\dfrac{x^2}{12x^3}$ 16. $\dfrac{(x-5)^2}{(x-5)^3}$

PART 2

In Part 1 of this section we examined the situation where a common factor of the numerator and denominator of a fraction can be a binomial. Unfortunately, the binomial common factors may not always be as obvious as they were in the exercises above. Consider, for instance, the fraction where the numerator

$$\dfrac{x^2-9}{x^2-x-12},$$

and denominator of this fraction do have a factor in common. To determine what this common binomial factor is, we must rewrite each of the polynomials in factored form.

$$\dfrac{x^2-9}{x^2-x-12}=\dfrac{(x+3)(x-3)}{(x+3)(x-4)}$$

By examining the fraction now, we can see that the common factor is $(x+3)$. Hence, when the above fraction is reduced to lowest terms it will be

$$\dfrac{x-3}{x-4}.$$

As we know, the process of reducing a fraction to lowest terms involves finding the G.C.F. of the numerator and denominator. Apparently, then, if we are to find the greatest common factor, the numerator and denominator of the fraction will have to be expressed in *factored* form.

Example 1

Reduce the following fraction to lowest terms.

$$\frac{25x^2 - 49}{5x^2 - 12x + 7}$$

Solution $\frac{25x^2 - 49}{5x^2 - 12x + 7} = \frac{(5x-7)(5x+7)}{(5x-7)(x-1)} = \frac{(5x+7)}{(x-1)} \cdot \frac{(5x-7)}{(5x-7)} = \frac{5x+7}{x-1}$

$\quad\quad\quad\quad\quad\quad\quad\;\;\uparrow\quad\quad\quad\quad\quad\uparrow\quad\quad\quad\quad\quad\;\;\uparrow\quad\quad\quad\quad\quad\uparrow$
$\quad\quad\quad\quad\quad\quad\quad\;\;(1)\quad\quad\quad\quad\;\;(2)\quad\quad\quad\quad\quad\;(3)\quad\quad\quad\quad\;\;(4)$

Explanation As suggested earlier, step (3) above may be omitted.

Example 2

Reduce the following fraction to lowest terms.

$$\frac{3x^2 - 15x}{6x^3 + 12x^2}$$

Solution $\frac{3x^2 - 15x}{6x^3 + 12x^2} = \frac{3x(x-5)}{6x^2(x+2)} = \frac{(x-5)}{2x(x+2)} \cdot \frac{3x}{3x} = \frac{(x-5)}{2x(x+2)}$

Example 3

Reduce the following fraction to lowest terms.

$$\frac{2x - 2y}{3y - 3x}$$

Explanation Upon factoring, we discover that the binomial factor in the numerator and the denominator are the same except for the fact that the terms of one are the additive inverses of the terms of the other. The $x - y$ in the

$$\frac{2(x-y)}{3(y-x)},$$

numerator is the additive inverse of $y - x$ in the denominator for their sum is 0.

$$(x + y) + (y - x) = x - y + y - x = 0$$

In this situation if -2 rather than $+2$ is used as the common factor in the numerator the binomial factors would be identical. Since addition is commutative,

$$\frac{-2(-x+y)}{3(y-x)},$$

$-x + y$ can be written as $y - x$. Thus the fraction above becomes

$$\frac{-2(y-x)}{3(y-x)}.$$

This fraction reduces to $\frac{-2}{3}$.

    Solution $\dfrac{2x-2y}{3y-3x} = \dfrac{-2(-x+y)}{3(y-x)} = \dfrac{-2(y-x)}{3(y-x)} = \dfrac{-2}{3}$

Example 4

Reduce the following fraction to lowest terms.

$$\frac{y-x}{x^2-y^2}$$

    Solution $\dfrac{y-x}{x^2-y^2} = \dfrac{y-x}{(x+y)(x-y)} = \dfrac{-1(x-y)}{(x+y)(x-y)} = \dfrac{-1}{(x+y)}$

    Explanation After factoring the numerator and denominator we discover that the factor $(y - x)$ in the numerator is the additive inverse of the factor $x - y$ in the denominator. To eliminate this difficulty, we use -1 as a common factor of the two terms in the numerator. Hence, $y - x$ becomes $-1(-y + x)$, which we rewrite as $-1(x - y)$.

EXERCISES 7.1 (Part 2)

A

Reduce each of the following fractions to lowest terms.

1. $\dfrac{a-b}{b-a}$ 6. $\dfrac{b-a}{(a+b)(a-b)}$

2. $\dfrac{x-3}{3-x}$ 7. $\dfrac{(x-2)(x+2)}{2-x}$

3. $\dfrac{5-y}{y-5}$ 8. $\dfrac{3+2x}{(2x+3)(2x-3)}$

4. $\dfrac{4a-4b}{5b-5a}$ 9. $\dfrac{5a-10b}{(2b+a)(2b-a)}$

5. $\dfrac{6x-6}{7-7x}$ 10. $\dfrac{(3x-4)(3x+2)}{4-3x}$

7.1/Reducing Fractions to Lowest Terms

B

Reduce each of the following fractions to lowest terms.

1. $\dfrac{2a + 2b}{5a + 5b}$

2. $\dfrac{6x - 12y}{8x - 16y}$

3. $\dfrac{x^2 - 4}{2x + 4}$

4. $\dfrac{7x + 21}{x^2 + 7x + 12}$

5. $\dfrac{x^2 + x - 30}{x^2 - 3x - 10}$

6. $\dfrac{x^2 + 5x - 6}{x^2 - 36}$

7. $\dfrac{2x^2 - 18}{x^2 - 5x + 6}$

8. $\dfrac{x^2 + 4x - 5}{x^2 + 10x + 25}$

9. $\dfrac{ax^2 + 10ax + 21a}{ax^2 + 5ax - 14a}$

10. $\dfrac{6a^2(x^2 - 3x + 2)}{12a^2(x^2 + x - 6)}$

11. $\dfrac{x^3 - x^2}{x^3 + x^2 - 2x}$

12. $\dfrac{3(x + 2)}{a(x + 2) + b(x + 2)}$

13. $\dfrac{3a(x - y) + 4b(x - y)}{5(x - y)}$

14. $\dfrac{2x + 2y}{ax + ay + bx + by}$

15. $\dfrac{3x - 3}{ax - a + bx - b}$

16. $\dfrac{4x^2 + 4x - 3}{4x^2 - 1}$

17. $\dfrac{3x^2 - 2x - 5}{9x - 15}$

18. $\dfrac{a - b}{b^2 - a^2}$

19. $\dfrac{6 - 3x}{x^2 + 3x - 10}$

20. $\dfrac{1 - x}{x^2 + 4x - 5}$

21. $\dfrac{3b - 2a}{4a^2 - 12ab + 9b^2}$

22. $\dfrac{x^4 + 2x^2 - 3}{x^2 + 6x + 5}$

23. $\dfrac{a^4 - 8a^2 - 9}{a^2 + 2a - 3}$

24. $\dfrac{x^4 + x^2 - 20}{10x^2 + 50}$

25. $\dfrac{x^4 - 5x^2 + 4}{x^2 + x - 2}$

26. $\dfrac{9a^2 + 3a - 2}{3a^2 - 22a + 7}$

27. $\dfrac{6a^2x + 11ax + 3x}{2axy + 3xy}$

28. $\dfrac{25 - y^2}{2y - 10}$

29. $\dfrac{x^2 + x - 12}{6 + x - x^2}$

30. $\dfrac{x^2 - 2x - 35}{28 + 3x - x^2}$

7.2 MULTIPLICATION OF FRACTIONS

Finding the product of two fractions is a direct application of the multiplication principle of fractions. As a result of accepting this principle, we know that

$$\frac{a}{b} \cdot \frac{c}{d} = \frac{ac}{bd}.$$

This same principle is applied when finding the product of two fractions such as

$$\frac{8}{15} \text{ and } \frac{5}{4}.$$

Thus,

$$\underset{(1)}{\frac{8}{15} \cdot \frac{5}{4}} = \underset{(2)}{\frac{8 \cdot 5}{15 \cdot 4}} = \underset{(3)}{\frac{40}{60}} = \underset{(4)}{\frac{2}{3} \cdot \frac{20}{20}} = \underset{(5)}{\frac{2}{3}}$$

In going from step (1) to step (2) the product of the fractions is determined by finding the product of the numerators and writing this over the product of the denominators. Once step (3) is found, the work is completed by reducing the fraction

$$\frac{40}{60}$$

to lowest terms.

Example 1

Find the product of the two fractions below.

$$\frac{3x}{2y} \cdot \frac{10xy}{9x^2}$$

Solution

$$\underset{(1)}{\frac{3x}{2y} \cdot \frac{10xy}{9x^2}} = \underset{(2)}{\frac{3x \cdot 10xy}{2y \cdot 9x^2}} = \underset{(3)}{\frac{30x^2y}{18x^2y}} = \underset{(4)}{\frac{5}{3} \cdot \frac{6x^2y}{6x^2y}} = \underset{(5)}{\frac{5}{3}}$$

Explanation Since step (2) is just about the same as step (1) it may be omitted.

7.2/Multiplication of Fractions

Based on Example 1, finding the product of two fractions involves expressing the two fractions as a single fraction that is then reduced to lowest terms. If the numerators and denominators of the fractions happen to be binomials or trinomials, as they are below, then the multiplication becomes very cumbersome.

$$\frac{ax-a}{bx+b} \cdot \frac{x^2-1}{2x^2-x-1}$$

In addition, after the product is found, as shown here,

$$\frac{ax^3 - ax^2 - ax + a}{2bx^3 + bx^2 - 2bx - 1}$$

it is now necessary to try to factor the numerator and denominator in order to determine which factors they have in common! Needless to say, factoring the polynomials in this fraction would challenge the best of us.

However, were we to reverse the process and factor the *original* numerators and denominators we would find the work far, far easier. Thus,

(1) $\quad \dfrac{ax-a}{bx+b} \cdot \dfrac{x^2-1}{2x^2-x-1} = \dfrac{a(x-1)}{b(x+1)} \cdot \dfrac{(x+1)(x-1)}{(2x+1)(x-1)} \quad$ (2)

$$= \frac{a(x-1)(x-1)(x+1)}{b(x-1)(x+1)(2x+1)} \quad (3)$$

$$= \frac{a(x-1)}{b(2x+1)} \cdot \frac{(x-1)(x+1)}{(x-1)(x+1)} \quad (4)$$

$$= \frac{a(x-1)}{b(2x+1)} \quad (5)$$

Notice that *at no time* do we actually find any products between the various factors. The products are merely *expressed* between the factors.

Mathematicians feel that the work above is somewhat too time consuming. They know that the only purpose for steps (3) and (4) above is to express the product of the fractions as a single fraction. This is then followed by eliminating *factors* that are common to the numerator and denominator. In view of this, once they have written the original polynomials in factored form, they simply cross off **factors that are common to the numerators and denominators.** Thus,

$$\frac{ax-a}{bx+b} \cdot \frac{x^2-1}{2x^2-x-1} = \frac{a(x-1)}{b\cancel{(x+1)}} \cdot \frac{\cancel{(x+1)}\cancel{(x-1)}}{(2x+1)\cancel{(x-1)}}$$

$$= \frac{a(x-1)}{b(2x+1)}$$

Example 2

Find the product of the two fractions below.

$$\frac{x^2 - 4x - 5}{3 - 3x} \cdot \frac{5x - 10}{x^2 - 7x + 10}$$

Explanation Rewrite each of the polynomials in factored form. Follow this by drawing lines through those factors common to the numerators and denominators.

Solution

$$\frac{x^2 - 4x - 5}{3x - 3} \cdot \frac{5x - 10}{x^2 - 7x + 10} = \frac{\cancel{(x-5)}(x+1)}{3(x-1)} \cdot \frac{5\cancel{(x-2)}}{\cancel{(x-5)}\cancel{(x-2)}}$$

$$= \frac{5(x+1)}{3(x-1)}$$

Example 3

Find the product of the two fractions below.

$$\frac{2x^2}{9x - 9} \cdot \frac{6x^2 - 6}{x^3}$$

Solution

$$\frac{2x^2}{9x - 9} \cdot \frac{6x^2 - 6}{x^3} = \frac{2x^2}{9(x - 1)} \cdot \frac{6(x^2 - 1)}{x^3}$$

$$= \frac{2x^2}{9\cancel{(x-1)}} \cdot \frac{6(x+1)\cancel{(x-1)}}{x^3}$$

$$= \frac{12x^2 (x + 1)}{9x^3} \qquad\qquad (4)$$

$$= \frac{4(x + 1)}{3x} \cdot \frac{3x^2}{3x^2}$$

$$= \frac{4(x + 1)}{3x}$$

Explanation After examining step (4) it was found that the numerator and denominator had the monomial factor $3x^2$ in common. Hence, it was necessary to reduce that fraction to lowest terms.

Example 4

Find the product of the fractions below.

$$\frac{3x^2 - 12x}{x^3 - x^2 - 12x} \cdot \frac{x^2 - 9}{2x^2} \cdot \frac{6x^5}{x^2 - 6x + 9}$$

7.2/Multiplication of Fractions

Solution

$$\frac{3x^2-12x}{x^3-x^2-12x} \cdot \frac{x^2-9}{2x^2} \cdot \frac{6x^5}{x^2-6x+9} = \frac{3x(x-4)}{x(x^2-x-12)} \cdot \frac{(x+3)(x-3)}{2x^2} \cdot \frac{6x^5}{(x-3)(x-3)}$$

$$= \frac{3x\cancel{(x-4)}}{x\cancel{(x-4)}\cancel{(x+3)}} \cdot \frac{\cancel{(x+3)}\cancel{(x-3)}}{2x^2} \cdot \frac{6x^5}{\cancel{(x-3)}(x-3)}$$

$$= \frac{18x^6}{2x^3(x-3)}$$

$$= \frac{9x^3}{x-3} \cdot \frac{2x^3}{2x^3}$$

$$= \frac{9x^3}{x-3}$$

EXERCISES 7.2

A

Find the product in each of the following exercises.

1. $\dfrac{a}{b} \cdot \dfrac{b^3}{a^2}$

2. $\dfrac{5x^2}{2y} \cdot \dfrac{6y^2}{10x^2}$

3. $\dfrac{12}{a^3} \cdot \dfrac{a^2 b}{4b}$

4. $\dfrac{6xy^2}{4y} \cdot \dfrac{2y^3}{3x^4}$

5. $\dfrac{2x}{3y} \cdot \dfrac{5y^2}{4xy} \cdot \dfrac{6x^2}{5x}$

6. $\dfrac{4abc}{3a^2} \cdot \dfrac{6b^3}{ac^2} \cdot \dfrac{c^3}{8ab}$

7. $\dfrac{a}{(x+y)} \cdot \dfrac{(x+y)}{b}$

8. $\dfrac{(x-1)(x+2)}{a} \cdot \dfrac{b}{(x-1)}$

9. $\dfrac{x^3}{(a-5)(a+1)} \cdot \dfrac{(a-5)(a+2)}{x^2}$

10. $\dfrac{(x+3)(x+4)}{(x-5)} \cdot \dfrac{(x-5)}{(x+3)(x+1)}$

11. $\dfrac{2x(a-b)}{x^2} \cdot \dfrac{5x^2(a+2b)}{(a-b)(a+2b)}$

12. $\dfrac{a-5}{a+3} \cdot \dfrac{(a+3)(a+2)}{5-a}$

B

Find the product in each of the following exercises.

1. $\dfrac{x+2}{x-3} \cdot \dfrac{x^2-4x+3}{x+2}$

2. $\dfrac{ax+3a}{bx-4b} \cdot \dfrac{x-4}{x+3}$

3. $\dfrac{x^2-4}{x+2} \cdot \dfrac{x-3}{x^2+3x-10}$

4. $\dfrac{x^2+2x-15}{x^2+7x+10} \cdot \dfrac{x+2}{x+7}$

262 7/The Operations with Fractions

5. $\dfrac{7x}{x-4} \cdot \dfrac{x^2-7x+12}{x^2-x-6}$

6. $\dfrac{(x+y)a+(x+y)b}{x^2-y^2} \cdot \dfrac{5x-5y}{a+b}$

7. $\dfrac{a(x-3)-b(x-3)}{a-b} \cdot \dfrac{7}{2x-6}$

8. $\dfrac{3a+3b}{10} \cdot \dfrac{2x+2y}{ax+ay+bx+by}$

9. $\dfrac{ax-bx}{5x^2} \cdot \dfrac{7x+7y}{ax+ay-bx-by}$

10. $\dfrac{2x^2+x-3}{x^2-1} \cdot \dfrac{x^2+3x+2}{2x+3}$

11. $\dfrac{5x^2+13x-6}{4x+12} \cdot \dfrac{6x-6}{5x^2-7x+2}$

12. $\dfrac{x^2-2x-15}{x^2+x-30} \cdot \dfrac{x^2+3x-18}{x^2-5x+6}$

13. $\dfrac{x^2+6x-7}{x^2+5x+6} \cdot \dfrac{x^2-5x-24}{x^2-9x+8}$

14. $\dfrac{6x^2-24}{2x+4} \cdot \dfrac{15x-9}{5x^2-13x+6}$

15. $\dfrac{2a^2x-2a^2}{3b^2x+9b^2} \cdot \dfrac{6abx+18ab}{5a^3x-5a^3}$

16. $\dfrac{x^4-x^2-12}{2x^2+6} \cdot \dfrac{5x-15}{x^2-x-6}$

17. $\dfrac{x+2}{x-3} \cdot \dfrac{x-3}{x+5} \cdot \dfrac{x+5}{x-7}$

18. $\dfrac{5a^2b}{x^2-1} \cdot \dfrac{ax+a}{3a^4} \cdot \dfrac{bx-b}{10b^3}$

19. $\dfrac{x^2-9}{x^2-2x-8} \cdot \dfrac{x^2-5x-14}{x^2+2x-15} \cdot \dfrac{x^2+x-20}{x^2-11x+28}$

20. $\dfrac{x^3-25x}{2x^2+10x} \cdot \dfrac{3xy^2+12y^2}{2x^2-11x+5} \cdot \dfrac{8x^2+2x-3}{xy+4y}$

7.3 DIVISION OF FRACTIONS

When we examined the operation of division in Chapter 3 we discovered that

"a divided by b" means "a times the multiplicative inverse of b."

This is expressed with symbols as

$$a \div b \quad \text{means} \quad a \cdot \dfrac{1}{b}.$$

Before we can apply this information to division of fractions we will first have to determine what the multiplicative inverse of a fraction is. You may recall that the relation between a number and its multiplicative inverse is such that the product of the two is 1. Thus,

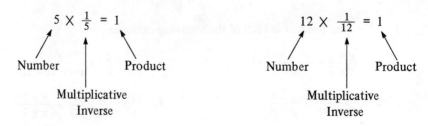

7.3/Division of Fractions

In view of this, the multiplicative inverse of $\frac{2}{3}$ will be the number that when multiplied by $\frac{2}{3}$ will yield a product of 1. That number is $\frac{3}{2}$.

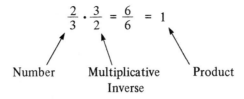

In general we can say that the multiplicative inverse of the fraction

$$\frac{a}{b} \text{ is } \frac{b}{a}.$$

Notice that the numerator of a fraction becomes the denominator of its multiplicative inverse and the denominator becomes the numerator.

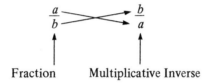

We frequently call the multiplicative inverse of a fraction the *reciprocal* of the fraction.

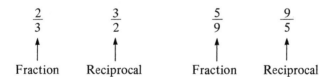

In view of what we have just shown, we can interpret the meaning of division when applied to fractions as,

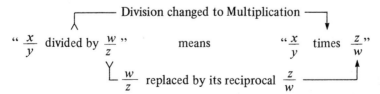

Using symbols we express this as

$$\frac{x}{y} \div \frac{w}{z} = \frac{x}{y} \cdot \frac{z}{w}$$

Example 1

Find the quotient in the exercise below.

$$\frac{5a^2}{3ab} \div \frac{10a^3b}{9b^4}$$

Solution

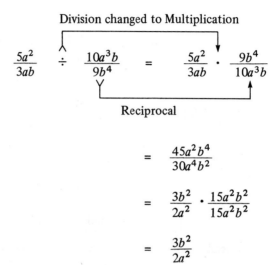

$$= \frac{45a^2b^4}{30a^4b^2}$$

$$= \frac{3b^2}{2a^2} \cdot \frac{15a^2b^2}{15a^2b^2}$$

$$= \frac{3b^2}{2a^2}$$

For our needs at the moment we have to recall that whenever any number is divided by 1 the quotient is that same number. Thus,

$$6 \div 1 = 6 \quad \text{and} \quad 2{,}475 \div 1 = 2{,}475.$$

In general, we express this principle as

$$\frac{x}{1} = x \quad \text{or} \quad x = \frac{x}{1}$$

The following illustration applies this principle.

Example 2

Find the quotient in the following exercise.

$$\frac{4a^2}{5b} \div 6a^3$$

7.3/Division of Fractions

Solution

$$\frac{4a^2}{5b} \div 6a^3 = \frac{4a^2}{5b} \div \frac{6a^3}{1}$$

$$= \frac{4a^2}{5b} \cdot \frac{1}{6a^3}$$

$$= \frac{4a^2}{30a^2 b}$$

$$= \frac{2}{15ab} \cdot \frac{2a^2}{2a^2}$$

$$= \frac{2}{15ab}$$

Example 3

Find the quotient in the exercise below.

$$\frac{15x^5}{x^3 - 9x} \div \frac{5x^3}{x^2 + 3x - 18}$$

Solution Division changed to Multiplication

$$\frac{15x^5}{x^3 - 9x} \div \frac{5x^3}{x^2 + 3x - 18} = \frac{15x^5}{x^3 - 9x} \cdot \frac{x^2 + 3x - 18}{5x^3}$$

Reciprocal

$$= \frac{15x^5}{x(x^2 - 9)} \cdot \frac{(x+6)(x-3)}{5x^3}$$

$$= \frac{15x^5}{x(x+3)(x-3)} \cdot \frac{(x+6)(x-3)}{5x^3}$$

$$= \frac{15x^5 (x+6)}{5x^4 (x+3)}$$

$$= \frac{3x(x+6)}{(x+3)} \cdot \frac{5x^4}{5x^4}$$

$$= \frac{3x(x+6)}{(x+3)}$$

Example 4

Perform the operations indicated in the exercise below.

$$\frac{ax - a}{x^2 + 4x - 5} \cdot \frac{x^2 + 7x + 10}{a^2 b} \div \frac{b^2 x + 2b^2}{3x + 3}$$

Solution

$$\frac{ax-a}{x^2+4x-5} \cdot \frac{x^2+7x+10}{a^2b} \div \frac{b^2x+2b^2}{3x+3} = \frac{ax-a}{x^2+4x-5} \cdot \frac{x^2+7x+10}{a^2b} \cdot \frac{3x+3}{b^2x+2b^2}$$

Division changed to Multiplication; Reciprocal

$$= \frac{a(x-1)}{(x+5)(x-1)} \cdot \frac{(x+2)(x+5)}{a^2b} \cdot \frac{3(x+1)}{b^2(x+2)}$$

$$= \frac{3a(x+1)}{a^2b^3}$$

$$= \frac{3(x+1)}{ab^3} \cdot \frac{a}{a}$$

$$= \frac{3(x+1)}{ab^3}$$

EXERCISES 7.3

A

Perform the operations indicated in the exercises below.

1. $\dfrac{4a}{3b^2} \div \dfrac{6a^3}{b^3}$

2. $\dfrac{5a}{9b} \div \dfrac{10a}{3b}$

3. $\dfrac{8x^2y}{5y^2} \div \dfrac{4x^3}{15y}$

4. $\dfrac{12a^3bc^2}{9ab^3} \div \dfrac{8b^2c^5}{6a^4}$

5. $\dfrac{14x}{5} \div 7x^2$

6. $8a^3 \div \dfrac{6a^2}{5b}$

7. $9x^2y^3 \div \dfrac{12y^2}{5x}$

8. $\dfrac{16a^2c}{6a^3b} \div 4bc^2$

9. $\dfrac{a^2}{b} \cdot \dfrac{b^2c}{a} \div \dfrac{bc^2}{a^3}$

10. $\dfrac{3xy}{4y^5} \cdot \dfrac{6y^5}{x^2} \div \dfrac{9x^3y^2}{2x}$

B

Find the quotient in each of the following exercises.

1. $\dfrac{x+3}{x+2} \div \dfrac{x+3}{x+5}$

2. $\dfrac{a^2}{x-3} \div \dfrac{a^3b}{x-3}$

3. $\dfrac{x^2+7x+10}{4a} \div \dfrac{x^2+6x+5}{6a^2}$

4. $\dfrac{6x^2+5x+1}{10x+5} \div \dfrac{3x+1}{10x^2}$

7.4/Addition and Subtraction of Fractions

5. $\dfrac{10x^2 - 7x + 1}{4x^2 - 1} \div \dfrac{5xy - y}{2x}$

6. $(x - 3) \div \dfrac{x^2 - 4x + 3}{x^2 - 1}$

7. $\dfrac{3x^2 - x - 2}{x - 1} \div (3x + 2)$

8. $(x^2 - 11x + 30) \div \dfrac{2x - 10}{6x}$

9. $\dfrac{5x^2 - 11x + 6}{5x^2 - 6x} \div (x^2 - 3x + 2)$

10. $\dfrac{ax - ay + 3x - 3y}{x^2 + 4xy - 5y^2} \div \dfrac{2a + 6}{5a}$

11. $\dfrac{4x^2 - 12x + 9}{2x^2 - x - 3} \div \dfrac{4x^2 - 4x - 3}{x^2 + 2x + 1}$

12. $\dfrac{6x^2 + 13x - 5}{9x^2 - 3x} \div \dfrac{4x^2 + 20x + 25}{12x^2}$

C

Perform the operations indicated in the exercises below.

1. $\dfrac{x^2 - 2x - 15}{x^2 + 3x} \cdot \dfrac{4x^2 - 4x - 8}{x^2 - 4x - 5} \div \dfrac{6x - 12}{2x^2}$

2. $\dfrac{x^2 - 6x + 9}{x^2 - x - 6} \cdot \dfrac{4x - 20}{x^2 - 4x + 3} \div \dfrac{2x^2 - 10x}{x^2 + 2x}$

3. $(3x - 15) \cdot \dfrac{x^2 - x - 12}{x^2 - 25} \div (6x - 24)$

7.4 ADDITION AND SUBTRACTION OF FRACTIONS

PART 1

Our knowledge of division proves to be useful to us when we try to develop some way for finding the sum of two fractions. We will begin by examining the sum of two fractions where the denominators of the two fractions are the same.

$$\frac{a}{c} + \frac{b}{c}. \qquad (1)$$

Since each fraction implies the operation of division, then (1) can be rewritten as

$$\frac{a}{c} + \frac{b}{c} = a \div c + b \div c. \qquad (2)$$

However, $a \div c$ means $a \cdot 1/c$, while $b \div c$ means $b \cdot 1/c$. In view of this, the right member of (2) can be expressed as

$$a \div c + b \div c = a \cdot \frac{1}{c} + b \cdot \frac{1}{c}. \qquad (3)$$

At this point we recognize the fact that $1/c$ is a common factor of the terms in the right member of (3). Hence, we can factor that expression.

$$a \cdot \frac{1}{c} + b \cdot \frac{1}{c} = \frac{1}{c}(a+b) \qquad (4)$$

We now notice that the operation between $1/c$ and $(a+b)/1$ is multiplication. Since the product of these two quantities is $(a+b)/c$, we can rewrite the right member of (4) as

$$\frac{1}{c}(a+b) = \frac{a+b}{c} .$$

Thus it would seem that the original sum $a/c + b/c$ can be expressed as the single fraction $(a+b)/c$.

$$\frac{a}{c} + \frac{b}{c} = \frac{a+b}{c}$$

In view of this we can say, if the denominators of two fractions are the same, then their sum can be found by adding the numerators and writing that quantity over the "common" denominator.

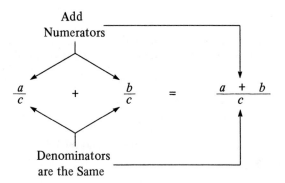

Example 1

Find the sum of the following fractions.

$$\frac{3x}{y} + \frac{5x}{y}$$

7.4 / Addition and Subtraction of Fractions

Explanation

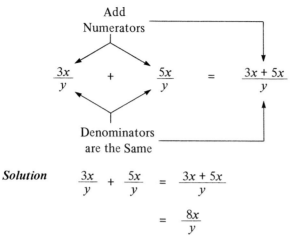

Solution

$$\frac{3x}{y} + \frac{5x}{y} = \frac{3x + 5x}{y}$$

$$= \frac{8x}{y}$$

Explanation Since the denominators of both fractions are the same, we merely add the numerators $3x$ and $5x$ and write their sum over y.

Example 2

Find the sum of the following two fractions.

$$\frac{4x + 2y}{3a} + \frac{5x - 7y}{3a}$$

Explanation

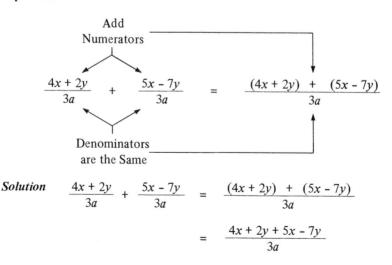

Solution

$$\frac{4x + 2y}{3a} + \frac{5x - 7y}{3a} = \frac{(4x + 2y) + (5x - 7y)}{3a}$$

$$= \frac{4x + 2y + 5x - 7y}{3a}$$

$$= \frac{9x - 5y}{3a}$$

7/The Operations with Fractions

Explanation continued Notice that the two binomials, $(4x + 2y)$ and $(4x - 7y)$, are initially written within parentheses. The reason for this will be explained very shortly. After removing the parentheses, like terms are collected.

Had we investigated the difference of two fractions earlier, rather than the sum, we would have discovered that

$$\frac{a}{c} - \frac{b}{c} = \frac{a-b}{c}$$

Thus, the difference of two fractions having the *same* denominator is found by determining the difference of their numerators and writing that quantity over the "common" denominator.

Example 3

Find the difference in the following exercise.

$$\frac{2x^2}{3a} - \frac{11x^2}{3a}$$

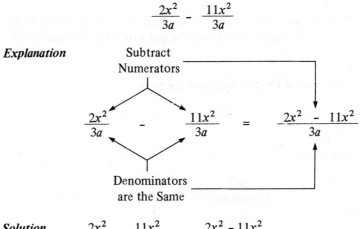

Solution

$$\frac{2x^2}{3a} - \frac{11x^2}{3a} = \frac{2x^2 - 11x^2}{3a} \quad (1)$$

$$= \frac{-9x^2}{3a} \quad (2)$$

$$= \frac{-3x^2}{a} \cdot \frac{3}{3} \quad (3)$$

$$= \frac{-3x^2}{a} \quad (4)$$

Explanation continued Normally, the work would have been completed in step (2). This fraction, though, is not reduced to lowest terms. When finding either the sum or difference of two fractions, it is best to examine your answer to make certain that the numerator and denominator are relatively prime.

7.4/Addition and Subtraction of Fractions

Example 4

Find the difference in the following exercise.

$$\frac{5a - 2}{4b} - \frac{7a - 5}{4b}$$

Solution

$$\frac{5a - 2}{4b} - \frac{7a - 5}{4b} = \frac{(5a - 2) - (7a - 5)}{4b}$$

$$= \frac{5a - 2 - 7a + 5}{4b}$$

$$= \frac{-2a + 3}{4b}$$

Explanation It is in an exercise such as this that we discover the need for the parentheses around the binomials $(5a - 2)$ and $(7a - 5)$. The symbol between these two quantities indicates the operation of subtraction. In view of this, the additive inverse of each of the terms in $7a - 5$ must be found. You might want to take a few moments to review the operation of subtraction. This particular point is discussed in Section 2.5.

Example 5

Perform the indicated operation in the following exercise.

$$\frac{x^2 + 5x}{x - 5} - \frac{6x + 20}{x - 5}$$

Solution

$$\frac{x^2 + 5x}{x - 5} - \frac{6x + 20}{x - 5} = \frac{(x^2 + 5x) - (6x + 20)}{x - 5}$$

$$= \frac{x^2 + 5x - 6x - 20}{x - 5}$$

$$= \frac{x^2 - x - 20}{x - 5}$$

$$= \frac{\cancel{(x - 5)}(x + 4)}{\cancel{(x - 5)}}$$

$$= x + 4$$

EXERCISES 7.4 (Part 1)

A

Perform the operations indicated in each of the following exercises.

1. $\dfrac{5x}{2} + \dfrac{4y}{2}$

2. $\dfrac{3a}{5} + \dfrac{7b}{5}$

3. $\dfrac{9x}{a} - \dfrac{4y}{a}$

4. $\dfrac{2a}{3b} + \dfrac{5a}{3b}$

5. $\dfrac{4x^2}{7y} + \dfrac{10x^2}{7y}$

6. $\dfrac{9a^2 b}{2} - \dfrac{5a^2 b}{2}$

7. $\dfrac{x^2}{2y} - \dfrac{6x^2}{2y}$

8. $\dfrac{4ab}{5c} - \dfrac{24ab}{5c}$

9. $\dfrac{3a}{2b} + \dfrac{5a}{2b} + \dfrac{a}{2b}$

10. $\dfrac{6x}{y} - \dfrac{x}{y} + \dfrac{3x}{y}$

11. $\dfrac{4a}{b} - \dfrac{5c}{b} + \dfrac{6a}{b}$

12. $\dfrac{3}{x^2} + \dfrac{5}{x^2} - \dfrac{2y}{x^2}$

B

Perform the operations indicated in each of the following exercises.

1. $\dfrac{x+2}{3} + \dfrac{x+5}{3}$

2. $\dfrac{3x-7}{a} + \dfrac{4x+6}{a}$

3. $\dfrac{3y}{x} + \dfrac{4y-9}{x}$

4. $\dfrac{2a-3b}{5} + \dfrac{a+5b}{5}$

5. $\dfrac{7a-4}{6x} + \dfrac{5a}{6x}$

6. $\dfrac{2a-3}{x^2} + \dfrac{-7a+5}{x^2}$

7. $\dfrac{3x+6}{5a} - \dfrac{x+8}{5a}$

8. $\dfrac{ab-2}{c^2} - \dfrac{ab-3}{c^2}$

9. $\dfrac{2x-4}{ab} - \dfrac{-5x-3}{ab}$

10. $\dfrac{4a+5}{c} - \dfrac{6a}{c}$

11. $\dfrac{7a}{b} - \dfrac{-4a+c}{b}$

12. $\dfrac{-2x+3}{y} - \dfrac{-5x}{y}$

C

Perform the operations indicated in each of the following exercises. You will find that each answer must be reduced to lowest terms.

1. $\dfrac{x^2}{x-2} - \dfrac{4}{x-2}$

2. $\dfrac{a^2}{a+b} - \dfrac{b^2}{a+b}$

3. $\dfrac{2x^2}{x-1} + \dfrac{x-3}{x-1}$

4. $\dfrac{x^2-5x}{x-2} + \dfrac{10-2x}{x-2}$

5. $\dfrac{x}{x-3} + \dfrac{x^2-12}{x-3}$

6. $\dfrac{a^2}{a+5} - \dfrac{a+30}{a+5}$

7. $\dfrac{2a^2+a}{2a-5} - \dfrac{10a-10}{2a-5}$

8. $\dfrac{5x^2-7x}{x-1} - \dfrac{2x^2-4}{x-1}$

9. $\dfrac{x^2}{y-x} + \dfrac{2xy-3y^2}{y-x}$

10. $\dfrac{a^2-5ab}{2b-a} - \dfrac{2ab-10b^2}{2b-a}$

7.4/Addition and Subtraction of Fractions

PART 2

At this stage the only fractions we can either add or subtract are those whose denominators are the same. Our next move, apparently, should be to discover some method by which we can add fractions whose denominators are *not* the same.

As an example, let's consider how we might search out some way that will enable us to add the fractions below.

$$\frac{3x}{4} + \frac{5x}{8}$$

In keeping with what we learned in Part 1, our approach will be to rename these fractions so that their denominators will be the same. However, there is but one way we have for renaming a fraction. Were we to multiply a fraction by some form of the number 1, we know that the value of the fraction will remain unchanged. Thus,

$$\frac{2}{3} \times 1 = \frac{2}{3}; \quad \frac{5}{7} \times 1 = \frac{5}{7}; \quad \frac{29}{36} \times 1 = \frac{29}{36}.$$

The number 1, though, has many forms. A few of these are:

$$\frac{7}{7}, \quad \frac{5}{5}, \quad \frac{a}{a}, \quad \frac{x^2}{x^2}, \quad \frac{2a^3 b}{2a^3 b}.$$

In fact, any quantity over itself names the number 1. Hence, if we multiply a fraction by any one of the forms of the number 1, the value of the fraction remains unchanged.

Let's return now to the original exercise,

$$\frac{3x}{4} + \frac{5x}{8}.$$

We are faced here with finding some form of the number 1 by which to multiply $3x/4$ so as to change the denominator 4 to an 8. Were that denominator an 8, the sum of the two fractions could be found.

Quite apparently the form of the number 1 that fits our needs is $\frac{2}{2}$. When $\frac{2}{2}$ is multiplied by $3x/4$, the new name for $3x/4$ will be $6x/8$.

$$\frac{3x}{4} + \frac{5x}{8} = \frac{2}{2} \cdot \frac{3x}{4} + \frac{5x}{8}$$

$$= \frac{6x}{8} + \frac{5x}{8}$$

$$= \frac{11x}{8}$$

274   7/The Operations with Fractions

Example 1

Find the sum of the following fractions.

$$\frac{7x}{18} + \frac{5x}{6}$$

Explanation In this situation we are looking for a form of 1 by which we can multiply $5x/6$ so that the denominator of the new fraction will be 18. It can readily be seen that the form of 1 needed here is $\frac{3}{3}$.

Solution

$$\frac{7x}{18} + \frac{5x}{6} = \frac{7x}{18} + \frac{3}{3} \cdot \frac{5x}{6}$$

$$= \frac{7x}{18} + \frac{15x}{18}$$

$$= \frac{22x}{18}$$

$$= \frac{11x}{9}$$

Before going any further, it is necessary to recall what is meant by **a multiple of a number**. For instance, the numbers

$$3, 6, 9, 12, 15, 18, \ldots$$

are all multiples of 3. Quite apparently, a multiple of 3 is a number in the "3 times table." Formally, we express this by saying that

---

A multiple of 3 is a number of the form $3n$ where n is a positive integer.

In Example 1 where we were asked to determine the sum for

$$\frac{7x}{18} + \frac{5x}{6}$$

the denominators were so designed that the larger denominator was a multiple of the smaller denominator. Because of this it was quite easy to spot the form of 1 by which $5x/6$ had to be multiplied. Simply dividing the 6 into the 18 gave us the number 3. In turn, the number 3 led to the fraction $\frac{3}{3}$ which is the form of 1 that had to be used. Whenever the larger denominator is a multiple of the smaller denominator, then dividing the larger denominator by the smaller one will lead us to the form of "1" we need.

7.4/Addition and Subtraction of Fractions

Example 2

Perform the operation indicated in the following exercise.

$$\frac{5x}{6} + \frac{7x}{24}$$

Explanation Since 24 is a multiple of 6, we divide 24 by 6 to obtain a quotient of 4. The fraction $\frac{4}{4}$ is the form of the number 1 that we need.

Solution
$$\frac{5x}{6} + \frac{7x}{24} = \frac{4}{4} \cdot \frac{5x}{6} + \frac{7x}{24}$$

$$= \frac{20x}{24} + \frac{7x}{24}$$

$$= \frac{20x + 7x}{24}$$

$$= \frac{27x}{24}$$

$$= \frac{9x}{8}$$

Example 3

Perform the operation indicated in the following exercises.

$$\frac{3x-2}{15} - \frac{2x-7}{3}$$

Solution
$$\frac{3x-2}{15} - \frac{2x-7}{3} = \frac{3x-2}{15} - \frac{5}{5} \cdot \frac{2x-7}{3} \quad (1)$$

$$= \frac{3x-2}{15} - \frac{10x-35}{15} \quad (2)$$

$$= \frac{(3x-2) - (10x-35)}{15} \quad (3)$$

$$= \frac{3x - 2 - 10x + 35}{15} \quad (4)$$

$$= \frac{-7x + 33}{15} \quad (5)$$

Explanation Notice that the fraction $\frac{5}{5}$ in step (1) is placed immediately before the fraction $(2x - 7)/3$.

EXERCISES 7.4 (Part 2)

A

Perform the operation indicated in each of the following exercises.

1. $\dfrac{5a}{3} + \dfrac{7a}{6}$

2. $\dfrac{x}{2} + \dfrac{5x}{8}$

3. $\dfrac{3b}{8} + \dfrac{7b}{2}$

4. $\dfrac{9x}{16} - \dfrac{3x}{8}$

5. $\dfrac{7a}{20} - \dfrac{2a}{5}$

6. $\dfrac{9x^2}{4} - \dfrac{3x^2}{28}$

7. $\dfrac{5ab}{36} + \dfrac{2ab}{9}$

8. $\dfrac{x^2}{2} - \dfrac{x^2}{12}$

9. $\dfrac{a^2 b}{8} - \dfrac{3a^2 b}{40}$

10. $\dfrac{-5y}{21} + \dfrac{-2y}{7}$

11. $\dfrac{3x}{10} - \dfrac{-2x}{5}$

12. $\dfrac{5a}{3x} + \dfrac{7a}{6x}$

B

Perform the operation indicated in each of the following exercises.

1. $\dfrac{x+3}{2} + \dfrac{x-5}{4}$

2. $\dfrac{2x-1}{5} + \dfrac{3x+7}{15}$

3. $\dfrac{4x-2}{9} + \dfrac{2x+1}{3}$

4. $\dfrac{7x+5}{12} + \dfrac{3x-2}{3}$

5. $\dfrac{5x-4}{2} - \dfrac{x+7}{6}$

6. $\dfrac{x-3}{4} - \dfrac{x-7}{20}$

7. $\dfrac{3x-1}{16} - \dfrac{2x+5}{4}$

8. $\dfrac{5x-3y}{30} - \dfrac{x-y}{5}$

9. $\dfrac{3x+4y}{4} + \dfrac{2x+3y}{8}$

10. $\dfrac{6x-5y}{15} - \dfrac{3x-y}{3}$

PART 3

The next question would seem to be pretty much the obvious one, "What do we do in the event the larger denominator is not a multiple of the smaller denominator?" The answer would seem to be equally as apparent—why not seek out a multiple that is common to both denominators.

7.4/Addition and Subtraction of Fractions

Consider the situation where we must determine the sum of the two fractions below:

$$\frac{5a}{6} + \frac{7a}{9}.$$

A multiple of both 6 and 9 is the number 18. By multiplying the first fraction by $\frac{3}{3}$ we can rename it with a denominator of 18. Similarly, by multiplying the second fraction by $\frac{2}{2}$ we can rename it with a denominator of 18. Thus,

$$\frac{5a}{6} + \frac{7a}{9} = \frac{3}{3} \cdot \frac{5a}{6} + \frac{2}{2} \cdot \frac{7a}{9}$$

$$= \frac{15a}{18} + \frac{14a}{18}$$

$$= \frac{29a}{18}.$$

There are other multiples that are common to both 6 and 9. Some of these are 36, 54, and 72. If we use any of the numbers rather than 18, the sum of the two fractions will be the same as the one above. However, the work would be somewhat more difficult for the numbers are larger.

Although the fractions can be renamed with any common multiple of their denominators, it would be best to use the *lowest common multiple* (L.C.M.).

Now we want to face the problem of finding the L.C.M. under conditions where the denominators involve variables with exponents. To do this it would be best if we first examine two fractions such as the following.

$$\frac{7a}{8} + \frac{11a}{32}.$$

The numbers 8 and 32 can be factored as:

$$8 = 2^3 \quad \text{and} \quad 32 = 2^5.$$

Thus, if we are seeking the L.C.M. of 8 and 32 we are in reality seeking the L.C.M. of 2^3 and 2^5. Quite apparently, a number such as 2^2 is *not* a multiple of 2^3 for there is no integer by which 2^3 can be multiplied to give 2^2 as an answer. Perhaps a clearer way of saying this is

"There is no integer by which 8 can be multiplied to give 4 as an answer."

For much the same reason, 2^2 cannot be a multiple of 2^5. On the other hand, 2^4 is a multiple of 2^3 for

"2^3 times 2 is 2^4."

That is to say,

"8 times 2 is 16."

However, 2^4 is not a multiple of 2^5 for once again,

"There is no number by which 32 can be multiplied to give 16 as an answer."

It appears then that the smallest number that is a multiple of both 2^3 and 2^5 will have to be 2^5. The integer by which we can multiply 2^3 to obtain 2^5 is 2^2. That is,

"8 times 4 is 32."

Also, the integer by which we can multiply 2^5 to obtain 2^5 is 1. That is,

"32 times 1 is 32."

In view of this,

the L.C.M. of 2^3 and 2^5 is 2^5.

Thus it would seem that if two numbers have the *same* base, their L.C.M. will be that base with the larger of the two exponents.

$$\text{L.C.M. of } 2^3 \text{ and } 2^5 \text{ is } 2^5.$$

(Same Base; Larger Exponent)

Example 1

Find the L.C.M. of 3^7 and 3^2.

Explanation Since the bases are the same we look for the larger exponent.

Solution

$$\text{L.C.M. of } 3^7 \text{ and } 3^2 \text{ is } 3^7.$$

(Same Base; Larger Exponent)

Example 2

Find the L.C.M. of 25 and 625.

Explanation The first step is to rewrite each of these numbers in factored form:

$$25 = 5^2 \quad \text{and} \quad 625 = 5^4$$

Solution Factored form of 25 is 5^2.
Factored form of 625 is 5^4.

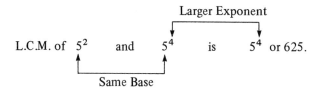

L.C.M. of 5^2 and 5^4 is 5^4 or 625.

Example 3
Find the L.C.M. of 48 and 72.

Solution Factored form of 48 is $2^4 \cdot 3$.
Factored form of 72 is $2^3 \cdot 3^2$.

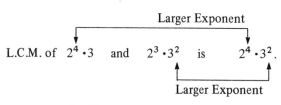

L.C.M. of $2^4 \cdot 3$ and $2^3 \cdot 3^2$ is $2^4 \cdot 3^2$.

L.C.M. = $2^4 \cdot 3^2$ = 144

Explanation Notice that the L.C.M. has as its factors every prime factor that appears in either number. The exponent of that factor is the larger exponent to which it appears in the two numbers.

Example 4
Find the L.C.M. for 96 and 120.

Solution $96 = 2^5 \cdot 3$ $120 = 2^3 \cdot 3 \cdot 5$

L.C.M. = $2^5 \cdot 3 \cdot 5$ = 480

Explanation All the different prime factors that appear in 96 and 120 are 2, 3, and 5. The larger exponent for the 2 is 5; the larger for 3 is 1; and the larger for 5 is 1.

Example 5
Find the L.C.M. for $a^3 b^2$ and $ab^5 c^2$.

Solution L.C.M. = $a^3 b^5 c^2$

Explanation Since each of the monomials is already written in factored form we know immediately that the different prime factors are a, b, and c. The larger exponent for a is 3; the larger for b is 5; and the larger for c is 2.

Example 6

Find the L.C.M. for $8x^2y^5$ and $12x^4w^3$.

Solution $\quad 8x^2y^5 = 2^3x^2y^5 \qquad\qquad 12x^4w^3 = 2^2 \cdot 3x^4w^3$

$\qquad\qquad$ L.C.M. $= 2^3 \cdot 3 \cdot x^4 \cdot y^5 \cdot w^3 = 24x^4y^5w^3$

Explanation This example is simply a combination of the preceding two. In many cases the L.C.M. of the coefficients can be found mentally. Rather than writing the 8 and 12 in factored form as 2^3 and $2^2 \cdot 3$, we often find it just as simple to look at the numbers 8 and 12 and determine their L.C.M. mentally.

EXERCISES 7.4 (Part 3)

Find the L.C.M. in each of the following exercises.

1. x^5 and x^3
2. x and x^5
3. x^2y^3 and x^5y^2
4. xy^2 and x^2y
5. x and y
6. x^2y^3w and x^3w^2
7. $6x^2$ and $4x^2$
8. $4x^3y$ and $9xy^2w$
9. 12 and 15
10. 16 and 20
11. 24 and 30
12. x^3, x^2, and x^5
13. x^2y^3, x^4w^2, and y^2w^3
14. $4x^5$, $3x^4$, and $6x^6$
15. $9xy^2$, $12x^5y$, and $4w^2$
16. 36, 48, and 54

PART 4

With the knowledge of how to determine the L.C.M. of any two monomials as a background we are now prepared to tackle Example 1.

Example 1

Perform the operation indicated in the following exercise.

$$\frac{5}{3x^2} + \frac{9}{4x^3}$$

Solution L.C.M. $= 12x^3$

$$\frac{5}{3x^2} + \frac{9}{4x^3} = \frac{4x}{4x} \cdot \frac{5}{3x^2} + \frac{3}{3} \cdot \frac{9}{4x^3}$$

$$= \frac{20x}{12x^3} + \frac{27}{12x^3}$$

$$= \frac{20x + 27}{12x^3}$$

7.4/Addition and Subtraction of Fractions

Explanation The first step is to determine the L.C.M. of the denominators. In this situation it is $12x^3$. The form of 1 by which each fraction is multiplied is found as before. For the first fraction it is found by dividing $12x^3$ by $3x^2$. The quotient $4x$ tells us that the form of 1 will be $4x/4x$. The first fraction $5/3x^2$ is then multiplied by $4x/4x$. Similarly, by dividing $12x^3$ by $4x^3$, the quotient 3 is found. In turn, this gives us the number $\frac{3}{3}$ by which the second fraction is multiplied.

Example 2

Perform the operation indicated in the following exercise.

$$\frac{2x-3}{6} + \frac{5x+4}{8}$$

Solution L.C.M. = 24

$$\frac{2x-3}{6} + \frac{5x+4}{8} = \frac{4}{4} \cdot \frac{2x-3}{6} + \frac{3}{3} \cdot \frac{5x+4}{8}$$

$$= \frac{8x-12}{24} + \frac{15x+12}{24}$$

$$= \frac{(8x-12)+(15x+12)}{24}$$

$$= \frac{8x-12+15x+12}{24}$$

$$= \frac{23x}{24}$$

Example 3

Perform the operation indicated in the following exercise.

$$\frac{3x^2-5}{6x^4} - \frac{2x-1}{4x^3}$$

Solution L.C.M. = $12x^4$

$$\frac{3x^2-5}{6x^4} - \frac{2x-1}{4x^3} = \frac{2}{2} \cdot \frac{3x^2-5}{6x^4} - \frac{3x}{3x} \cdot \frac{2x-1}{4x^3}$$

$$= \frac{6x^2-10}{12x^4} - \frac{6x^2-3x}{12x^4}$$

$$= \frac{(6x^2-10)-(6x^2-3x)}{12x^4}$$

$$= \frac{6x^2-10-6x^2+3x}{12x^4}$$

$$= \frac{3x-10}{12x^4}$$

7/The Operations with Fractions

EXERCISES 7.4 (Part 4)

A

Perform the operations indicated in each of the following exercises.

1. $\dfrac{3x}{8} + \dfrac{5x}{6}$

2. $\dfrac{4x}{9} - \dfrac{7x}{6}$

3. $\dfrac{2}{x^2} + \dfrac{4}{x^3}$

4. $\dfrac{7}{x^5} - \dfrac{5}{x}$

5. $\dfrac{3}{x} + \dfrac{4}{y}$

6. $\dfrac{a}{b} + \dfrac{c}{d}$

7. $\dfrac{5}{6x} + \dfrac{3}{5y}$

8. $\dfrac{7}{ab} - \dfrac{4}{a^2}$

9. $\dfrac{3}{4a^2} + \dfrac{5}{6ab^2}$

10. $\dfrac{3}{x} + \dfrac{4}{y} - \dfrac{5}{z}$

11. $\dfrac{a}{xy} - \dfrac{b}{xz} + \dfrac{c}{yz}$

12. $\dfrac{2x}{y} - \dfrac{3y}{x} + \dfrac{1}{xy}$

B

Perform the operations indicated in each of the following exercises.

1. $\dfrac{x+5}{3} + \dfrac{x+2}{4}$

2. $\dfrac{2x-1}{5} + \dfrac{3x-1}{3}$

3. $\dfrac{5x-4}{6} + \dfrac{2x-5}{9}$

4. $\dfrac{x-2}{4} - \dfrac{x-3}{6}$

5. $\dfrac{4x-3}{9} - \dfrac{x+5}{4}$

6. $\dfrac{3}{4} + \dfrac{2x-3}{5x}$

7. $\dfrac{4x-1}{12x} - \dfrac{5}{8}$

8. $\dfrac{7}{x} + \dfrac{2y-3}{xy}$

9. $\dfrac{5}{xy^2} - \dfrac{4x-3}{x^2y^2}$

10. $\dfrac{3x-2y}{x} + \dfrac{2x-3y}{y}$

11. $\dfrac{x-y}{x} - \dfrac{x-y}{y}$

12. $\dfrac{x-y}{xy} + \dfrac{y-z}{yz}$

13. $\dfrac{x+y}{x^2} + \dfrac{2}{xy} + \dfrac{x-y}{y^2}$

14. $\dfrac{3x-3y}{4x} - \dfrac{5}{3xy} - \dfrac{2x-y}{2y}$

PART 5

Our investigation of the sum and difference of fractions has thus far involved only those fractions whose denominators are monomials. However, our approach

7.4/Addition and Subtraction of Fractions

to these operations is the same whether the denominators have one term or more than one term. No matter what the nature of the denominators is, the L.C.M. must always be found before either addition or subtraction can take place. And since determining the L.C.M. of the denominators requires that they be expressed in factored form, factoring will have to be our first step. Let's examine several examples in which this is the case.

Example 1

Perform the operation indicated below.

$$\frac{5x}{x^2 - 5x + 6} + \frac{2x}{x^2 - 9}$$

Solution
$x^2 - 5x + 6 = (x - 3)(x - 2)$
$x^2 - 9 = (x + 3)(x - 3)$
L.C.M. $= (x + 3)(x - 3)(x - 2)$

$$\frac{5x}{x^2 - 5x + 6} + \frac{2x}{x^2 - 9} = \frac{5x}{(x - 3)(x - 2)} + \frac{2x}{(x + 3)(x - 3)}$$

$$= \frac{(x + 3)}{(x + 3)} \cdot \frac{5x}{(x - 3)(x - 2)} + \frac{(x - 2)}{(x - 2)} \cdot \frac{2x}{(x + 3)(x - 3)}$$

$$= \frac{5x^2 + 15x}{(x + 3)(x - 3)(x - 2)} + \frac{2x^2 - 4x}{(x + 3)(x - 3)(x - 2)}$$

$$= \frac{(5x^2 + 15x) + (2x^2 - 4x)}{(x + 3)(x - 3)(x - 2)}$$

$$= \frac{5x^2 + 15x + 2x^2 - 4x}{(x + 3)(x - 3)(x - 2)}$$

$$= \frac{7x^2 + 11x}{(x + 3)(x - 3)(x - 2)}$$

Explanation Once each of the denominators is factored, the L.C.M. is found as before. All the different prime factors that appear in either denominator are $(x - 3)$, $(x - 2)$, and $(x + 3)$. Each appears to the first power. Hence, their product is the L.C.M.

Example 2

Perform the operation indicated below.

$$\frac{3}{x^2 - 6x + 9} - \frac{5}{x^2 - 3x}$$

Solution
$$x^2 - 6x + 9 = (x - 3)^2$$
$$x^2 - 3x = x(x - 3)$$
$$\text{L.C.M.} = x(x - 3)^2$$

$$\frac{3}{x^2 - 6x + 9} - \frac{5}{x^2 - 3x} = \frac{3}{(x-3)^2} - \frac{5}{x(x-3)}$$

$$= \frac{x}{x} \cdot \frac{3}{(x-3)^2} - \frac{(x-3)}{(x-3)} \cdot \frac{5}{x(x-3)^2}$$

$$= \frac{3x}{x(x-3)^2} - \frac{5x - 15}{x(x-3)^2}$$

$$= \frac{3x - (5x - 15)}{x(x-3)^2}$$

$$= \frac{3x - 5x + 15}{x(x-3)^2}$$

$$= \frac{-2x + 15}{x(x-3)^2}$$

Explanation The larger exponent to which the factor $(x - 3)$ appears in both denominators is 2. Therefore, the exponent to which it appears in the L.C.M. is also 2. The only other point where you may encounter difficulty is in dividing $x(x - 3)$ into $x(x - 3)^2$. This can be done in the same manner as when reducing a fraction to lowest terms:

$$\frac{x(x-3)^2}{x(x-3)} = \frac{(x-3)}{1} \cdot \frac{x(x-3)}{x(x-3)} = \frac{x-3}{1} = x - 3.$$

EXERCISES 7.4 (Part 5)

A

Perform the operations indicated in each of the exercises below.

1. $\dfrac{3}{x-y} + \dfrac{4}{x+y}$

2. $\dfrac{5}{2x-1} + \dfrac{6}{3x+1}$

3. $\dfrac{2}{x-5} + \dfrac{3}{x-6}$

4. $\dfrac{7}{3x-1} - \dfrac{3}{2x+1}$

5. $\dfrac{9}{5x-3} - \dfrac{7}{4x-1}$

6. $\dfrac{x}{x-1} + \dfrac{x}{x+1}$

7. $\dfrac{x}{x+3} - \dfrac{x}{x-3}$

8. $\dfrac{5}{2x+2} + \dfrac{4}{3x+3}$

9. $\dfrac{7}{5x-10} - \dfrac{4}{4x-8}$

10. $\dfrac{3}{ab+b} + \dfrac{2}{ac+c}$

7.5/Solution of Equations Containing Fractions

B

Perform the operations indicated in each of the exercises below.

1. $\dfrac{5}{x^2 + 5x + 6} + \dfrac{6}{x^2 + 4x + 3}$

2. $\dfrac{3}{x^2 - 25} + \dfrac{2}{x^2 + 3x - 10}$

3. $\dfrac{2x}{x^2 + 2x} + \dfrac{5x}{x^2 - 4}$

4. $\dfrac{4}{x^2 - 3x} - \dfrac{2}{x^2 - x - 6}$

5. $\dfrac{5}{2x^2 + x - 1} + \dfrac{1}{x^2 - 1}$

6. $\dfrac{3}{x^2 + 5x + 6} - \dfrac{4}{x^2 - 2x - 15}$

7. $\dfrac{3x}{6x^2 - 7x + 2} - \dfrac{2x}{3x^2 - 5x + 2}$

8. $\dfrac{4}{x^2 + 6x + 9} + \dfrac{7}{x^2 + 7x + 12}$

9. $\dfrac{2}{x^2 - 5x} - \dfrac{3}{x^2 - 10x + 25}$

10. $\dfrac{4}{2x^2 - 3x} + \dfrac{2x + 1}{2x^2 - 5x + 3}$

11. $\dfrac{3x - 4}{x^2 - 1} - \dfrac{5}{2x^2 - 2x}$

12. $\dfrac{3x}{8x^2 + 2x - 1} - \dfrac{5x}{2x^2 - 9x - 5}$

C

Perform the operations indicated in the exercises below.

1. $\dfrac{2}{x^2 + x - 2} + \dfrac{5}{x^2 + 2x - 3} + \dfrac{3}{x^2 + 5x + 6}$

2. $\dfrac{4}{x^2 - 1} + \dfrac{3}{x^2 - 5x + 4} - \dfrac{7}{x^2 - 3x - 4}$

3. $\dfrac{1}{x^2 + 2x - 15} - \dfrac{5}{x^2 - 3x} - \dfrac{4}{x^2 + 5x}$

4. $\dfrac{2x}{x^2 - 4x - 21} - \dfrac{3x}{x^2 - 9x + 14} - \dfrac{x}{x^2 + x - 6}$

5. $\dfrac{5}{6 - 3x} + \dfrac{4}{x^2 + x - 6}$

7.5 SOLUTION OF EQUATIONS CONTAINING FRACTIONS

PART 1

Having learned the fundamental operations relative to fractions, it would seem only natural that we would now apply this information to the solution of equations that contain fractions. The mathematician makes a distinction between two different types of equations that contain fractions. If the denominators of the fractions consist of *numerical quantities only*, the equation containing them is called an **equation with fractions**. An equation such as this is

$$\frac{3x}{2} + \frac{5x-4}{3} = \frac{2x-7}{5}.$$

In this situation, the denominators are the numbers 2, 3, and 5.

On the other hand, if any of the denominators of the fractions contains a variable, then that equation is called a **fractional equation**. In the equation below

$$\frac{5x-4}{2x} - \frac{3x}{5x-2} = \frac{1}{2}$$

the denominators of the first and second fractions contain variables. Hence, this equation is a fractional equation.

To understand the approach we take when solving an equation with fractions, it would be best to review the method by which we multiply 6 by $x/3$. Earlier in this chapter we learned that the number 6 can be expressed as $\frac{6}{1}$. Therefore, after 6 is changed to $\frac{6}{1}$, the steps below are the usual ones for finding the product of any two fractions.

$$6 \cdot \frac{x}{3} = \frac{6}{1} \cdot \frac{x}{3}$$

$$= \frac{6x}{3}$$

$$= \frac{3}{3} \cdot \frac{2x}{1}$$

$$= \frac{2x}{1}$$

$$= 2x$$

Were we to write the answer immediately to the right of the original exercise we would discover that the $2x$ could have been found far more easily than is done above. If we simply divide the denominator 3 into the 6 and multiply that number by x, we obtain the answer $2x$ with far less effort.

$$6 \cdot \frac{x}{3} = 2x$$

7.5/Solution of Equations Containing Fractions

$$\overset{2}{\cancel{6}} \cdot \frac{x}{\cancel{3}} = 2x$$

Similarly, 8 times $\frac{y}{2}$ is found as follows.

$$\overset{4}{\cancel{8}} \cdot \frac{y}{\cancel{2}} = 4y$$

Example 1

Find the product of the quantities below.

$$5 \cdot \frac{2x}{5}$$

Solution $\overset{1}{\cancel{5}} \cdot \frac{2x}{\cancel{5}} = 2x$

Example 2

Find the product of the quantities below.

$$12 \cdot \frac{2x - 7}{4}$$

Solution $\overset{3}{\cancel{12}} \cdot \frac{(2x - 7)}{\cancel{4}} = 6x - 21$

Explanation There is only one slight difference between the solution above and the one in Example 1. In this one it is necessary to multiply the monomial 3 by the binomial $2x - 7$.

EXERCISES 7.5 (Part 1)

Find the product of the quantities below.

1. $8 \cdot \frac{x}{4}$
2. $12 \cdot \frac{x}{2}$
3. $-6 \cdot \frac{a}{2}$
4. $-18 \cdot \frac{2x}{3}$

5. $15 \cdot \frac{4a}{5}$
6. $-24 \cdot \frac{3a}{4}$
7. $12 \cdot \frac{-2a}{3}$
8. $6 \cdot \frac{-5x}{6}$

9. $4 \cdot \dfrac{x+3}{4}$

10. $16 \cdot \dfrac{x-4}{2}$

11. $24 \cdot \dfrac{3x-2}{6}$

12. $-14 \cdot \dfrac{2x-5}{7}$

13. $-20 \cdot \dfrac{3x+7}{5}$

14. $18 \cdot \dfrac{-2a-5}{6}$

15. $28 \cdot \dfrac{7-3a}{4}$

16. $-24 \cdot \dfrac{-5a-1}{6}$

PART 2

With the background of the material in Part 1, we are now prepared to solve equations with fractions. The equations that we know how to solve in general resemble the one here.

$$3x - 5 = 7x + 12. \qquad (1)$$

Hence, if it were possible for us to take an equation such as

$$\dfrac{5x}{3} + 10 = \dfrac{x}{2} + 17 \qquad (2)$$

and somehow eliminate the fractions, equation (2) would be of the same form as (1). Of course, once it resembles (1) we will be able to solve it.

From our study of equations we know that there are but two principles that can aid us in finding the roots of an equation.

1. The multiplication principle of equality
2. The addition principle of equality

Our approach will be to find the L.C.M. of the denominators in equation (2). Then by applying the multiplication principle we can eliminate the fractions from that equation. Thus,

The L.C.M. of 3 and 2 is 6.

$$6\left(\dfrac{5x}{3} + 10\right) = 6\left(\dfrac{x}{2} + 17\right) \qquad (3)$$

$$6 \cdot \dfrac{5x}{3} + 6 \cdot 10 = 6 \cdot \dfrac{x}{2} + 6 \cdot 17 \qquad (4)$$

$$10x + 60 = 3x + 102 \qquad (5)$$

$$10x = 3x + 102 + (-60) \qquad (6)$$

$$10x + (-3x) = 102 + (-60) \qquad (7)$$

$$7x = 42 \qquad (8)$$

$$x = 6 \quad \text{(Root)}$$

7.5/Solution of Equations Containing Fractions

Once we were able to eliminate the fractions in step (5), that equation was solved in the same manner as in the past.

Example 1

Solve and check the following equation.

$$\frac{4x}{3} - \frac{7x}{4} = -5$$

Explanation The L.C.M. of 3 and 4 is 12. It is this number by which both members of the equation are multiplied.

Solution

$$\frac{4x}{3} - \frac{7x}{4} = -5$$

$$12 \cdot \left(\frac{4x}{3} - \frac{7x}{4}\right) = 12 \cdot (-5)$$

$$12\left(\frac{4x}{3}\right) - 12\left(\frac{7x}{4}\right) = -60$$

$$16x - 21x = -60$$

$$-5x = -60$$

$$x = 12 \quad \text{(Root)}$$

Check for $x = 12$:

$$\frac{4x}{3} - \frac{7x}{4} = -5$$

$$\frac{4 \cdot 12}{3} - \frac{7 \cdot 12}{4} \stackrel{?}{=} -5$$

$$\frac{48}{3} - \frac{84}{4} \stackrel{?}{=} -5$$

$$16 - 21 \stackrel{?}{=} -5$$

$$-5 = -5$$

Example 2

Solve and check the following equation.

$$\frac{7x-3}{4} - \frac{5x+3}{6} = \frac{4x}{3}$$

7/The Operations with Fractions

Solution

$$\frac{7x-3}{4} - \frac{5x+3}{6} = \frac{4x}{3} \qquad (1)$$

$$12\left(\frac{7x-3}{4} - \frac{5x+3}{6}\right) = 12\left(\frac{4x}{3}\right) \qquad (2)$$

$$\overset{3}{\cancel{12}} \cdot \frac{(7x-3)}{\cancel{4}} - \overset{2}{\cancel{12}} \cdot \frac{(5x+3)}{\cancel{6}} = \overset{4}{\cancel{12}}\left(\frac{4x}{\cancel{3}}\right) \qquad (3)$$

$$21x - 9 - 10x - 6 = 16x \qquad (4)$$

$$11x - 15 = 16x$$

$$11x = 16x + (+15)$$

$$11x + (-16x) = +15$$

$$-5x = +15$$

$$x = -3 \quad \text{(Root)}$$

Check for $x = -3$:

$$\frac{7x-3}{4} - \frac{5x+3}{6} = \frac{4x}{3}$$

$$\frac{7(-3)-3}{4} - \frac{5(-3)+3}{6} \overset{?}{=} \frac{4(-3)}{3}$$

$$\frac{-21-3}{4} - \frac{-15+3}{6} \overset{?}{=} \frac{-12}{3}$$

$$\frac{-24}{4} - \frac{-12}{6} \overset{?}{=} -4$$

$$-6 - (-2) \overset{?}{=} -4$$

$$-6 + 2 \overset{?}{=} -4$$

$$-4 = -4$$

Explanation We must be particularly careful in going from step (3) to step (4). Notice in step (3) that 6 is divided into a *negative* 12. Hence, it is the quotient of *negative* 2 that is multiplied by $(5x + 3)$ to give the product of $-10x - 6$.

EXERCISES 7.5 (Part 2)

A

Solve each of the following equations.

1. $\frac{x}{3} + \frac{2x}{3} = 6$
2. $\frac{2x}{5} = \frac{7x}{5} - 5$

3. $\dfrac{2x}{3} = \dfrac{5x}{6} - 1$

4. $\dfrac{5x}{4} - \dfrac{x}{2} = \dfrac{x}{4} + 4$

5. $\dfrac{x}{2} - 5 = \dfrac{3x}{5} - 4$

6. $\dfrac{x}{4} + \dfrac{x}{2} = \dfrac{7x}{10} + 8$

7. $\dfrac{x}{3} - \dfrac{4x}{8} = \dfrac{x}{2} - 6$

8. $\dfrac{3x}{8} - \dfrac{5}{4} = 4 + \dfrac{5x}{4}$

9. $\dfrac{7x}{3} + \dfrac{3x}{5} = \dfrac{4x}{15} - 4$

10. $\dfrac{4x}{3} - \dfrac{5}{4} = \dfrac{5}{6} - \dfrac{3x}{4}$

B

Solve each of the following equations.

1. $\dfrac{x+4}{2} + \dfrac{x}{3} = 7$

2. $\dfrac{2x-1}{5} + \dfrac{5x-1}{3} = 16$

3. $\dfrac{x+5}{4} + \dfrac{x+2}{6} = \dfrac{6x-1}{6}$

4. $\dfrac{3x-2}{2} - 2 = x - \dfrac{x}{4}$

5. $4x + \dfrac{2x+3}{3} = 5x - \dfrac{x}{6}$

6. $\dfrac{x+2}{5} - \dfrac{x-3}{3} = \dfrac{1}{3}$

7. $\dfrac{3x+2}{4} - \dfrac{2x+1}{3} = \dfrac{1}{2}$

8. $\dfrac{4x+3}{6} - \dfrac{10-x}{4} = \dfrac{7}{2}$

9. $\dfrac{5-x}{8} - \dfrac{x+3}{6} = -\dfrac{x}{3}$

10. $\dfrac{2x+3}{2} - \dfrac{x+5}{4} = \dfrac{2x-3}{4}$

11. $\dfrac{2x-1}{3} + \dfrac{5x-4}{8} = \dfrac{3x}{2}$

12. $\dfrac{4x-2}{9} - \dfrac{-x-3}{6} = \dfrac{2x-1}{2}$

C

Solve and check each of the following equations.

1. $\dfrac{x}{3} + \dfrac{3x}{4} = 13$

2. $\dfrac{5x+1}{3} + \dfrac{7x-3}{2} = 4$

3. $\dfrac{2x-2}{5} - \dfrac{4x-3}{3} = -\dfrac{x}{3}$

4. $\dfrac{x-2}{6} - \dfrac{x}{2} = \dfrac{x+8}{4}$

7.6 SOLUTION OF INEQUALITIES CONTAINING FRACTIONS

You may recall that the solution of an inequality followed the same pattern as that of an equality. There was only one step in the solution of an inequality that was of any concern. That step was the last one.

1. If both members of an inequality are multiplied by a *positive* number, the order of inequality *remains the same*.
2. If both members of an inequality are multiplied by a *negative* number, the order of inequality is *reversed*.

The solution of an inequality with fractions—no variables in any denominators—again follows the same pattern as the solution of an equality.

Example 1

Solve the following inequality.

$$\frac{x}{3} > \frac{5x}{6} + \frac{7}{2}$$

Solution

$$\frac{x}{3} > \frac{5x}{6} + \frac{7}{2}$$

$$6 \cdot \frac{x}{3} > 6\left(\frac{5x}{6} + \frac{7}{2}\right)$$

$$6 \cdot \frac{x}{3} > 6 \cdot \frac{5x}{6} + 6 \cdot \frac{7}{2}$$

$$2x > 5x + 21$$

$$2x + (-5x) > 21$$

$$-3x > 21 \qquad (6)$$

$$x < -7 \quad \text{(Roots)} \qquad (7)$$

Explanation Both members of the inequality in step (6) are multiplied by $-\frac{1}{3}$. Hence, the order of inequality in step (7) is the reverse of that in step (6).

Example 2

Solve the following inequality.

$$\frac{5x-2}{3} - \frac{7x-3}{5} < \frac{x}{4}$$

Solution

$$\frac{5x-2}{3} - \frac{7x-3}{5} < \frac{x}{4}$$

$$60 \cdot \left(\frac{5x-2}{3} - \frac{7x-3}{5}\right) < 60 \cdot \frac{x}{4}$$

$$\overset{20}{\cancel{60}} \cdot \frac{(5x-2)}{\cancel{3}} - \overset{12}{\cancel{60}} \cdot \frac{(7x-3)}{\cancel{5}} < \overset{15}{\cancel{60}} \cdot \frac{x}{\cancel{4}}$$

$$100x - 40 - 84x + 36 < 15x$$

$$16x - 4 < 15x$$

$$16x < 15x + (+4)$$
$$16x + (-15x) < +4$$
$$x < 4 \quad \text{(Roots)}$$

EXERCISES 7.6

Solve each of the following inequalities.

1. $\dfrac{3x}{4} + \dfrac{5x}{2} > \dfrac{39}{4}$

2. $\dfrac{2x}{3} - \dfrac{x}{2} < \dfrac{7}{2}$

3. $\dfrac{4x}{3} - \dfrac{3}{4} < x + \dfrac{9}{4}$

4. $\dfrac{x+3}{4} + \dfrac{4x}{3} > \dfrac{6x+5}{4}$

5. $\dfrac{2x+1}{5} + \dfrac{x-1}{2} > \dfrac{3x-1}{10}$

6. $\dfrac{2x}{3} + \dfrac{4x+2}{7} > x + 1$

7. $\dfrac{5x-1}{12} - \dfrac{3x+1}{4} < \dfrac{x-11}{3}$

8. $\dfrac{7x-1}{3} - x < \dfrac{7x+2}{6}$

9. $\dfrac{2x-2}{5} - \dfrac{3x}{4} > \dfrac{x}{4}$

10. $\dfrac{7x+1}{10} - \dfrac{x-1}{2} > \dfrac{3}{5}$

11. $\dfrac{6x+1}{4} - \dfrac{6-8x}{2} < 2x - 1$

12. $\dfrac{18x-1}{5} - \dfrac{20x}{3} > \dfrac{8x+4}{3} - 13$

7.7 SOLUTION OF FRACTIONAL EQUATIONS

It is but one short step from solving equations containing fractions to solving fractional equations. In both situations it is necessary to determine the lowest common multiple of the denominators. It is this quantity by which both members of the equation are multiplied. You will find that the solutions in the following examples are much the same as the solutions of equations containing fractions.

Example 1

Solve and check the following equation.

$$\frac{3}{x} + \frac{5}{2x} = \frac{11}{8}$$

Explanation The L.C.M. of x, $2x$, and 8 is $8x$. We must now multiply both members of the equation by $8x$.

Solution

$$\frac{3}{x} + \frac{5}{2x} = \frac{11}{8}$$

$$8x\left(\frac{3}{x} + \frac{5}{2x}\right) = 8x\left(\frac{11}{8}\right)$$

$$\overset{8}{\cancel{8x}} \cdot \frac{3}{\cancel{x}} + \cancel{8x} \cdot \frac{\overset{4}{5}}{\cancel{2x}} = \overset{x}{\cancel{8x}} \cdot \frac{11}{\cancel{8}}$$

$$24 + 20 = 11x$$
$$44 = 11x$$
$$x = 4$$

Check for $x = 4$:

$$\frac{3}{x} + \frac{5}{2x} = \frac{11}{8}$$

$$\frac{3}{4} + \frac{5}{8} \overset{?}{=} \frac{11}{8}$$

$$\frac{6}{8} + \frac{5}{8} \overset{?}{=} \frac{11}{8}$$

$$\frac{11}{8} = \frac{11}{8} \qquad \text{Hence, 4 is a root.}$$

Example 2
Solve and check the following equation.

$$\frac{2}{x-2} + \frac{1}{3} = \frac{5}{x}$$

Solution

$$\frac{2}{x-2} + \frac{1}{3} = \frac{5}{x}$$

L.C.M.: $3x(x-2)$

$$3x(x-2)\left(\frac{2}{x-2} + \frac{1}{3}\right) = 3x(x-2) \cdot \frac{5}{x} \qquad (1)$$

$$\overset{3x}{\cancel{3x(x-2)}} \cdot \frac{2}{\cancel{x-2}} + \frac{x(x-2)}{\cancel{3x(x-2)}} \cdot \frac{1}{\cancel{3}} = \frac{3(x-2)}{\cancel{3x(x-2)}} \cdot \frac{5}{\cancel{x}} \qquad (2)$$

$$6x + x(x-2) = 15(x-2) \qquad (3)$$
$$6x + x^2 - 2x = 15x - 30 \qquad (4)$$
$$x^2 + 6x - 2x + (-15x) + (+30) = 0 \qquad (5)$$
$$x^2 - 11x + 30 = 0 \qquad (6)$$
$$(x-5)(x-6) = 0$$

7.7/Solution of Fractional Equations

Check for $x = 5$:

$$x - 5 = 0$$
$$x = 5$$

$$\frac{2}{x-2} + \frac{1}{3} = \frac{5}{x}$$

$$\frac{2}{5-2} + \frac{1}{3} \stackrel{?}{=} \frac{5}{5}$$

$$\frac{2}{3} + \frac{1}{3} \stackrel{?}{=} 1$$

$$\frac{3}{3} \stackrel{?}{=} 1$$

$$1 = 1$$

Hence, 5 is a root.

Check for $x = 6$:

$$x - 6 = 0$$
$$x = 6$$

$$\frac{2}{x-2} + \frac{1}{3} = \frac{5}{x}$$

$$\frac{2}{6-2} + \frac{1}{3} \stackrel{?}{=} \frac{5}{6}$$

$$\frac{2}{4} + \frac{1}{3} \stackrel{?}{=} \frac{5}{6}$$

$$\frac{1}{2} + \frac{1}{3} \stackrel{?}{=} \frac{5}{6}$$

$$\frac{3}{6} + \frac{2}{6} \stackrel{?}{=} \frac{5}{6}$$

$$\frac{5}{6} = \frac{5}{6}$$

Hence, 6 is a root.

Explanation The process of eliminating the fractions in step (2) led to the quadratic equation in step (4). Recall that in order to solve a quadratic equation it is important that the right member of the equation be zero. You may want to review Section 6.6, where we examined the solution of the quadratic equation.

Example 3

Solve and check the following equation.

$$\frac{40}{x^2 - 25} + 1 = \frac{4}{x - 5}$$

Solution

$$\frac{40}{x^2 - 25} + 1 = \frac{4}{x - 5}$$

$$\frac{40}{(x+5)(x-5)} + 1 = \frac{4}{x-5}$$

L.C.M.: $(x+5)(x-5)$

$$(x+5)(x-5) \cdot \left[\frac{40}{(x+5)(x-5)} + 1\right] = (x+5)(x-5) \cdot \frac{4}{x-5}$$

$$\cancel{(x+5)(x-5)} \cdot \frac{40}{\cancel{(x+5)(x-5)}} + (x+5)(x-5) \cdot 1 = (x+5)\cancel{(x-5)} \cdot \frac{4}{\cancel{x-5}}$$

$$40 + (x^2 - 25) = 4x + 20$$
$$40 + x^2 - 25 = 4x + 20$$
$$x^2 + 15 = 4x + 20$$
$$x^2 + 15 + (-4x) + (-20) = 0$$
$$x^2 - 4x - 5 = 0$$
$$(x + 1)(x - 5) = 0$$

$x + 1 = 0$	$x - 5 = 0$
$x = -1$	$x = 5$
Check for $x = -1$	Check for $x = 5$
$\dfrac{40}{x^2 - 25} + 1 = \dfrac{4}{x - 5}$	$\dfrac{40}{x^2 - 25} + 1 = \dfrac{4}{x - 5}$
$\dfrac{40}{(-1)^2 - 25} + 1 \stackrel{?}{=} \dfrac{4}{-1 - 5}$	$\dfrac{40}{5^2 - 25} + 1 \stackrel{?}{=} \dfrac{4}{5 - 5}$
$\dfrac{40}{1 - 25} + 1 \stackrel{?}{=} \dfrac{4}{-6}$	$\dfrac{40}{25 - 25} + 1 \stackrel{?}{=} \dfrac{4}{0}$
$\dfrac{40}{-24} + 1 \stackrel{?}{=} -\dfrac{2}{3}$	$\dfrac{40}{0} + 1 \neq \dfrac{4}{0}$
$-\dfrac{5}{3} + \dfrac{3}{3} \stackrel{?}{=} -\dfrac{2}{3}$	Hence, 5 is not a root.
$-\dfrac{2}{3} = -\dfrac{2}{3}$	

Hence, −1 is a root.

Explanation Notice that when we checked to determine whether 5 is a root of the equation, the denominator of the fraction $4/(x - 5)$ turned out to be zero. The same was true for the fraction $4/(x^2 - 25)$. In view of the fact that division by zero is not defined (see explanation at the introduction of this chapter), then 5 cannot be a root of the equation. The number 5 does not make the equation a true sentence.

EXERCISES 7.7

A

Solve each of the following equations.

1. $\dfrac{2y + 8}{y} = \dfrac{16}{y}$ 2. $\dfrac{3x - 5}{2x} = \dfrac{2x - 5}{x}$

7.7/Solution of Fractional Equations

3. $1 - \dfrac{1}{2w} = \dfrac{3}{4} + \dfrac{1}{w}$

4. $\dfrac{2}{3x} + \dfrac{3}{2} = 2 - \dfrac{5}{6x}$

5. $\dfrac{5}{4} - \dfrac{1}{2a} = 2 + \dfrac{1}{a}$

6. $\dfrac{16}{9} - \dfrac{1}{3x} = \dfrac{4}{3} - \dfrac{5}{3x}$

7. $\dfrac{8z + 1}{z} = 9$

8. $3 - \dfrac{5}{x} = \dfrac{x + 9}{x}$

9. $\dfrac{7w - 4}{5w} + \dfrac{4}{w} = \dfrac{9}{5}$

10. $\dfrac{x + 1}{x} - \dfrac{2}{3x} = \dfrac{x - 3}{2x} - \dfrac{1}{6x}$

11. $\dfrac{2b - 1}{3b} + \dfrac{1}{9} = \dfrac{b + 2}{b} + \dfrac{1}{9b}$

12. $\dfrac{3y - 1}{3y} - \dfrac{17}{15y} = \dfrac{5 - y}{5y} - \dfrac{y - 1}{15y}$

B

Solve each of the following equations.

1. $\dfrac{3}{x^2} = \dfrac{7}{4x} - \dfrac{1}{4}$

2. $\dfrac{1}{4} = \dfrac{1}{2y^2} + \dfrac{1}{4y}$

3. $\dfrac{1}{20} - \dfrac{1}{5z} = \dfrac{1}{4z} - \dfrac{1}{z^2}$

4. $\dfrac{36}{(c - 1)(c + 1)} = \dfrac{4}{c - 1} + \dfrac{3}{c + 1}$

5. $\dfrac{32}{m^2 - 4} - \dfrac{5}{m - 2} = \dfrac{7}{m + 2}$

6. $\dfrac{10}{d^2 - 5d + 6} = \dfrac{5}{d - 3} - \dfrac{4}{d - 2}$

7. $\dfrac{x - 5}{2x - 9} = \dfrac{2}{x}$

8. $\dfrac{3}{a} = \dfrac{a - 4}{a - 2}$

9. $\dfrac{4}{x(x - 6)} + 1 = \dfrac{1}{x}$

10. $\dfrac{24}{x^2 + 5x} = \dfrac{6}{x} - 1$

11. $4 + \dfrac{5}{y} = \dfrac{3y}{y - 1} + \dfrac{7}{y^2 - y}$

12. $1 + \dfrac{8}{a^2 - 5a + 4} = \dfrac{3}{a - 4}$

13. $\dfrac{20}{x^2 - 3x + 2} + \dfrac{2}{x - 1} = 1$

14. $\dfrac{a - 4}{a + 2} + \dfrac{8}{a + 2} = \dfrac{5}{a - 2}$

15. $\dfrac{y + 1}{y - 4} = \dfrac{3}{y - 4} + \dfrac{6}{y - 3}$

16. $\dfrac{x + 2}{x - 2} = \dfrac{-4}{x + 7} - \dfrac{6}{x - 2}$

C

Solve and check each of the following equations.

1. $\dfrac{3x + 4}{x} = \dfrac{10}{x}$

2. $3 - \dfrac{1}{y} = \dfrac{4}{y} + 2$

3. $\dfrac{5}{x} + 1 = \dfrac{1}{2x} - \dfrac{1}{2}$

4. $\dfrac{x}{x + 1} + \dfrac{1}{x - 3} = 0$

5. $\dfrac{x + 2}{x - 6} = \dfrac{5x + 1}{x - 3}$

6. $\dfrac{x + 3}{x - 3} = \dfrac{8}{x - 1} + \dfrac{13}{x^2 - 4x + 3}$

7.8 SOLUTION OF LITERAL EQUATIONS

Earlier in our work we examined a number of narrative problems that involved the use of the formula for the perimeter of a rectangle.

$$P = 2l + 2w$$

This equation is said to be **solved for P**. A statement such as this implies that the variable P appears on one side of the equation and all other variables and constants appear on the other side of the equation.

Under this arrangement should we be given a number of different pairs of values for l and w, we can readily find the value of P. For instance,

If $l = 4$ and $w = 3$, (1)
then $P = 2 \cdot 4 + 2 \cdot 3 = 8 + 6 = 14$

or   If $l = 53$ and $w = 42$ (2)
then $P = 2 \cdot 53 + 2 \cdot 42 = 106 + 84 = 190$

There are times, though, when we are given values for P and w and asked to determine the value of l in this same formula.

$$P = 2l + 2w.$$

Thus, if $P = 24$ and $w = 5$

then $24 = 2l + 2 \cdot 5$ (1)

At this point it would be necessary to solve this linear equation for l.

Were we given a great number of different pairs of values for P and w, and asked to find each corresponding value for l we would have to solve an equation such as (1) over and over again. Rather than continually repeating this process, we find it easier to **solve the formula for l**. This implies that l will appear by itself on one side of the equation and all other variables or constants will appear on the other side of the equation.

The method for solving the equation

$$P = 2l + 2w$$

for l is the same as we always use for solving any linear equation:

1. Use the addition principle of equality to isolate the *terms* containing the variable for which we are solving the equation.
2. Multiply both members of the equation by the multiplicative inverse of the coefficient of the variable.

7.8/Solution of Literal Equations

Let us apply this procedure to solve the equation $P = 2l + 2w$ for l.

$$P = 2l + 2w$$
$$2l + 2w = P$$
$$2l = P + (-2w) \quad \text{Addition Principle}$$
$$2l = P - 2w$$
$$l = \frac{1}{2}(P - 2w) \quad \text{Multiplication Principle}$$
$$l = \frac{1}{2}P - w$$

An equation that contains more than one variable is called a **literal equation**.

Example 1

Solve the following equation for b.

$$rb + c = d$$

Solution
$$rb + c = d$$
$$rb = d + (-c) \quad \text{Addition Principle}$$
$$rb = d - c$$
$$b = \frac{1}{r}(d - c) \quad \text{Multiplication Principle}$$

Explanation Just as 2 is the coefficient of x in the expression $2x$, so, too, is the r the coefficient of b in the expression rb. In the expression $2x$, there are 2 "x's"; in the expression rb, there are r "b's."

The value of b in Example 1 can be written in two ways other than the manner given in the solution. These are,

$$b = \frac{d}{r} - \frac{c}{r} \quad \text{and} \quad b = \frac{d-c}{r}.$$

Can you justify why each of these is correct?

Example 2

Solve the following equation for t.

$$r = \frac{A - P}{Pt}$$

Explanation The equation above is a fractional equation. Hence, the first step is to find the L.C.M. of the denominators. Since there is only one denominator, Pt, the L.C.M. will have to be Pt.

Solution

$$r = \frac{A-P}{Pt} \quad (1)$$

$$Pt \cdot r = Pt \cdot \frac{A-P}{Pt} \quad (2)$$

$$Ptr = A - P \quad (3)$$

$$Prt = A - P \quad (4)$$

$$t = \frac{1}{Pr}(A-P) \quad (5)$$

Explanation continued The term containing the variable t was isolated in step (3). At that point it was necessary to determine the coefficient of t. To do this, we recall that t times r is equal to r times t. Hence, we can rewrite *Ptr* as *Prt* in step (4). When written in this form, we realize the *Pr* is the coefficient of t. The solution is completed by multiplying both members of the equation by $1/Pr$.

The situation in both Examples 1 and 2 calls our attention to our greatest difficulty when solving a literal equation—that is the problem of determining the coefficient of the variable for which we are solving.

In expressions such as those below it is quite simple to spot the coefficient of the variable x for each coefficient is the multiplier of x. Each coefficient indicates the number of x's under consideration.

$3x$ $7x$ $295x$ $\frac{4}{5}x$

Three x's   Seven x's   Two Hundred Ninety-five x's   Four-fifths x's

The same is true when the coefficient of x happens to be another variable. The manner in which the coefficient of x is determined is by recognizing the quantity that is the multiplier of x. It may appear under a number of different conditions similar to those below.

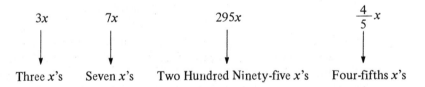

On occasion, however, the coefficient may be somewhat "disguised" as it was in Example 2. In that situation we were looking for the coefficient of t in the term *Ptr*. When this occurred we applied the commutative principle of

7.8/Solution of Literal Equations

multiplication and rearranged the factors so that the t appeared as the last factor:

$$Ptr = Prt$$

Similarly, to find the coefficient of x in the term xm we would again rearrange the factors so that x appears as the last factor. Thus,

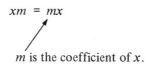

m is the coefficient of x.

Example 3

What is the coefficient of n in the expression ncb?

Explanation Rearrange the factors of ncb so the n is the last factor.

Solution $$ncb = cbn$$

cb is the coefficient of n.

In one of the examples above we saw that $a + b$ was the coefficient of x in the expression

$$(a + b)x. \qquad (1)$$

We will find, though, that there are many times when this same expression is written as

$$ax + bx. \qquad (2)$$

When this occurs and we are seeking the coefficient of x, it will be necessary to factor (2) as

$$x(a + b). \qquad (3)$$

At this point, if we write x as the **last** factor, (3) will become

$$(a + b)x. \qquad (4)$$

Now it is evident that $a + b$ is the coefficient of x.

Example 4

What is the coefficient of P in the expression

$$Pr + Prt\ ?$$

Solution
$$Pr + Prt = P(r + rt)$$
$$= (r + rt)P$$
$(r + rt)$ is the coefficient of P.

Explanation Notice that the greatest common factor of the terms in $Pr + Prt$ is Pr. However, when the expression $Pr + Prt$ was factored, the common factor that was used was P and **not** Pr. This was so for we were searching only for the coefficient of P and, hence, wanted the factor P to stand alone.

Example 5
Solve the following equation for d.
$$b = \frac{dc - v}{d}$$

Solution
$$b = \frac{dc - v}{d} \tag{1}$$
$$d \cdot b = \overset{1}{d} \cdot \frac{dc - v}{\underset{1}{d}} \tag{2}$$
$$d \cdot b = dc - v \tag{3}$$
$$db - dc = -v \tag{4}$$
$$d(b - c) = -v \tag{5}$$
$$(b - c)d = -v \tag{6}$$
$$d = \frac{1}{b - c} \cdot (-v) \tag{7}$$
$$d = \frac{-v}{b - c} \tag{8}$$

Explanation In going from step (6) to step (7), both members of equation (6) were multiplied by the reciprocal (multiplicative inverse) of the coefficient of d. That reciprocal is $1/(b - c)$. Recall that $-v$ can be written as $-v/1$. When $1/(b - c)$ is multiplied by $-v/1$ the product is $-v/(b - c)$.

EXERCISES 7.8

A

Determine the coefficient of the variable indicated in each of the exercises below.

	Expression	Variable		Expression	Variable
1.	ab	b	3.	rst	t
2.	ab	a	4.	rst	r

7.8/Solution of Literal Equations

	Expression	Variable		Expression	Variable
5.	rst	s	13.	$(c+d)a$	a
6.	$3ab$	b	14.	$b(c+d)$	b
7.	$5mn$	m	15.	$(2a-b)c$	c
8.	$\frac{2}{3}bc$	b	16.	$ab+ac$	a
9.	a^2c	c	17.	$2y-ay$	y
10.	cb^2	c	18.	$mg-mk$	m
11.	ab^2c	c	19.	$ab-a$	a
12.	ab^2c	a	20.	$6ac-4bc$	c

B

Solve each of the following equations for x.

1. $ax = b$
2. $x + a = b$
3. $x - a = b$
4. $a - x = b$
5. $b = x + c$
6. $ax + b = c$
7. $ax - b = c$
8. $\frac{x}{a} + b = c$
9. $ax + \frac{b}{c} = d$
10. $\frac{ax}{b} + \frac{c}{d} = 3$
11. $\frac{b}{c} + \frac{ax}{2} = d$
12. $a(x + b) = c$
13. $a(b - x) = c$
14. $\frac{x + a}{b} = c$
15. $\frac{4x - a}{3} = \frac{c}{2}$
16. $\frac{x - a}{b + 2} = 5$
17. $5(x + a) = 3(x - b)$
18. $6x = a - 4(x - b)$
19. $3x - a^2 = 2x - b^2$
20. $a^2x - b = c$
21. $(a + b)x = c$
22. $(a - b)x = -c$
23. $rxs = m$
24. $xbc - t = m$
25. $x(b - c) = a$
26. $bx + cx = m$
27. $ax = bx + c$
28. $a(x + c) = mx$
29. $\frac{x}{a} + \frac{b}{a} = x$
30. $\frac{x - a}{b} = \frac{x}{d}$
31. $\frac{a - 2x}{3} = \frac{x}{b}$
32. $b(x - a) = c(x - b)$
33. $abx + cbx = 7$
34. $\frac{p - qx}{r^2} = \frac{x}{r}$
35. $\frac{ax}{b + c} - x = \frac{a}{b + c}$
36. $a^2x + b = x$
37. $ax = ab$
38. $c^2x = c^5$
39. $xa - xb = a^2 - b^2$
40. $ax + 3a = 2x + a^2 + 2$

C

Solve each of the following equations for the variable indicated.

	Equation	Variable
1.	$A = lw$	w
2.	$A = lw$	l
3.	$P = 4s$	s
4.	$P = 2l + 2w$	l
5.	$C = \pi D$	D
6.	$C = 2\pi R$	R
7.	$A = P + Prt$	t
8.	$A = P + Prt$	r
9.	$A = P + Prt$	P
10.	$S = \dfrac{gt^2}{2}$	g
11.	$V = \dfrac{Bh}{3}$	B
12.	$V = \dfrac{\pi r^2 h}{3}$	h
13.	$A = \pi R^2 + \pi Rl$	l
14.	$A = 2\pi R^2 + 2\pi Rh$	h
15.	$\dfrac{A}{a} = \dfrac{B}{b}$	b
16.	$A = \dfrac{h(a+b)}{2}$	a
17.	$S = \dfrac{n(a+l)}{2}$	l
18.	$S = \dfrac{ar^3 - a}{r - 1}$	a
19.	$A = a + ab$	a
20.	$V = 3rh^2 - rh$	r

7.9 PROBLEM SOLVING INVOLVING FRACTIONS

PART 1

Once again we are going to return to the topic of narrative problem solving. In view of the subject area we have just been examining, the narrative problems

7.9/Problem Solving Involving Fractions   305

we intend to investigate are those that will lead to equations that contain fractions. Part 1 of this section will be concerned only with "number problems" that are similar to those encountered in the past. Following these, we will take a look at several narratives that require a somewhat different approach than used before to determine their translation to mathematical sentences.

Example 1

The numerator of a fraction is 2 less than the denominator. If 5 is added to the numerator and 4 is added to the denominator, the fraction will be $\frac{10}{11}$. What are the original numerator and denominator of the fraction?

Explanation As we learned earlier, in a narrative problem that involves two numbers, two pieces of information will be given.

1. One of the pieces of information will relate the size of one of the numbers to the other number.
2. The other piece of information will enable us to determine the equation.

In this problem the two pieces of information are:

1. The numerator is 2 less than the denominator.
2. If 5 is added to the numerator and 4 is added to the denominator, the fraction will be $\frac{10}{11}$.

Based on the first piece of information:

$$\text{Denominator is } x$$
$$\text{Numerator is } x - 2$$

Based on second piece of information:

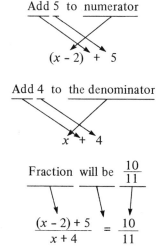

$$\frac{(x-2)+5}{x+4} = \frac{10}{11}$$

Solution Let x represent the denominator.
Then $x - 2$ represents the numerator.

Therefore,
$$\frac{(x-2)+5}{x+4} = \frac{10}{11}$$

$$\frac{x-2+5}{x+4} = \frac{10}{11}$$

$$\frac{x+3}{x+4} = \frac{10}{11}$$

$$11(x+4)\frac{x+3}{x+4} = 11(x+4)\frac{10}{11}$$

$$11(x+3) = 10(x+4)$$

$$11x + 33 = 10x + 40$$

$$11x - 10x = 40 - 33$$

$$x = 7 \quad \text{(Denominator)}$$

$$x - 2 = 5 \quad \text{(Numerator)}$$

Check: $\dfrac{5+5}{7+4} = \dfrac{10}{11}$

Example 2

The sum of two numbers is 25. If the first number is divided by the second number the quotient is 1 less than the second number. What are the numbers?

Explanation If the sum of two numbers is 25, and one of the numbers is 3, the other number will have to be 22. The number 22 is found by subtracting 3 from 25:

$$25 - 3.$$

Similarly, if the sum of the two numbers is 25 and one is x, then the other is found by subtracting x from 25:

$$25 - x.$$

Solution Let x represent the first number.
Then $25 - x$ represents the second number.

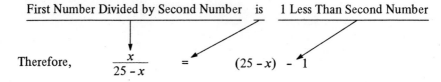

Therefore, $\dfrac{x}{25-x} = (25-x) - 1$

7.9/Problem Solving Involving Fractions

$$\frac{x}{25-x} = 25 - x - 1$$

$$\frac{x}{25-x} = 24 - x$$

$$(25-x) \cdot \frac{x}{25-x} = (25-x)(24-x)$$

$$x = 600 - 49x + x^2$$

$$x^2 - 49x + 600 = x$$

$$x^2 - 49x + 600 - x = 0$$

$$x^2 - 50x + 600 = 0$$

$$(x-30)(x-20) = 0$$

$x - 30 = 0$	$x - 20 = 0$
$x = 30$ (First Number)	$x = 20$ (First Number)
$25 - x = 25 - 30 = -5$ (Second Number)	$25 - x = 25 - 20 = 5$ (Second Number)
Check: $\frac{30}{-5} = -6$	Check: $\frac{20}{5} = 4$
-6 is 1 less than -5.	4 is 1 less than 5.

Explanation   The second piece of information in the narrative problem led to a quadratic equation which, in turn, led to *two* values for the *first number*. Hence, there are two values also for the second number. The two numbers must always be tied to one another. That is when the first number is 30, then the second *must be* -5 **not** 5. Also, when the first number is 20, then the second *must be* 5 **not** -5.

EXERCISES 7.9 (Part 1)

Find the number requested in each of the following problems.

1. One number is 15 more than another. When the larger number is divided by the smaller number the quotient is 4. What are the numbers?

2. One number is 5 more than another. When the smaller number is divided by 2 and the larger number is divided by 5, their sum will be 8. What are the two numbers?

3. The denominator of a fraction is 5 more than the numerator. If 2 is added to the numerator and 3 is added to the denominator, the new fraction will be $\frac{3}{5}$. What are the numerator and denominator of the original fraction?

4. The numerator of a fraction is 2 less than the denominator. When 4 is added to the numerator and 1 is subtracted from the denominator, the new fraction is $\frac{13}{10}$. What are the numerator and denominator of the original fraction?

5. Mr. Johnson is 30 years older than his son Tommy. When Mr. Johnson's age is divided by Tommy's age, the quotient is 7. How old is each of them?

6. The numerator of a fraction is 4 more than the denominator. When $\frac{1}{2}$ is added to the fraction, the sum is 2. What are the numerator and denominator of the fraction?

7. The denominator of a fraction is 1 less than the numerator. If $\frac{3}{4}$ is subtracted from the fraction, the difference is $\frac{3}{8}$. What are the numerator and denominator of the fraction?

8. A larger number is 4 more than a smaller number. When 2 more than 3 times the larger number is divided by the smaller number, the quotient is 10. What are the numbers?

9. A smaller number is 3 less than a larger number. When the smaller number is divided by 15 less than twice the larger number, the quotient is 2. What are the numbers?

10. The sum of two numbers is 9. When the first number is divided by the second number the quotient is 1 less than the second number. What are the numbers?

11. The sum of two numbers is 12. If the larger number is divided by the smaller number, the quotient is the smaller number. What are the numbers?

12. The denominator of a fraction is 1 more than the numerator. The sum of the fraction and its reciprocal is $\frac{13}{6}$. What are the numerator and denominator of the fraction?

13. Two numbers are consecutive integers. When the reciprocal of the first is added to the reciprocal of the second, the sum is $\frac{5}{6}$. What are the numbers? (Hint: Remember to discard answers that are fractions.)

14. Two numbers are consecutive even numbers. The sum of the reciprocals of the two numbers is $\frac{5}{12}$. What are the two numbers?

15. The sum of Mary's age and Sarah's age is 9. The sum of the reciprocals of their ages is $\frac{1}{2}$. How old is each of them if Mary is the older?

PART 2

There is a collection of narrative problems called "work problems" that students of algebra usually find of interest to them. These problems are concerned with situations where several men may be working at the same job but all work at different speeds. Our goal in some problems may be to determine how long it

7.9/Problem Solving Involving Fractions

will take the men to complete the work. In other cases there may be two different sized drain pipes that can be used to empty a tank. Under this condition we will very likely be asked to determine the time it will take to drain the tank if both pipes are opened.

The interest—and difficulty, unfortunately—can be heightened by having some pipes fill a pool while others are emptying the pool at the very same time. Questions arise such as, "Can the pool ever be filled?" and if so, "How long will it take to fill it?"

In each of these situations the immediate objective is to determine how much of the job is completed in a unit of time. Thus, if it takes a man *5 days* to paint an apartment, how much of the apartment can he complete in *1 day*? Similarly, if a water pipe can fill a tank in *40 minutes*, how much of the tank will it fill in *1 minute*? In the first situation the unit of time is 1 day, while in the second it is 1 minute.

Let's continue our investigation of the man who spends 5 days painting an apartment. Quite apparently, in 1 day he will complete $\frac{1}{5}$ of the work. In two days he will do twice as much, or $\frac{2}{5}$ of the work. In four days he will do $\frac{4}{5}$ of the work. And in 5 days he will do $\frac{5}{5}$ of the work which, of course, we can represent as **1**. Thus, the *entire job* can be represented by the number **1**.

In the same way, the water pipe that fills the tank in 40 minutes will fill $\frac{15}{40}$ of the tank in 15 minutes. In 40 minutes it will fill $\frac{40}{40}$ of the tank. The $\frac{40}{40}$ here again implies that the *entire job* can be represented by the number 1.

Example 1

It will take a person 12 days to complete a job. What fraction of the job will he complete in *y* days?

Explanation In 1 day he will complete $\frac{1}{12}$ of the job.
3 days: $\frac{3}{12}$ of the job.
y days: *y*/12 of the job.

Quite apparently, *y* cannot be greater than 12 for the person cannot complete more than $\frac{12}{12}$ of the job.

Solution Fraction of the job completed in *y* days: *y*/12.

Example 2

It takes *m* minutes for a drainpipe to empty a tank. What fraction of the tank will it empty in 5 minutes?

Explanation If it took the drainpipe 20 minutes to empty the tank, then in 5 minutes it would empty

$$\frac{5}{20} \text{ of the tank.}$$

Since it takes the drainpipe m minutes to empty the tank, then in 5 minutes it will empty

$$\frac{5}{m} \text{ of the tank.}$$

Solution Fraction of the tank emptied in 5 minutes: $5/m$

Example 3

Working alone Fred can paint an apartment in 6 days. Working alone, Tom can paint the same apartment in 8 days. If the two men worked together, how long will it take them to complete the job?

Explanation As usual, we allow some variable such as x to represent the number of days it will take both men working together to complete the job. Since Fred can complete the entire job in 6 days, then in the x days he works he will complete

$$x/6\text{ths of the job.}$$

Similarly, during the x days that Tom works he will complete

$$x/8\text{ths of the job.}$$

However, the sum of these two fractions must be the **entire job** and this, we learned, is represented by the number 1.

Solution Let x represent the number of days required for both men to complete the work. Then $x/6$ represents the fraction of work completed by Fred. And $x/8$ represents the fraction of work completed by Tom.

Therefore,
$$\frac{x}{6} + \frac{x}{8} = 1$$

$$\overset{4}{\cancel{24}} \cdot \frac{x}{\cancel{6}} + \overset{3}{\cancel{24}} \cdot \frac{x}{\cancel{8}} = 24 \cdot 1$$

$$4x + 3x = 24$$
$$7x = 24$$
$$x = 3\frac{4}{7} \text{ days}$$

Example 4

A water pipe can fill a pool in 8 hours while a drain pipe can empty the pool in 20 hours. If the pool is empty and both pipes are opened at the same time, how long will it take to fill the pool?

7.9/Problem Solving Involving Fractions

Solution   Let x represent the number of hours it will take to fill the pool. Then $x/8$ represents fraction of pool *filled* in x hours. And $x/20$ represents fraction of pool *drained* in x hours.

Therefore,
$$\frac{x}{8} - \frac{x}{20} = 1$$

$$\overset{5}{\cancel{40}} \cdot \frac{x}{\cancel{8}} - \overset{2}{\cancel{40}} \cdot \frac{x}{\cancel{20}} = 40 \cdot 1$$

$$5x - 2x = 40$$

$$3x = 40$$

$$x = 13\tfrac{1}{3} \text{ hours}$$

Explanation   Notice that when writing the equation it was necessary to subtract $x/20$ from $x/8$. This was brought about by the fact that during every hour $\frac{1}{20}$ of the pool was *drained off* as $\frac{1}{8}$ was *filled up*. Therefore, in the x hours it took to fill the pool, $x/20$ were drained off while $x/8$ were filled.

Example 5

If John were working alone, he could cement a walk in 60 hours. If Nick were working alone, he could cement the same walk in 40 hours. John spends 15 hours working on the walk before he is joined by Nick. How many hours will it take them to complete the walk?

Explanation   John will complete $\frac{15}{60}$ of the job before Nick appears for work. If we let x represent the number of hours they work together to complete the job, then John will complete another $x/60$ of the job after Nick reports for work. Nick, on the other hand, will do $x/40$ of the work during the x hours he is on the job.

The amount John does is: $\frac{15}{60} + \frac{x}{60}$.

The amount Nick does is: $\frac{x}{40}$.

But John's amount added to Nick's amount is the **entire job**.

Solution   Let x represent the number of hours they work together.

Therefore,
$$\left(\frac{15}{60} + \frac{x}{60}\right) + \frac{x}{40} = 1$$

$$\frac{15}{60} + \frac{x}{60} + \frac{x}{40} = 1$$

$$\overset{2}{\cancel{120}} \cdot \frac{15}{\cancel{60}} + \overset{2}{\cancel{120}} \cdot \frac{x}{\cancel{60}} + \overset{3}{\cancel{120}} \cdot \frac{x}{\cancel{40}} = 120 \cdot 1$$

$$30 + 2x + 3x = 120$$

$$2x + 3x = 120 - 30$$

$$5x = 90$$

$$x = 18 \text{ hours they worked together.}$$

EXERCISES 7.9 (Part 2)

A

Express the fraction in each of the following situations.

1. A man can paint a room in 12 hours. What fraction of the room can he paint in 1 hour?

2. A boy can mow a lawn in 200 minutes. What fraction of the lawn can be mowed in 40 minutes.

3. A water pipe can fill a tank in 6 hours. What fraction of the tank will it fill in 2 hours?

4. Wall-to-wall carpeting can be laid in a large room in 8 hours. What fraction of the carpeting will be laid in 5 hours?

5. A field can be plowed in 16 hours. What fraction of the field will be plowed in h hours?

6. An oil burner operating constantly will consume a tank of oil in d days. What fraction of the tank of oil will it consume in 12 days ($12 < d$)?

7. A water pipe can fill a pool in 10 hours. A drain pipe can empty the pool in 25 hours.
    a. If the drain pipe is turned off, what fraction of the pool is filled in 2 hours?
    b. If the water pipe is turned off, what fraction of the pool is drained in 2 hours?
    c. If both pipes are open, what fraction of the pool is filled in 2 hours?

8. Betty can type a set of letters in 200 minutes. Alice will take 300 minutes to type the same set of lettters.
    a. What fraction of the letters will Betty type in 60 minutes?
    b. What fraction of the letters will Alice type in 60 minutes?
    c. What fraction of the letters will they both type together in 60 minutes?
    d. What fraction of the letters will they both type together in x minutes?

B

1. It will take Paul 15 days working alone to paint the outside of a house. It will take Ben 10 days working alone to do the same job. If they work together, how many days will it take them to do the job?

2. When Pat mows Mrs. Tyler's lawn, it takes him 4 hours. When Fred does the job, it takes him 6 hours. They become partners and mow the lawn together. How long will it take them?

3. There are two water pipes that lead into a pool. One can fill the pool in 12 hours, while it only takes 8 hours for the other to fill the pool. If both are opened at the same time, how long will it take to fill the pool?

4. A rocket engine can drain a tank of fuel in 90 seconds. A second rocket engine can drain the same tank of fuel in 60 seconds. Both rockets are turned on at the same moment and both drain the fuel from the same tank. In how many seconds will the tank be empty?

5. A water pipe can fill a pool in 5 hours while a drain pipe can empty the pool in 15 hours. If both are left open, how long will it take to fill the pool?

6. A swimming pool has two water pipes that are used to fill the pool and one drain pipe to empty it. One of the water pipes can fill the pool in 8 hours while it takes the other 12 hours to fill the pool. The drain pipe can empty the pool in 24 hours. All three pipes are opened simultaneously on an empty pool. How long will it take to fill the pool?

7. It will take a water pipe 10 hours to fill a tank. It will take the drain pipe for that tank 25 hours to drain the tank. After the water pipe has been open for 4 hours, the drain pipe is also opened. Both pipes remain open until the tank is filled. How many hours will this take?

8. Betty can type a set of letters in 200 minutes. Alice will take 300 minutes to type the same set of letters. After Betty has worked on the letters for 60 minutes, Alice joins her and together they complete the typing. For how many minutes will they work together?

9. A farmer has two tractors—a new one that will plow one of his fields in 12 hours and an old one that will take 18 hours to plow the same field. He uses the old tractor for 6 hours and then switches over to the new one. How many hours will it take him to complete the plowing with the new tractor?

10. It takes Jane 16 days to check over the company records at the close of each year. If Linda does this same work it takes her 18 days. After working 4 days on the records Jane was transferred to other duties and Linda was called in to complete the checking. How many days did it take Linda to finish the work?

11. It would take Carl 12 days to wallpaper the rooms of a house. After working on the job 3 days, he is joined by Mario and together they complete the job in 5 days. How many days would the job have taken Mario if he were working alone?

12. A water pump is brought in to drain the water from a larger excavation. Operating alone the pump would have drained off all the water in 30 hours. However, after 5 hours a second pump is brought in and together the two pumps empty the excavation in 13 hours. In how many hours would the second pump have drained the entire excavation had it been working alone?

CHAPTER REVIEW

A

1. Reduce each of the following fractions to lowest terms.

   a. $\dfrac{5a}{10b}$

   b. $\dfrac{9x^3 y}{6x^2 y^3}$

   c. $\dfrac{10x}{x}$

   d. $\dfrac{12a^5 (a-2)}{8a^2 (a-2)}$

2. Reduce each of the following fractions to lowest terms.

   a. $\dfrac{x-7}{7-x}$

   b. $\dfrac{(5+a)(5-a)}{a+5}$

   c. $\dfrac{x+4}{x^2 + 6x + 8}$

   d. $\dfrac{3-x}{x^2 - 9}$

   e. $\dfrac{y^3 - y}{y^3 + 2y^2 - 3y}$

   f. $\dfrac{2x(a-3) + 3y(a-3)}{5(a-3)}$

3. Find the product in each of the following exercises.

   a. $\dfrac{x^2}{y} \cdot \dfrac{y^3}{x^4}$

   b. $\dfrac{(x+1)(x-2)}{(x-7)} \cdot \dfrac{(x-7)}{(x-2)(x-1)}$

   c. $\dfrac{3a^2 b}{c^3} \cdot \dfrac{c^2}{6b^3} \cdot \dfrac{8bc}{a}$

   d. $\dfrac{a-3}{a+5} \cdot \dfrac{a^2 - 25}{a^2 - 2a - 3}$

   e. $\dfrac{x^2 + 2x - 8}{x^3 - 2x^2} \cdot \dfrac{x^2 + 5x}{x^2 + 9x + 20}$

   f. $\dfrac{2y^2 + y - 6}{y^2 - 3y - 10} \cdot \dfrac{2y - 2}{2y^2 - 5y + 3}$

Chapter Review

4. Find the quotient in each of the following exercises.

   a. $\dfrac{5a^3}{6b^4} \div \dfrac{3a}{10b^2}$

   b. $8a^2b^3 \div \dfrac{4ab^5}{3c}$

   c. $\dfrac{9x^3}{6x^2y} \div 2xy^2$

   d. $\dfrac{x^2 - 7x + 10}{6a^2} \div \dfrac{x^2 - 2x - 15}{10a^5}$

   e. $(y - 4) \div \dfrac{y^2 - 5y + 4}{y^2 - 1}$

   f. $\dfrac{4a^2 - 12a + 9}{2ax - 3x} \div (2a^2 - 5a + 3)$

5. Perform the operation indicated in each of the following exercises.

   a. $\dfrac{6x}{7} + \dfrac{5x}{7}$

   b. $\dfrac{3a^2}{2x} - \dfrac{8a^2}{2x}$

   c. $\dfrac{3x - 4y}{a} + \dfrac{x + 3y}{a}$

   d. $\dfrac{a - 3b}{x - y} - \dfrac{a + 3b}{x - y}$

6. Perform the operation indicated in each of the following exercises.

   a. $\dfrac{6c}{5} + \dfrac{3c}{20}$

   b. $\dfrac{3b}{4x} - \dfrac{7b}{2x}$

   c. $\dfrac{3x - 2}{6} + \dfrac{x - 5}{2}$

   d. $-\dfrac{4a - 3}{3} + \dfrac{5a + 1}{9}$

7. Perform the operations indicated in each of the following exercises.

   a. $\dfrac{5}{a} + \dfrac{4}{b}$

   b. $\dfrac{6}{a} + \dfrac{3}{ab} + \dfrac{2}{b}$

   c. $\dfrac{4}{3a^2b} - \dfrac{5}{6ab^2}$

   d. $\dfrac{5}{x} + \dfrac{4 - 2y}{xy}$

   e. $\dfrac{a - 2b}{a} - \dfrac{2a - b}{b}$

   f. $\dfrac{a - b}{2a} + \dfrac{b - a}{4b} - \dfrac{3}{2ab}$

8. Perform the operation indicated in each of the following exercises.

   a. $\dfrac{5}{x + 2} + \dfrac{3}{x - 5}$

   b. $\dfrac{3x}{5x + 5} + \dfrac{2x}{7x + 7}$

   c. $\dfrac{2}{x^2 - 9} + \dfrac{4}{x^2 + 4x + 3}$

   d. $\dfrac{5}{x^2 - 2x} - \dfrac{4}{x^2 - x - 2}$

   e. $\dfrac{3}{x^2 - 7x + 6} - \dfrac{2}{x^2 - 5x - 6}$

   f. $\dfrac{x + 1}{x^2 - 4} - \dfrac{4}{3x^2 - 6x}$

9. Solve each of the following equations.

a. $\dfrac{x}{2} + \dfrac{4x}{3} = 11$

b. $\dfrac{7a}{4} - \dfrac{8a}{5} = \dfrac{a}{2} - \dfrac{7}{5}$

c. $\dfrac{7x-4}{3} + \dfrac{3x+1}{5} = \dfrac{10x-1}{3}$

d. $\dfrac{5y-1}{3} - \dfrac{7y+1}{4} = \dfrac{3-y}{2}$

10. Solve each of the following equations.

a. $\dfrac{3a+5}{a} = \dfrac{-10}{a}$

b. $6 + \dfrac{5}{x} = \dfrac{2x-7}{x}$

c. $\dfrac{x+1}{x} - \dfrac{x-2}{5x} = 1 + \dfrac{2x-17}{5x}$

d. $\dfrac{y-5}{y+6} = \dfrac{1}{y+6} - \dfrac{2}{y+3}$

11. Solve each of the following inequalities.

a. $\dfrac{3x}{2} - \dfrac{5x}{6} > \dfrac{8}{3}$

b. $\dfrac{4x}{3} < \dfrac{6x}{5} - \dfrac{4}{15}$

c. $\dfrac{x+10}{2} - \dfrac{x}{10} < \dfrac{x}{5}$

d. $\dfrac{x+1}{2} - \dfrac{x}{12} > \dfrac{2x-2}{3} - \dfrac{1}{3}$

12. Solve each of the following equations for x.

a. $b + cx = d$

b. $ax + bx = c$

c. $ax = bx + c$

d. $A = rx$

e. $\dfrac{x}{a} + \dfrac{b}{a} = c$

f. $a(x+b) = c(x+d)$

g. $xab = a^2$

h. $P = 2x + w$

B

Find the numbers requested in each of the following problems.

1. The denominator of a fraction is 8 more than the numerator. If 1 is added to the denominator and 3 is subtracted from the numerator, the new fraction will be $\dfrac{2}{3}$. What are the numerator and denominator of the original fraction?

2. The sum of two numbers is 8. If the larger number is divided by the smaller number, the quotient is 1 more than the smaller number. What are the numbers?

3. There are two faucets that can be used to fill a bathtub. If only the first is used, the tub can be filled in 15 minutes. If only the second is used, the tub can be filled in 9 minutes. How long will it take to fill the tub if both faucets are used?

4. Betty can knit a sweater in 40 hours. Elsie can knit the same sweater in 30 hours. Betty starts the sweater and works for 8 hours. She then asks Elsie to complete the sweater for her. How many hours will it take Elsie to finish the sweater?

The Graph of an Open Sentence in Two Variables

At the very outset of our work we developed a method for naming the points of a line. We did this by establishing a matching between the real numbers and the points of the line. You may recall that once we had decided upon the point to be named by the number 0 and the point to be named by the number 1, the name for every other point was locked into place. Thus, in Figure 8-1 where

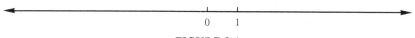

FIGURE 8-1

the "0" point and the "1" point have already been named, the point to be named by the number 4 will have to be to the *right* of the origin, "0" point, and four times as far from the origin as the point "1" is from the origin (Figure 8-2).

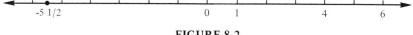

FIGURE 8-2

In the same way, the point named by the number $-5\frac{1}{2}$ is five and one-half units to the left of the origin. Positive numbers name points to the right of the origin while negative numbers name points to the left of the origin.

We now try to establish some way by which we might name the points of a plane rather than the points of a line.

8.1 NAMING POINTS IN A PLANE

PART 1

As we have just indicated, the naming of a point of a line involves a direction given with reference to a specific point of the line called the origin.

In much the same way, when naming points of a plane, we establish two lines in the plane that intersect at right angles. These lines are called *axes* (Figure 8-3). The directions for naming a point are given with reference to each of the axes.

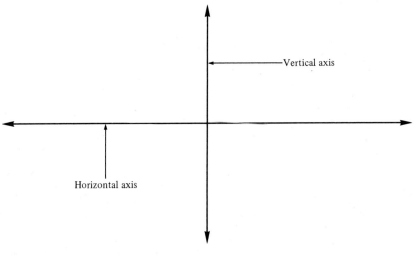

FIGURE 8-3

However, now a single direction is not sufficient for naming a point. For instance, if we say that we would like to locate the point that is

4 units to the right of the vertical axis

we find that there are infinitely many points that fit this description. Each of these points is shown in Figure 8-4. Notice that each of these points is on a line parallel to the vertical axis.

Similarly if we are given the single direction that the point is to be

2 units below the horizontal axis,

we would find that here again there are infinitely many points that fit this description. Each of these points is on a line that is parallel to the horizontal axis, as seen in Figure 8-5.

8.1/Naming Points in a Plane

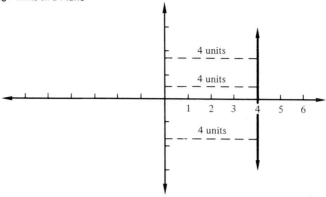

FIGURE 8-4

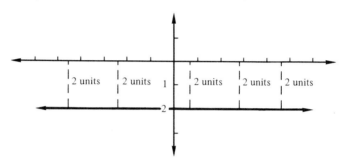

FIGURE 8-5

However, if we name the point by giving both directions, that is, the point is

1. 4 units to the right of the vertical axis
2. 2 units below the horizontal axis

then there will be but one point named by this description. This is the point where the bold line in Figure 8-4 intersects the bold line in Figure 8-5. This situation is pictured in Figure 8-6.

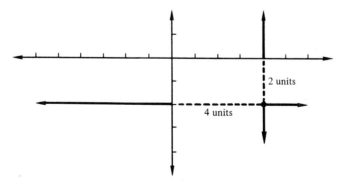

FIGURE 8-6

Thus, the naming of a point in a plane requires two directions.

1. The *first* direction is given with reference to the vertical axis.
2. The *second* direction is given with reference to the horizontal axis.

Rather than state the directions in sentence form as we have above, we will simply express them as a pair of numbers written within parentheses. Thus, the directions for naming the point in Figure 8-6 are given as

$$(+4, -2).$$

The *first* number gives us a direction with reference to the *vertical* axis; the positive sign before the 4 indicates that the point is to the *right* of that axis. Had the sign been *negative* we would know that the point is to the *left* of the vertical axis. The second number indicates a direction with reference to the *horizontal* axis, the fact the sign of the 2 is *negative* informs us that the point is *below* the *horizontal* axis. Had the sign of the 2 been *positive* it would indicate that the point is *above* the *horizontal* axis.

The pair of numbers (+4, -2) is called an **ordered pair of numbers**. The *first* of these numbers *must* give the direction with reference to the *vertical axis*; the second *must* give the direction with reference to the *horizontal axis*.

Example 1

Express the following ordered pair of numbers as a pair of directions.

$$(-5, +2)$$

Explanation The *negative* sign before the first direction implies that the point will be to the *left* of the vertical axis. The positive sign before the second direction implies the point will be *above* the horizontal axis.

Solution 1. The point is 5 units to the left of the vertical axis.
and 2. The point is 2 units above the horizontal axis.

Example 2

Express the following ordered pair of numbers as a pair of directions.

$$(0, 5)$$

Solution 1. The point is 0 units from the vertical axis.
and 2. The point is 5 units above the horizontal axis.

Explanation The first direction states that the point is 0 units from the vertical axis. This implies that since the point is neither to the left nor right of the vertical axis it must be *on* the vertical axis itself. Also, no sign of direction before the 5 implies that the sign is positive. Hence, the point is *above* the horizontal axis. The point named by (0, 5) is shown in Figure 8-7.

8.1/Naming Points in a Plane

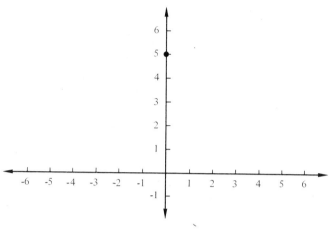

FIGURE 8-7

Example 3

Express the following two directions as an ordered pair of numbers.

    1. The point is 4 units to the left of the vertical axis.

and 2. The point is 9 units below the horizontal axis.

Solution   (-4, -9)

EXERCISES 8.1 (Part 1)

A

Express each of the following ordered pairs of numbers as a pair of directions.

1. (+4, +5)
2. (+7, +3)
3. (+6, 2)
4. (5, +3)
5. (-4, +8)
6. (-7, 5)
7. (+3, -4)
8. (+6, -9)
9. (10, -4)
10. (-3, -5)
11. (-6, -11)
12. (0, +4)
13. (+7, 0)
14. (0, 0)

B

Express each of the following pairs of directions as an ordered pair of numbers.

1. The point is
        3 units to the right of the vertical axis
   and 2 units above the horizontal axis.

2. The point is
    12 units to the right of the vertical axis
    and   9 units above the horizontal axis.

3. The point is
    7 units to the left of the vertical axis
    and   6 units above the horizontal axis.

4. The point is
    10 units to the right of the vertical axis
    and   1 unit below the horizontal axis.

5. The point is
    15 units to the left of the vertical axis
    and   20 units below the horizontal axis.

6. The point is
    0 units from the vertical axis
    and   16 units above the horizontal axis.

7. The point is
    8 units to the left of the vertical axis
    and   0 units from the the horizontal axis.

8. The point is
    0 units from the vertical axis
    and   0 units from the horizontal axis.

PART 2

The pair of numbers that name the point in a plane is called the **coordinates** of the point. The first of the numbers is called the **first coordinate** while the second is the **second coordinate**.

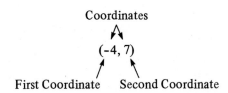

Another term frequently used in reference to the *first* coordinate of a point is **abscissa of a point**. In the same way, the term **ordinate of a point** is used when speaking of the *second* coordinate of a point.

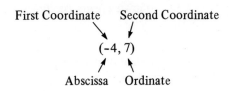

8.1/Naming Points in a Plane

Information that the abscissa of a point is negative 4 will imply that the point is 4 units to the left of the vertical axis. Similarly, if the ordinate of a point is positive 7, it will imply that the point is 7 units above the horizontal axis. The term *abscissa* is really but another name for the first *direction*. The term *ordinate* is another name for the *second direction*.

Since the points of the plane are named by pairs of coordinates, the plane itself is called a **coordinate plane**. Similarly, the vertical and horizontal reference axes are called the **coordinate axes** (Figure 8-8).

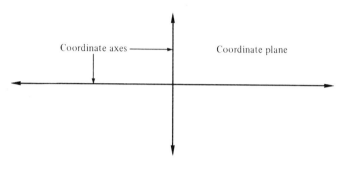

FIGURE 8-8

Example 1

Plot the point whose abscissa is 3 and whose ordinate is −4.

Explanation The term "plot the point" is the mathematical way of asking that we locate the point named by a pair of directions. Since the abscissa is 3, the point is 3 units to the right of the vertical axis. Second, the ordinate being −4 implies that the point is 4 units below the horizontal axis.

Solution

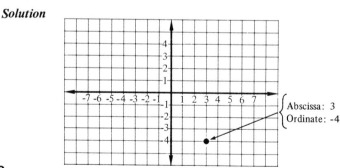

Example 2

Plot the point whose abscissa is −5.

Explanation In view of the fact that only *one* direction is given, there will be infinitely many points that are named by this information. Since the abscissa is −5 the points will be 5 units to the left of the vertical axis. In view of this they will be points of a line that is parallel to the vertical axis.

Solution

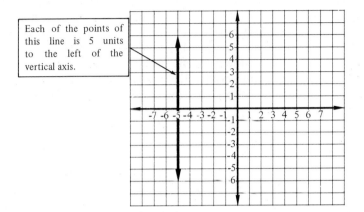

Each of the points of this line is 5 units to the left of the vertical axis.

We have already examined three different ways by which a point can be named. We have yet to examine the fourth and final method. The pair of directions, (5, -3), can be expressed as the pair of equations

$$x = 5$$

$$\text{and } y = -3.$$

The equation $x = 5$ is but another way of saying that

"The point is 5 units to the right of the vertical axis."

Similarly, the equation $y = -3$ is but another way of saying that

"This point is also 3 units below the horizontal axis."

Example 3

Plot the point named by the equations below.

$$x = -1$$
$$y = -2$$

Explanation The direction $x = -1$ states that the point is 1 unit to the left of the vertical axis. The direction $y = -2$ states that the point is 2 units below the horizontal axis.

Solution

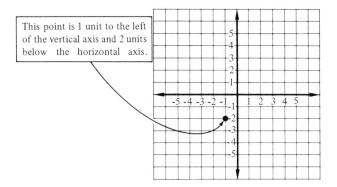

This point is 1 unit to the left of the vertical axis and 2 units below the horizontal axis.

When a point is named by a pair of equations such as

$$x = -1$$
$$y = -2$$

we give special names to the vertical and horizontal axes. The vertical axis is usually called the Y-axis while the horizontal axis is called the X-axis. There are some mathematicians, though, who prefer to use the small letters y and x rather than the capital letters.

The X and Y axes separate the plane into 4 regions that are called **quadrants**. Each of these quadrants has its own special name. The **first quadrant** is that section of the plane in the upper right corner. Each of the other quadrants are then named by moving in a counterclockwise direction (⤴).

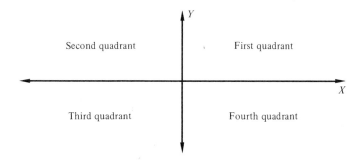

Example 4

What are the signs of the abscissas and the ordinates of points in the second quadrant?

Explanation Since the points in the second quadrant are to the *left* of the vertical axis then the abscissas of these points will have to be negative. Since the points are *above* the horizontal axis, the ordinates will have to be positive.

Solution Abscissas are negative. Ordinates are positive.

Example 5

What are the coordinates of point A shown below?

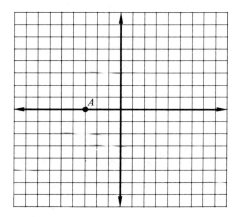

Explanation As point A is 3 units to the left of the vertical axis, its abscissa must be -3. Since it is neither above nor below the horizontal axis, its ordinate must be 0.

Solution The coordinates of A are: $(-3, 0)$.

EXERCISES 8.1 (Part 2)

A

Write each of the following ordered pairs of numbers as a pair of equations.

1. $(5, 2)$
2. $(+4, 6)$
3. $(-2, +7)$
4. $(+5, -2)$
5. $(-9, -7)$
6. $(0, 8)$

B

Write each of the following pairs of equations as an ordered pair of numbers.

1. $x = 5, y = 7$
2. $x = 9, y = -3$
3. $x = -2, y = -4$
4. $x = 0, y = -7$
5. $x = -6, y = 0$
6. $x = 0, y = 0$

C

Write each of the following points as an ordered pair of numbers.

1. Abscissa is 5; ordinate is -4.
2. Abscissa is -3; ordinate is $+6$.

8.1/Naming Points in a Plane

3. Ordinate is –7; abscissa is 0.
4. Ordinate is –9; abscissa is –2.
5. Abscissa is 2; ordinate is 6.
6. Ordinate is 0; abscissa is –1.

D

Plot each of the following points.

1. (+6, +3)
2. (+5, +1)
3. (+4, –2)
4. (+7, –4)
5. (–3, +5)
6. (–6, +6)
7. (–4, –2)
8. (–3, –8)
9. (5, –1)
10. (–6, 3)
11. (4, 4)
12. (7, –3)
13. (+5, 0)
14. (0, +6)
15. (–4, 0)
16. (0, –5)
17. (0, 0)
18. (–2, –2)
19. (–8, +7)
20. (+10, –5)

E

Plot the points named by each of the following pairs of equations.

1. $x = 1$, $y = 5$
2. $x = 0$, $y = -3$
3. $x = -5$, $y = 7$
4. $x = +8$, $y = 0$
5. $x = -8$, $y = 0$
6. $x = -3$, $y = -3$

F

Plot the points named in each of the following exercises.

1. Abscissa is –4; ordinate is +5.
2. Abscissa is 7; ordinate is 2.
3. Abscissa is –11; ordinate is –6.
4. Ordinate is –3; abscissa is –8.
5. Ordinate is +2; abscissa is 0.
6. Abscissa is +2; ordinate is 0.

G

Using the coordinate plane below, give the coordinates of each of the points in the exercises that follow. Exercise 1 is completed for you.

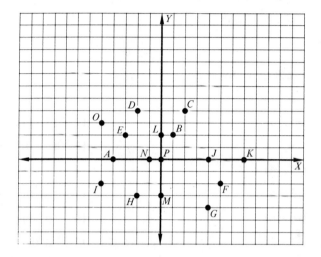

1. $A(-4, 0)$
2. $B(\;,\;)$
3. $C(\;,\;)$
4. $D(\;,\;)$
5. $E(\;,\;)$
6. $F(\;,\;)$
7. $G(\;,\;)$
8. $H(\;,\;)$
9. $I(\;,\;)$
10. $J(\;,\;)$
11. $K(\;,\;)$
12. $L(\;,\;)$
13. $M(\;,\;)$
14. $N(\;,\;)$
15. $O(\;,\;)$
16. $P(\;,\;)$

H

Using the coordinate plane below, write the pair of equations that names each of the points listed. Exercise 1 is completed for you.

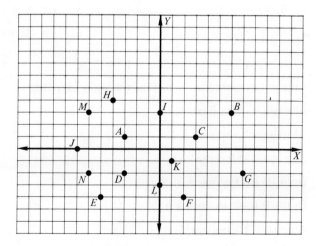

8.1/Naming Points in a Plane

1. $A: x = -3, y = 1$
2. $B:$
3. $C:$
4. $D:$
5. $E:$
6. $F:$
7. $G:$
8. $H:$
9. $I:$
10. $J:$
11. $K:$
12. $L:$
13. $M:$
14. $N:$

I

Using the coordinate plane below, write the abscissa and ordinate of each of the points in the exercises that follow. Exercise 1 is completed for you.

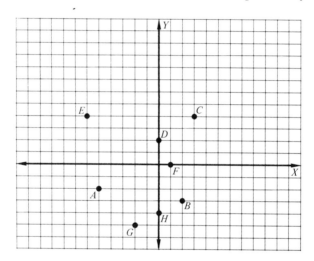

1. $A:$ abscissa $= -5$, ordinate $= -2$
2. $B:$
3. $C:$
4. $D:$
5. $E:$
6. $F:$
7. $G:$
8. $H:$

J

Answer each of the following.
1. What is the sign of the abscissa of each point in the first quadrant?
2. What is the sign of the ordinate of each point in the second quadrant?
3. What is the sign of the ordinate of each point in the fourth quadrant?
4. What is the sign of the abscissa of each point in the third quadrant?

5. If the signs of both the abscissa and the ordinate of a point are positive, in which quadrant is the point?
6. If the sign of the abscissa of a point is positive and the sign of the ordinate is negative, in which quadrant is the point?
7. If the abscissa of a point is 7, describe the position of the point.
8. If the ordinate of a point is -3, describe the position of the point.
9. If the ordinate of a point is 0, describe the position of the point.
10. The abscissa of a point is 0. Describe the position of the point.

K

Plot the set of points given by the single direction in each of the following exercises.

1. $x = 6$
2. $x = 9$
3. $y = 4$
4. $y = 7$
5. $x = -8$
6. $x = -1$
7. $y = -6$
8. $y = -2$
9. $x = 0$
10. $y = 0$

8.2 THE GRAPH OF A LINEAR EQUATION IN TWO VARIABLES

PART 1

In our study of a linear equation in one variable we discovered that there was but one number with which the variable can be replaced so that the equation will be a true sentence. Thus, in the equation

$$x + 5 = 7$$

the only number that can replace x so as to make this sentence true is the number 2.

At this time we are going to investigate equations in two variables and try to search out those numbers that will make these sentences true. Linear equations in two variables are equations whose form is similar to the one below.

$$3x + 9y = 5 \qquad (1)$$

Notice that this equation contains a single term in x to the first power, a single term in y to the first power, and a constant term. The general form of a linear equation in two variables is usually written as

$$ax + by = c. \qquad (2)$$

8.2/The Graph of a Linear Equation in Two Variables

When written in this manner, we recognize the a as the coefficient of x, the b as the coefficient of y, and the c as the constant term. Using equation (1) as an example, we see that

$$a = 3, \quad b = 9, \quad c = 5.$$

Let us begin our study of the linear equation in two variables with a relatively simple situation, such as

$$x + y = 6. \tag{3}$$

Here we notice that if 5 replaces x and 1 replaces y, this equation will be a true sentence. Hence we can say that the pair of numbers, 5 for x and 1 for y, is an element in the *solution set* of the equation $x + y = 6$.

Recall that the solution set of an equation consists of all replacements for the variables that will make the equation true. Since there are two variables in equation (3), the solution set of that equation will consist of *pairs* of numbers. Hence, the pair of numbers

$$x = 5 \quad \text{and} \quad y = 1$$

is an element in the solution set of equation (3). From what we learned earlier in this chapter, though, this pair of equations can be written as the ordered pair of numbers

$$(5, 1).$$

Not only is the ordered pair (5, 1) an element in the solution set of the equation

$$x + y = 6 \tag{3}$$

but so, too, is the pair of numbers (4, 2). When x is replaced by 4 and y by 2, we discover that this pair of numbers also satisfies equation (3).

$$x + y = 6$$
$$4 + 2 \stackrel{?}{=} 6$$
$$6 = 6$$

In fact, there are infinitely many pairs of numbers that will satisfy equation (3). Some of these are,

$$(5, 1), \ (4, 2), \ (3, 3), \ (2, 4), \ (1, 5), \ (0, 6), \ (6, 0), \ (-1, 7), \ (-2, 8).$$

In addition, there are infinitely many pairs of mixed numbers not listed above that will also satisfy equation (3). Some of these are,

$(4\frac{1}{2}, 1\frac{1}{2}), (3\frac{1}{2}, 2\frac{1}{2}), (5\frac{1}{4}, \frac{3}{4}), (7\frac{1}{8}, -1\frac{1}{8}), (-6\frac{2}{3}, +12\frac{2}{3})$.

It is quite apparent that an entire lifetime can be spent in just listing those pairs of numbers that satisfy this single equation. Rather than devote their time to this, mathematicians decided to plot several of the pairs of values that satisified this equation. When they did, they discovered a rather startling thing. Let us do the same by plotting in Figure 8-9 some of the pairs of integers we found above.

(5, 1), (4, 2), (3, 3), (2, 4), (1, 5), (0, 6), (6, 0), (-1, 7), (-2, 8)

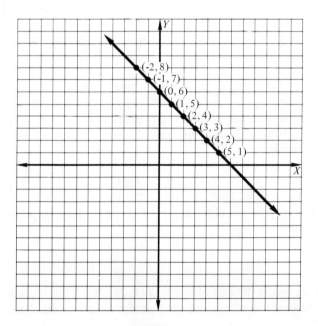

FIGURE 8-9

We are immediately struck by the fact that every point named by these pairs of numbers lies on the same line. Not only that, but if we plot the pairs of mixed numbers, they too name points on this line. In fact, every pair of numbers that satisfies the equation

$$x + y = 6$$

names a point on the line in Figure 8-9.

8.2/The Graph of a Linear Equation in Two Variables

In general, it is true that the set of points named by the solution set of a linear equation in two variables names points that lie on a line. The set of points named by the solution set of *any equation* is called the *graph* of that equation. Hence, what we have shown above implies that

"The graph of a linear equation in two variables is a straight line."

Example

Find 4 pairs of numbers in the solution set of the following equation and then graph the equation.

$$x + y = 10$$

Explanation At this stage, the easiest way to find pairs of values that satisfy this equation is to guess at them. If in doubt, merely replace x and y with the 2 numbers to determine if they do satisfy the equation. For instance, the test for the pair of numbers $(6, 4)$ is:

$$x + y = 10$$

$$6 + 4 \stackrel{?}{=} 10$$

$$10 = 10$$

Therefore, $(6, 4)$ is an element in the solution set.

Solution Four pairs of numbers in the solution set are

$$(6, 4), \quad (8, 2), \quad (4, 6), \quad (2, 8).$$

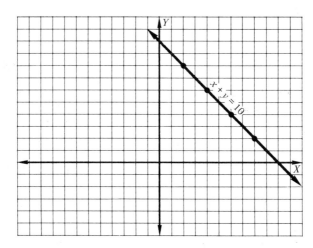

Explanation continued Although we have found only four points of the graph, all the points of the graph will lie on the line containing these four points.

EXERCISES 8.2 (Part 1)

Find 4 pairs of values in the solution set of each of the following equations and then draw the graph of each equation.

1. $x + y = 4$
2. $x + y = 8$
3. $x + y = 12$
4. $x - y = 5$
5. $x - y = 2$
6. $2x - y = 3$

PART 2

We have examined only the first of four relations that exist between an equation and its graph.

---

1. If a pair of numbers satisfies an equation, then it will name a point on the graph of the equation.
2. If a pair of numbers does not satisfy an equation, then it will name a point that is *not* on the graph of the equation.
3. If a point lies on the graph of an equation, then its coordinates will satisfy the equation.
4. If a point does *not* lie on the graph of an equation, then its coordinates will *not* satisfy the equation.

The following figures picture these four relations between the equation $x + y = 5$ and its graph.

Relation 1

$(3, 2)$ satisfies $x + y = 5$.
$(3, 2)$ is on the graph.

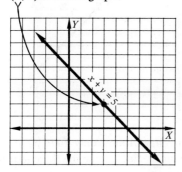

Relation 2

$(4, 7)$ does *not* satisfy $x + y = 5$.
$(4, 7)$ is *not* on the graph.

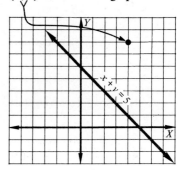

8.2/The Graph of a Linear Equation in Two Variables

Relation 3

(7, -2) is on the graph.
(7, -2) satisfies $x + y = 5$.

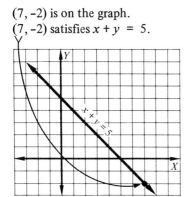

Relation 4

(-4, 2) is *not* on the graph.
(-4, 2) does *not* satisfy $x + y = 5$.

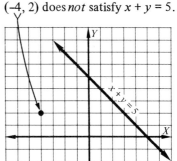

Example 1
Will the pair of numbers (6, -3) name a point that falls on the graph of the equation below?

$$2x = 9 - 2y$$

Solution

$$2x = 9 - 2y$$
$$2(6) \stackrel{?}{=} 9 - (-3)$$
$$12 \stackrel{?}{=} 9 + 6$$
$$12 \neq 15$$

Since the pair of numbers (6, -3) does not satisfy the equation $2x = 9 - 2y$, it does not name a point on the graph of this equation.

Example 2

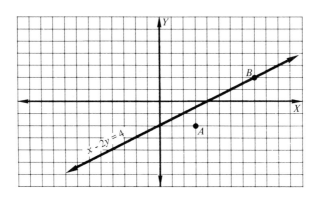

In the example figure will the coordinates of A satisfy the equation $x - 2y = 4$, the graph of which is shown?

Solution Point A does not lie on the graph of the equation $x - 2y = 4$, hence its coordinates will not satisfy that equation.

Example 3
Using the diagram in Example 2, what conclusion can be drawn concerning the coordinates of B? Justify your answer.

Solution Conclusion—the coordinates of B will satisfy the equation $x - 2y = 4$. Point B lies on the graph of the equation $x - 2y = 4$, hence its coordinates will satisfy the equation.

EXERCISES 8.2 (Part 2)

A

Next to each of the following equations, an ordered pair of numbers is given. Will the pair of numbers name a point that lies on the graph of the equation? Justify your answer.

1. $x + y = 12$ (7, 5)
2. $x - y = 15$ (12, 2)
3. $2x + 3y = 6$ (1, 2)
4. $5x - 2y = 7$ (3, 4)
5. $2x - 6y = 3$ (5, 2)
6. $3x + 5y = -4$ (2, -2)
7. $4x = 2y - 3$ (-2, 6)
8. $3y = x - 7$ (13, 2)
9. $x = 7y + 15$ (1, -2)
10. $y = 3x + 5$ (2, -1)
11. $x - y - 3 = 0$ (5, 2)
12. $3x + 2y - 7 = 0$ (4, -1)

B

In each of the following exercises an equation and its graph are shown. Will the point A in each exercise satisfy the equation of that exercise? Justify your answer.

1.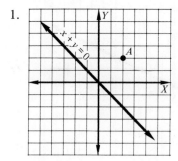

2.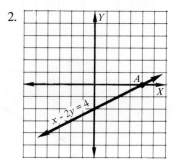

8.2/The Graph of a Linear Equation in Two Variables

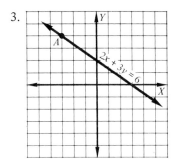
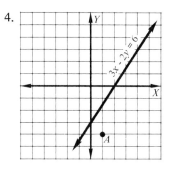

C

In each of the following exercises an equation and its graph are shown. What conclusion can be drawn concerning the coordinates of point B in each exercise? Justify your conclusion.

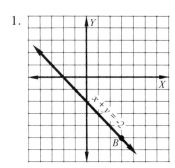
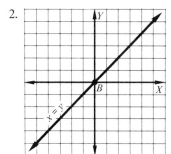

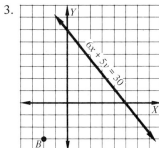
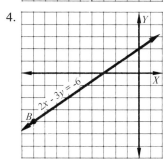

PART 3

The equations we were called upon to graph in Part 1 of this section were designed so that it was relatively easy to guess at pairs of numbers that satisfy these equations. When the coefficients of the variables are other than 1, the difficulty of guessing increases many times. In order to avoid the need for guessing we treat the equation as a literal equation and solve it for one of its variables.

As an illustration, consider the equation

$$3x + y = 15. \qquad (1)$$

This equation is solved for y by adding a $-3x$ to both of its members. By doing this, the equation above becomes

$$y = 15 - 3x. \qquad (2)$$

It now becomes a simple matter to replace x with any number we desire and thus find the corresponding value of y. That pair of numbers will be an element in the solution set of the original equation $3x + y = 15$.

Specifically, let us replace x in equation (2) with the number 4; equation (2) then becomes

$$y = 15 - 3 \cdot 4 \qquad \text{where } x = 4$$
$$y = 15 - 12$$
$$y = 3.$$

The pair of numbers (4, 3) is an element in the solution set of (1). This must be so since equation (2) is but another form of equation (1).

$$3x + y = 15 \qquad (1)$$
$$\text{and} \quad y = 15 - 3x \qquad (2)$$

These two equations are called **equivalent equations** for they have the *same* solution set. Out of curiosity, we will test to determine whether (4, 3) is actually an element in the solution set of both.

(1) $\quad 3x + y = 15 \qquad\qquad$ (2) $\quad y = 15 - 3x$
$\qquad 3 \cdot 4 + 3 \stackrel{?}{=} 15 \qquad\qquad\qquad 3 \stackrel{?}{=} 15 - 3 \cdot 4$
$\qquad 12 + 3 \stackrel{?}{=} 15 \qquad\qquad\qquad\; 3 \stackrel{?}{=} 15 - 12$
$\qquad\quad\; 15 = 15 \qquad\qquad\qquad\qquad\; 3 = 3$

As we had anticipated, not only is (4, 3) an element in the solution set of $y = 15 - 3x$, it is also an element in the solution set of $3x + y = 15$.

Thus far we have found one pair of numbers that is an element in the solution set of the equation

$$y = 15 - 3x.$$

8.2/The Graph of a Linear Equation in Two Variables

By replacing x with numbers other than 4, we can determine as many elements in the solution set as we desire. Actually, to draw the line that is the graph of the equation, we need only two pairs of numbers. These two pairs will name two points through which the line is drawn. Although we need but two points, we usually prefer four. The reason for this becomes apparent when we examine Figures 8-10, 8-11, and 8-12.

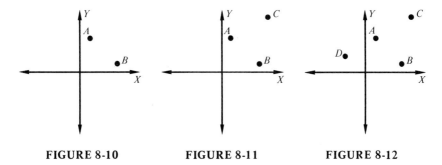

FIGURE 8-10   **FIGURE 8-11**   **FIGURE 8-12**

In Figure 8-10 after plotting points A and B we would not know whether either or neither has been plotted correctly. When we examine Figure 8-11, we definitely know that at least one of the points is incorrect for all three must lie on the same line. However, we do not know whether the incorrect point is A, B, or C. In Figure 8-12, though, we can see that the incorrect point is very likely B, for the remaining three lie on the same line.

When determining these four pairs of numbers we find it best to organize our work in a manner similar to that shown below.

$$3x + y = 15$$

The first step is to solve the equation for y.

$$y = 15 - 3x$$

We then arrange this equation within a framework similar to football goalposts.

$$15 - 3x = y$$

At this point, the heading "x" is placed at the top of the first column.

x	$15 - 3x =$	y

8/The Graph of an Open Sentence in Two Variables

Now, we simply list the replacements for x in the first column and find the corresponding values for y in the third column. In the situation below we replaced x with 2 and found the corresponding value of y to be 9.

x	$15 - 3x =$	y
2	$15 - 3 \cdot 2 =$	9

The same process is then repeated with three other replacements for x. This is shown below.

x	$15 - 3x =$	y	(x, y)
2	$15 - 3 \cdot 2 =$	9	(2, 9)
4	$15 - 3 \cdot 4 =$	3	(4, 3)
6	$15 - 3 \cdot 6 =$	-3	(6, -3)
7	$15 - 3 \cdot 7 =$	-6	(7, -6)

The pairs of numbers (2, 9), (4, 3), (6, -3), and (7, -6) are four elements in the solution set of the equation

$$3x + y = 15.$$

Using these four ordered pairs of numbers we graph this equation in the same manner as we graphed those in Part 1.

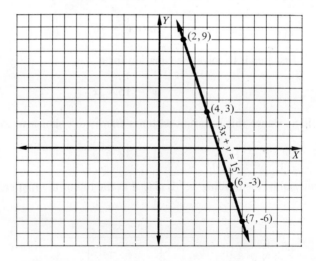

It is not necessary to name the four points as is done in the figure above. However, the equation is *always* written along the graph.

8.2/The Graph of a Linear Equation in Two Variables

Example 1

Draw the graph of the following equation.

$$y - 2x = 3$$

Explanation The initial step is to solve the equation for y. We then draw the "goalposts" and replace x with four different numbers.

Solution

$$y - 2x = 3$$
$$y = 3 + 2x$$

x	$3 + 2x$	=	y	(x, y)
-2	$3 + 2(-2) =$		-1	$(-2, -1)$
0	$3 + 2(0) =$		3	$(0, 3)$
2	$3 + 2(2) =$		7	$(2, 7)$
4	$3 + 2(4) =$		11	$(4, 11)$

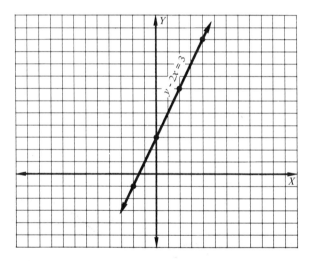

Explanation continued When selecting replacements for x try to avoid points that are "bunched" together. If at all possible use some negative numbers as replacements for x as well as positive numbers.

Example 2

Solve the following equation for y.

$$5x - y = -4$$

Solution

$$5x - y = -4 \quad (1)$$
$$5x = -4 + y \quad (2)$$
$$5x + 4 = y \quad (3)$$

Explanation Notice that equation (3) was solved for y by isolating y on the *right* side of the equation rather than the *left*.

Example 3

Draw the graph of the equation below.

$$4x - y = 1$$

Solution

$$4x - y = 1$$
$$4x = 1 + y$$
$$4x - 1 = y$$

x	$4x - 1$	$=$	y	(x, y)
-1	$4(-1) - 1 =$		-5	$(-1, -5)$
0	$4(0) - 1 =$		-1	$(0, -1)$
2	$4(2) - 1 =$		7	$(2, 7)$
3	$4(3) - 1 =$		11	$(3, 11)$

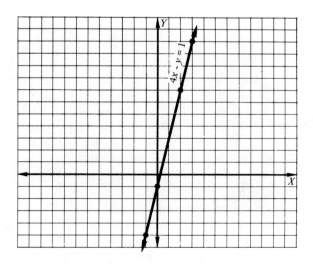

8.2/The Graph of a Linear Equation in Two Variables

EXERCISES 8.2 (Part 3)

A

Solve each of the following equations for y.

1. $y + 3x = 5$
2. $y + 5 = 6x$
3. $y - 7x = 8$
4. $6x + y = 3$
5. $9 + y = 5x$
6. $2x + y + 3 = 0$
7. $4x - y = 5$
8. $2x - y = -4$
9. $3 - y = 8x$
10. $x - y = 7$
11. $5 - x - y = 0$
12. $x - y - 3 = 0$

B

Graph each of the following equations.

1. $y = 2x + 5$
2. $y = 1 - 3x$
3. $y + 3x = 2$
4. $4x + y = 4$
5. $y - 3x = 1$
6. $2x - y = 1$
7. $3x - y = 3$
8. $y + 4 = 2x$
9. $y - 1 = 5x$
10. $2x = 2 - y$

PART 4

In Part 3 it was suggested that the first step in finding elements in the solution set of a linear equation in two variables is to solve the equation for y. However, should we follow this suggestion with the equation below, we create certain difficulties.

$$x + 3y = 5 \qquad (1)$$

Let us see what does occur when this equation is solved for y.

$$x + 3y = 5 \qquad (1)$$
$$3y = 5 - x \qquad (2)$$
$$y = \frac{1}{3}(5 - x) \qquad (3)$$

This is the first time we encountered fractions in the process of solving an equation for y. In order to avoid fractions, the exercises earlier in our work were designed so that the coefficient of y was either +1 or -1.

In finding pairs of values that satisfy an equation it does not matter whether that equation is solved for y or for x. Hence, to avoid fractions we will solve the equation for whichever variable has a coefficient of either +1 or -1. In the equation

$$x + 3y = 5$$

the x variable has a coefficient of 1; therefore this equation should be solved for x.

Example 1

Graph the following equation.

$$x + 2y = 5$$

Explanation Since the x variable has a coefficient of 1, the above equation is solved for x. This, though, will mean that we must interchange the positions of x and y when arranging the "goalposts."

Solution

$$x + 2y = 5$$
$$x = 5 - 2y$$

New "y" column	$5 - 2y =$	New "x" column	(x, y)
-1	$5 - 2(-1) =$	7	$(7, -1)$
0	$5 - 2(0) =$	5	$(5, 0)$
1	$5 - 2(1) =$	3	$(3, 1)$
4	$5 - 2(4) =$	-3	$(-3, 4)$

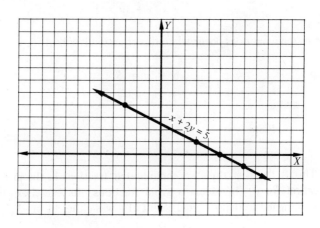

Example 2

Graph the following equation.

$$3y - x = 1$$

Solution

$$3y - x = 1$$
$$3y = 1 + x$$
$$3y - 1 = x$$

y	3y - 1 =	x	(x, y)
-2	3(-2) - 1 =	-7	(-7, -2)
0	3(0) - 1 =	-1	(-1, 0)
2	3(2) - 1 =	5	(5, 2)
3	3(3) - 1 =	8	(8, 3)

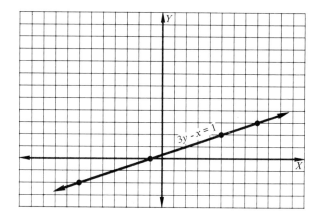

EXERCISES 8.2 (Part 4)

A

Solve each of the following equations for x.

1. $x + 4y = 7$
2. $x - 6y = 3$
3. $5y + x = 4$
4. $x + 8 = -y$
5. $9 + x = -3y$
6. $x + 2y - 3 = 0$
7. $3y - x = 5$
8. $6 - x = 2y$
9. $y - x = 9$
10. $4 - x = -3y$
11. $6 - x + 2y = 0$
12. $3y - x - 5 = 0$

B

Graph each of the following equations.

1. $x = 2y + 3$
2. $x = 3y - 4$
3. $x = 5 - 2y$
4. $x + 4y = 1$
5. $3y + x = 2$
6. $2y - x = 6$
7. $5y - x = -2$
8. $4 - x = 3y$
9. $7 - x = 2y$
10. $5 - x - 3y = 0$

PART 5

We now face the question of how to graph a linear equation if both variables have coefficients other than +1 and -1. Should this occur, we usually find it best to solve the equation for the variable that has the "smaller" coefficient—smaller, of course, in terms of its absolute value. In the following examples some suggestions are made to simplify the computations.

Example 1

Solve the following equation for x.

$$2x - 3y = 6$$

Solution

$$2x - 3y = 6 \quad (1)$$
$$2x = 6 + 3y \quad (2)$$
$$x = \frac{1}{2}(6 + 3y) \quad (3)$$
$$x = 3 + \frac{3y}{2} \quad (4)$$

Explanation For our purpose it would be best not to leave the value of x in the form in which it appears in (3) but actually multiply $\frac{1}{2}$ by $6 + 3y$. Doing this will make the arithmetic computation somewhat easier.

Example 2

Solve the following equation for the variable whose coefficient has the smaller absolute value.

$$5x - 3y = 12$$

Explanation Since the absolute value of -3 is smaller than the absolute value of 5, the equation is solved for the y variable. Since the coefficient of y is negative, it would be best to isolate the y term on the right side of the equation.

8.2/The Graph of a Linear Equation in Two Variables

Solution

$$5x - 3y = 12$$
$$5x = 12 + 3y$$
$$5x - 12 = 3y$$
$$\frac{1}{3}(5x - 12) = y$$
$$\frac{5x}{3} - 4 = y$$

Example 3

Find the corresponding value of y for each of the values of x that are shown below.

x	$\frac{3x}{4} - 5 =$	y
-12		
-4		
0		
8		
16		

Explanation When x is replaced with -12 in

$$\frac{3x}{4} - 5,$$

this expression becomes

$$\frac{3(-12)}{4} - 5.$$

Perhaps the easiest way to find the value of the fraction

$$\frac{3(-12)}{4}$$

is to divide 4 into -12 and then multiply that answer by 3. Thus,

$$\frac{3(\cancel{-12})^{-3}}{\cancel{4}} - 5$$
$$= -9 - 5$$
$$= -14$$

Solution

x	$\dfrac{3x}{4} - 5 =$	y
-12	$\dfrac{3(-12)}{4} - 5 =$	-14
-4	$\dfrac{3(-4)}{4} - 5 =$	-8
0	$\dfrac{3(0)}{4} - 5 =$	-5
8	$\dfrac{3(8)}{4} - 5 =$	1
16	$\dfrac{3(16)}{4} - 5 =$	7

Example 4

Graph the following equation.

$$3x - 4y = 6$$

Explanation Since the absolute value of 3 is less than the absolute value of -4, the equation is solved for x.

Solution

$$3x - 4y = 6$$
$$3x = 6 + 4y$$
$$x = \tfrac{1}{3}(6 + 4y)$$
$$x = 2 + \tfrac{4y}{3}$$

y	$2 + \dfrac{4y}{3} =$	x	(x, y)
-3	$2 + \dfrac{4(-3)}{3} =$	-2	(-2, -3)
0	$2 + \dfrac{4(0)}{3} =$	2	(2, 0)
3	$2 + \dfrac{4(3)}{3} =$	6	(6, 3)
6	$2 + \dfrac{4(6)}{3} =$	10	(10, 6)

8.2/The Graph of a Linear Equation in Two Variables

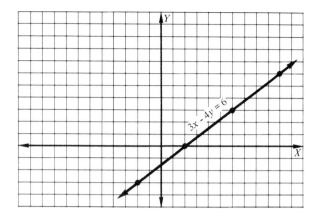

Explanation continued Notice that the replacements used for y are either 0 or positive and negative multiples of 3. This is done deliberately so that the value of the fraction $4y/3$ will turn out to be an integer. To make the computation as simple as possible, the replacements for the variable in the numerator should be multiples of the number in the denominator. Had the 3 in the denominator been a 5, then the replacements for y would have been positive and negative multiples of 5. Some of these numbers are: $-15, -10, -5, 5, 10, 15, 20, \ldots$.

Example 5

Graph the following equation.

$$5x - 4y = 8$$

Solution

$$5x - 4y = 8$$
$$5x = 8 + 4y$$
$$5x - 8 = 4y$$
$$\frac{1}{4}(5x - 8) = y$$
$$\frac{5x}{4} - 2 = y$$

x	$\frac{5x}{4} - 2 =$	y	(x,y)
-4	$\frac{5(-4)}{4} - 2 =$	-7	$(-4, -7)$
0	$\frac{5(0)}{4} - 2 =$	-2	$(0, -2)$
4	$\frac{5(4)}{4} - 2 =$	3	$(4, 3)$
8	$\frac{5(8)}{4} - 2 =$	8	$(8, 8)$

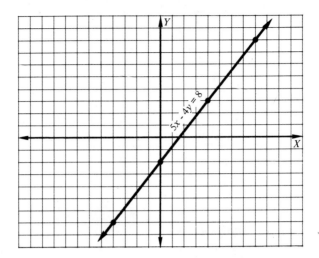

Explanation The replacements for x are multiples of 4 because the denominator of the fraction $5x/4$ is 4.

EXERCISES 8.2 (Part 5)

A

Solve each of the following equations for y.

1. $2y + 3x = 6$
2. $5x + 4y = 12$
3. $5x - 2y = 6$
4. $6x - 5y = 10$
5. $3x - 2y - 8 = 0$
6. $3y - 9 = 5x$

B

Solve each of the following equations for x.

1. $2x + 5y = 8$
2. $7y + 2x = 10$
3. $3y - 2x = 6$
4. $4y = 3x - 9$
5. $6 = 2x - 5y$
6. $7y - 2x - 8 = 0$

C

Solve each of the following equations for the variable whose coefficient has the smaller absolute value.

1. $5x + 2y = 20$
2. $4y + 3x = 12$
3. $-3x + 24 = 4y$
4. $7x = 2y - 14$

D

In each of these exercises, find the corresponding value of y for each of the values of x that are shown.

1.
x	$\frac{3x}{2} + 1 =$	y	(x,y)
-4			
-2			
0			
6			

2.
x	$\frac{5x}{3} + 4 =$	y	(x,y)
-9			
-3			
0			
6			

3.
x	$5 - \frac{2x}{3} =$	y	(x,y)
-6			
-3			
0			
12			

4.
x	$7 - \frac{6x}{5} =$	y	(x,y)
-15			
-5			
5			
10			

E

In each of the exercises, find the corresponding value of x for each of the values of y that are shown below.

1.
y	$\frac{4y}{3} + 1 =$	x	(x,y)
9			
6			
0			
-3			

2.
y	$\frac{5y}{2} - 3 =$	x	(x,y)
10			
6			
0			
-8			

3.
y	$8 - \frac{7y}{5} =$	x	(x,y)
15			
10			
-5			
-20			

4.
y	$-6 - \frac{5y}{4} =$	x	(x,y)
16			
4			
-8			
-12			

F

Graph each of the following equations.

1. $2y + 3x = 6$
2. $2y - 3x = 4$
3. $3x + 4y = 9$
4. $3x - 5y = 6$

5. $2y = 4 - 3x$
6. $2x + 3y - 8 = 0$
7. $5x - 2y = 10$
8. $3x = 5y - 9$
9. $4x - 3y + 6 = 0$
10. $8 - 2x = 3y$
11. $5y + 4x - 8 = 0$
12. $7x = 9 - 3y$

8.3 THE SLOPE OF A LINE

PART 1

We have often heard people refer to some hills as having a "gradual slope" while others are referred to as having a "steep slope." Usually, what is meant by the first of these terms is that the hill rises very slowly over a great distance (Figure 8-13).

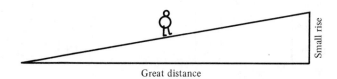

FIGURE 8-13

On the other hand, a hill with a "steep slope" would be one in which there is a great vertical change over a very small horizontal change (Figure 8-14).

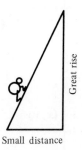

FIGURE 8-14

The mathematician uses this very same interpretation of the term "slope" when he refers to the "slope of a line." To him, slope of a line means the extent to which the vertical change between two points on the line compares with the horizontal change betweeen these same two points. Let us examine the meaning of this in terms of the line in Figure 8-15. In this situation we are interested in determining both the vertical change or "rise" from point A to point B and also the horizontal change or "run" that takes place in going from point A to point B.

8.3/The Slope of a Line

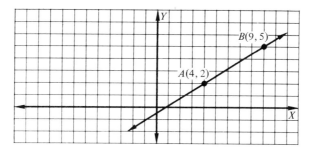

FIGURE 8-15

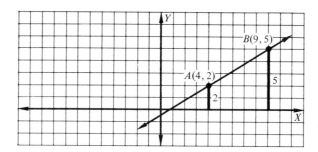

FIGURE 8-16

To determine the *rise* we make use of the fact that the ordinate of a point tells us the distance from that point to the horizontal axis. Hence, the distance from point A to the horizontal axis is 2 units (Figure 8-16). Similarly, since the ordinate of B is 5, then the distance from point B to the horizontal axis is 5 units. In view of this, a person going from point A to point B will move *vertically* 5 − 2 or 3 units. Another way of expressing this is to say that *the rise between these two points is 3 units.*

In general,

> The rise between two points is found by subtracting the ordinate of the first point from the ordinate of the second point.

Now let us examine what we will have to do to determine the *run* or horizontal change between the points A and B (Figure 8-17). To do this, we make use of the fact that the abscissa of a point tells us the distance from that point to the vertical axis. Hence, the distance from point A to the vertical axis is 4 units. Similarly, the abscissa of point B is 9; therefore the distance from point B to the vertical axis is 9 units. In view of this, a person going from

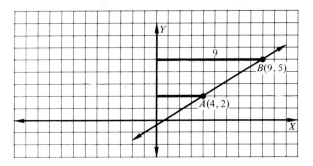

FIGURE 8-17

point A to point B will move *horizontally* 9 - 4 or 5 units. Another way of expressing this is to say that *the run between these two points is 5 units.*

In general,

The run between two points is found by subtracting the abscissa of the first point from the abscissa of the second point.

Figure 8-18 gathers all the information we have just found into a single diagram. Here we can see that between the two points, A and B, there is a vertical

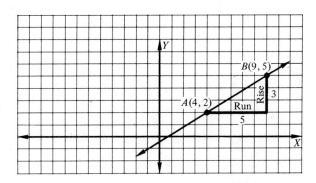

FIGURE 8-18

change of 3 units for a horizontal change of 5 units. This comparison is called the *slope of this line* and is expressed as follows:

$$\text{Slope} = \frac{\text{Rise}}{\text{Run}} = \frac{\text{Change in ordinates}}{\text{Change in abscissas}} = \frac{5-2}{9-4} = \frac{3}{5}.$$

8.3/The Slope of a Line

Example 1
What is the slope of the line such that two of its points are (-2, 3) and (5, -7)?

Explanation To determine the slope it is necessary to find both the rise and the run. Knowing these values, we then divide the rise by the run.

Solution
$$\text{Rise} = -7 - (3) = -7 - 3 = -10$$
$$\text{Run} = 5 - (-2) = 5 + 2 = 7$$
$$\text{Slope} = \frac{\text{Rise}}{\text{Run}} = \frac{-10}{7}$$

Notice in Example 1 that the rise turned out to be a negative quantity. This implies that we will be walking "downhill" when going from the first point to the second. When we plot the points (-2, 3) and (5, -7) we discover that this is so (Figure 8-19).

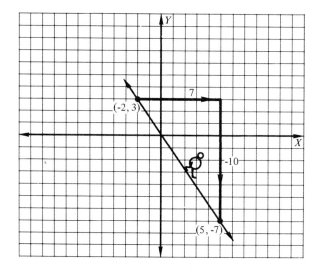

FIGURE 8-19

It is very important that we realize that when finding the slope we must *always* travel from *left to right*. This implies that the run must always be a positive number. In view of this, before finding the slope make certain that the abscissa of the first point is *smaller* than the abscissa of the second.

Example 2
Find the slope of the line whose equation is
$$3x + 2y = 8.$$

Explanation Before we can find the slope of the line, we must determine the coordinates of two points of that line. This is done in the same manner as we would when determining the coordinates of points for the graph of an equation. After coordinates of two points are determined, the slope is found in the same manner as in Example 1.

Solution

$$3x + 2y = 8$$

$$2y = 8 - 3x$$

$$y = \frac{1}{2}(8 - 3x)$$

$$y = 4 - \frac{3x}{2}$$

x	$4 - \frac{3x}{2} =$	y	(x, y)
-2	$4 - \frac{3(-2)}{2} =$	7	$(-2, 7)$
4	$4 - \frac{3(4)}{2} =$	-2	$(4, -2)$

$$\text{Rise} = -2 - (7) = -2 - 7 = -9$$

$$\text{Run} = 4 - (-2) = 4 + 2 = 6$$

$$\text{Slope} = \frac{\text{Rise}}{\text{Run}} = \frac{-9}{6} = \frac{-3}{2}$$

When examining the solution in Example 2 you may have wondered as to what might the slope have turned out to be had two points been found other than the two that were. Actually, it will always be true that

---

The slope found by using any two points of a line will always be the same as the slope found when using any other two points of that line.

Example 3

Find the slope of the line whose equation is

$$7x - 4y = 12$$

8.3/The Slope of a Line

Solution

$$7x - 4y = 12$$
$$7x = 4y - 12$$
$$7x + 12 = 4y$$
$$\frac{1}{4}(7x + 12) = y$$
$$\frac{7x}{4} + 3 = y$$

x	$\frac{7x}{4} + 3$	$=$	y	(x, y)
-8	$\frac{7(-8)}{4} + 3 =$		-11	$(-8, -11)$
4	$\frac{7(4)}{4} + 3 =$		10	$(4, 10)$

Rise $= 10 - (-11) = 10 + 11 = 21$

Run $= 4 - (-8) = 4 + 8 = 12$

Slope $= \frac{\text{Rise}}{\text{Run}} = \frac{21}{12} = \frac{7}{4}$

EXERCISES 8.3 (Part 1)

A

In each of the following exercises the coordinates of two of the points of a line are given. Find the slope of each of these lines.

1. $(2, 6)$ and $(6, 11)$
2. $(3, 7)$ and $(12, 9)$
3. $(-2, 4)$ and $(1, 7)$
4. $(-5, 3)$ and $(0, 9)$
5. $(1, -3)$ and $(5, 6)$
6. $(-3, -5)$ and $(0, 8)$
7. $(0, -4)$ and $(6, -3)$
8. $(-2, 7)$ and $(2, 0)$
9. $(-3, 1)$ and $(1, -7)$
10. $(-7, 3)$ and $(-2, 6)$
11. $(-10, -5)$ and $(-4, -1)$
12. $(0, 0)$ and $(4, 7)$
13. $(-4, 3)$ and $(0, 0)$
14. $(-6, 5)$ and $(6, -2)$
15. $(-1, -2)$ and $(1, 9)$
16. $(0, -3)$ and $(5, 3)$

B

Determine the slope of the line in each of the exercises below.

1.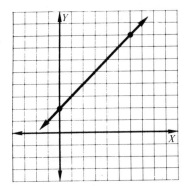

2.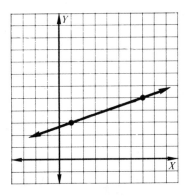

3.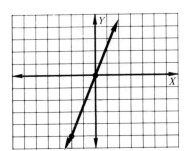

4.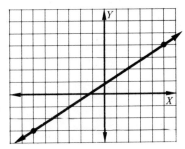

5.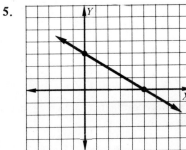

6.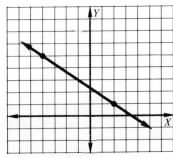

7.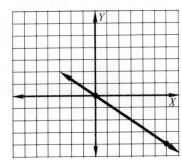

8.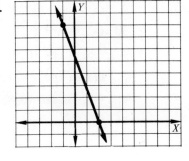

8.3/The Slope of a Line

9.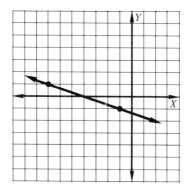

10.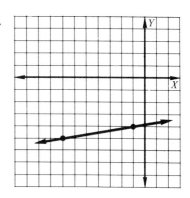

C

Determine the slope of each of the lines whose equations are given below.

1. $y = 3x + 1$
2. $y = 5 - 2x$
3. $x = 4y + 3$
4. $x = 5 - 3y$
5. $2y + x = 7$
6. $3x + y = 4$
7. $5x - y = 3$
8. $7y - x = 5$
9. $2x + 3y = 6$
10. $3x - 4y = 9$
11. $5x + 2y = 8$
12. $3x - 2y = 10$
13. $5y - 3x = 12$
14. $9y - 2x = 2$
15. $7x - 5y - 15 = 0$
16. $8y - 7x - 21 = 0$

PART 2

You may have noticed something rather interesting that occurred in Example 2 of Part 1 of this section. In that example we were asked to find the slope of the line whose equation is

$$3x + 2y = 8.$$

When the equation was solved for y, it turned out to be

$$y = 4 - \frac{3}{2}x.$$

And when we found the slope of this line, that turned out to be

$$\text{Slope} = \frac{-3}{2}.$$

Similarly, in Example 3 of Part 1 we were asked to find the slope of the line whose equation is

$$7x - 4y = 12.$$

When this equation was solved for y, it turned out to be

$$y = \frac{7}{4}x + 3,$$

while the slope of this line was

$$\text{Slope} = \frac{7}{4}.$$

In both of these situations the slope of each line was the very same number as the coefficient of x when the equation is solved for y. This does *not* happen to be a coincidence. In general it is true that

---

The slope of a line is the coefficient of x when the equation is solved for the y variable.

This discovery gives us a very short way for finding the slope of a line if we know the equation of that line. Rather than find the coordinates of two points of that line as we had in the past, we need simply solve the equation for y and then examine the coefficient of x. That number is the slope of the line.

Example 1

Find the slope of the line whose equation is

$$5x - 4y = 8.$$

Solution

$$5x - 4y = 8$$
$$5x = 8 + 4y$$
$$5x - 8 = 4y$$
$$\frac{1}{4}(5x - 8) = y$$
$$\frac{5}{4}x - 2 = y \quad \text{(Equation solved for } y\text{)}$$
$$\text{Slope} = \frac{5}{4} \quad \text{(Coefficient of } x\text{)}$$

8.3/The Slope of a Line

Knowing this short way for finding the slope of a line, we will be able to graph a linear equation far more rapidly than we had previously. Consider the equation $5x - 4y = 8$ in Example 1. We discovered that the slope of this line is

$$\frac{5}{4}.$$

This means that the rise from one point to another is 5 when the run is 4. Hence, if we know the position of any point on the graph, a second point can be found immediately by moving from that point 4 units horizontally and 5 units vertically.

The coordinates of a point can easily be found in the usual manner:

x	$\frac{5}{4}x - 2 =$	y	(x, y)
0	$\frac{5}{4} \cdot 0 - 2 =$	-2	$(0, -2)$

Zero is used as the replacement for x in order to make the computation as simple as possible. The point $(0, 2)$ is now plotted. Using the slope $\frac{5}{4}$, we find a second point as is shown in Figure 8-20. The line through these two points is the graph of the equation $5x - 4y = 8$.

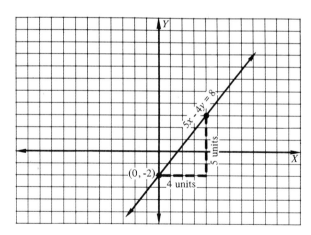

FIGURE 8-20

The method used in Example 1 for graphing a linear equation by using the slope of a line and a point of that line is called the **point-slope** method.

Example 2

Graph the following equation by the point-slope method.

$$2x + 3y = 15$$

Solution

$$2x + 3y = 15$$
$$3y = 15 - 2x$$
$$y = \frac{1}{3}(15 - 2x)$$
$$y = 5 - \frac{2}{3}x$$
$$\text{Slope} = \frac{-2}{3}$$

x	$5 - \frac{2}{3}x$	$=$	y	(x, y)
0	$5 - \frac{2}{3} \cdot 0$	$=$	5	$(0, 5)$

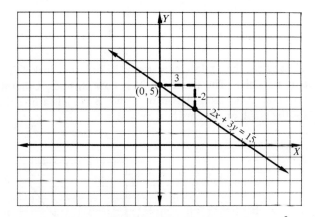

Explanation Notice that when the coefficient of x, the $-\frac{2}{3}$, is written for the slope, the negative sign is attached to the numerator. This is done simply to remind us that we are going "downhill" as we move from left to right.

Example 3

Graph the following two equations on the same set of axes. Use the point-slope method.

(1) $3x - 2y = 10$ (2) $2y - 3x = 6$

Solution

(1) $\quad 3x - 2y = 10$
$\quad\quad\quad 3x = 10 + 2y$
$\quad\quad 3x - 10 = 2y$
$\quad\quad \frac{3}{2}x - 5 = y$
$\quad\quad \text{Slope} = \frac{3}{2}$

(2) $\quad 2y - 3x = 6$
$\quad\quad\quad 2y = 6 + 3x$
$\quad\quad\quad y = 3 + \frac{3}{2}x$
$\quad\quad \text{Slope} = \frac{3}{2}$

8.3 / The Slope of a Line

x	$\frac{3}{2}x - 5$	=	y	(x, y)
0	$\frac{3}{2} \cdot 0 - 5$	=	-5	$(0, -5)$

x	$3 + \frac{3}{2}x$	=	y	(x, y)
0	$3 + \frac{3}{2} \cdot 0$	=	3	$(0, 3)$

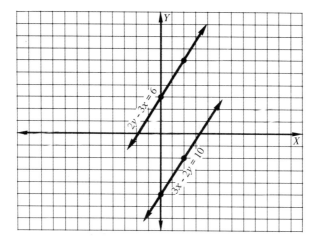

When we examine the graphs above, we notice something rather interesting—the two lines are parallel; that is, they do not intersect. Perhaps we should have suspected this when we saw that the "steepness" of the lines are the same. This, in turn, suggests that the lines will have to be parallel.

In general we can say that

If two different lines have the same slope, then the lines are parallel. It is also true that if they do *not* have the same slopes, then they will *not* be parallel.

Example 4

Are the graphs of the following two equations parallel lines? Justify your answer.

$$2y + 3x = 7 \quad \text{and} \quad 6x = 12 - 4y$$

Solution

$2y + 3x = 7$

$2y = 7 - 3x$

$y = \frac{7}{2} - \frac{3}{2}x$

Slope = $\frac{-3}{2}$

$6x = 15 - 4y$

$6x + 4y = 15$

$4y = 15 - 6x$

$y = \frac{15}{4} - \frac{6}{4}x$

Slope = $\frac{-6}{4} = \frac{-3}{2}$

Since the slopes are the same, the lines are parallel.

EXERCISES 8.3 (Part 2)

A

In each of the following exercises the equation of a line is given. Find the slope of each of these lines.

1. $y = \frac{2}{3}x + 7$
2. $y = \frac{-3}{5}x + 4$
3. $y = \frac{4}{1}x - 9$
4. $y = 5x - 6$
5. $y = 8 + \frac{6}{7}x$
6. $y = 6 - \frac{3}{4}x$
7. $y = 9 + 8x$
8. $y = 6 - 4x$
9. $y = 15 + x$
10. $y = 11 - x$
11. $3x + y = 5$
12. $5x - y = 7$
13. $7x + 2y = 6$
14. $2x + 7y = 14$
15. $3x - 4y = 8$
16. $8x + 6y = 5$
17. $2x = 3y + 9$
18. $4y - 7 = 2x$
19. $5y + 10 = x$
20. $x - 3y - 4 = 0$

B

In each of the following exercises the slope of a line and the coordinates of a point on the line are given. Graph the line in each exercise.

1. $(2, 3)$; slope $= \frac{2}{5}$
2. $(0, 2)$; slope $= \frac{3}{2}$
3. $(-3, 5)$; slope $= \frac{-2}{3}$
4. $(0, 0)$; slope $= \frac{4}{1}$
5. $(6, 0)$; slope $= 2$
6. $(-3, 0)$; slope $= -3$
7. $(0, -2)$; slope $= \frac{4}{5}$
8. $(0, 4)$; slope $= \frac{1}{3}$
9. $(-3, -5)$; slope $= \frac{2}{7}$
10. $(4, -3)$; slope $= \frac{-1}{2}$
11. $(3, 8)$; slope $= -\frac{5}{3}$
12. $(2, 6)$; slope $= \frac{0}{1}$

C

Graph the line of each of the following equations by using the point-slope method.

1. $y - 2x = 6$
2. $y + 3x = 5$
3. $5x + y = 4$
4. $4x - y = 2$
5. $2x + 3y = 12$
6. $3x - 5y = 20$
7. $5x - 2y = 14$
8. $1x + 3y = 9$
9. $x + 4y = 16$
10. $3x + 6y - 24 = 0$

D

In each of the following exercises the equations of two lines are given. You are to state whether the two lines are "parallel" or "not parallel" and justify your answer.

1. $y = 3x + 5$ and $y - 3x = 7$
2. $2x + y = 6$ and $2x = 4 - y$
3. $3x + 2y = 8$ and $17 - 6x = 4y$
4. $8x - 4y = 5$ and $10x = 5y - 7$
5. $3x - 4y = 6$ and $5x + 2y = 3$
6. $8x - 6y - 11 = 0$ and $15y - 20x + 17 = 0$
7. $9x + 4 = 6y$ and $18x + 12y - 5 = 0$

8.4 THE GRAPH OF A LINEAR EQUATION IN TWO VARIABLES WHERE ABSOLUTE VALUES ARE INVOLVED (OPTIONAL)

Our objective now is to determine a method for graphing an equation involving absolute values such as the one below.

$$|x + 2| = y - 4$$

If somehow we can express this equation in terms of linear equations that involve no absolute quantities, our problem will be solved for we know how to graph linear equations.

The approach, actually, is to show that the equation

$$|x + 2| = y - 4$$

can be replaced with two linear equations in two variables. Before doing this, though, it might be well to review the method by which we solved a linear equation in one variable where absolute values are involved.

8/The Graph of an Open Sentence in Two Variables

Consider the equation

$$|z| = 6.$$

If you recall, this equation implies that the numbers named by z are 6 units from the origin on the number line. Hence, the numbers named by z are either +6 or its *additive inverse* -6. Specifically,

z will name +6 when z is *greater* than 0

and

z will name the *additive inverse* of +6 when z is *less* than 0.

We can express this with symbols as follows:

1. $z = +6$ when $z > 0$

and 2. $z = -6$ when $z < 0$

In the same way, if

$$|x + 2| = y - 4,$$

then $x + 2$ also names two numbers:

1. The first of these is $+(y - 4)$ which occurs when $x + 2$ is *greater* than zero.
2. The second of these is the **additive inverse** of $+(y - 4)$ which occurs when $x + 2$ is *less* than zero.

In view of this, the equation

$$|x + 2| = y - 4$$

can be replaced by two linear equations:

(1) $x + 2 = +(y - 4)$ when $x + 2 > 0$
(2) $x + 2 = -(y - 4)$ when $x + 2 < 0$

At this point we examine each situation separately.

(2)		(1)	
$x + 2 = -(y - 4)$ when $x + 2 < 0$		$x + 2 = +(y - 4)$ when $x + 2 > 0$	
$x + 2 = -y + 4$ when $x < -2$		$x + 2 = y - 4$ when $x > -2$	
$x + 2 + y = 4$ when $x < -2$		$x + 2 + 4 = y$ when $x > -2$	
$y = 4 + (-2) + (-x)$ when $x < -2$		$x + 6 = y$ when $x > -2$	
$y = 2 - x$ when $x < -2$			

8.4/The Graph of a Linear Equation in Two Variables

Thus, what we have shown above is that the graph of the equation

$$|x + 2| = y - 4$$

consists of two sections. The first section is that part of the graph of the equation

$$x + 6 = y \qquad \text{where} \quad x > -2. \tag{1}$$

The second section is that part of the graph of the equation

$$y = 2 - x \qquad \text{where} \quad x < -2. \tag{2}$$

But what happens when $x = -2$? To find the corresponding value of y, we replace x with -2 in the original equation.

$$|x + 2| = y - 4$$
$$|-2 + 2| = y - 4$$
$$0 = y - 4$$
$$4 = y$$

Had we replaced x with -2 in either equation (1) or equation (2) the value of y would have again turned out to be 4. Hence, any one of the three equations can be used to find the value of y when $x = -2$.

Since we already have one point of the graph, the point $(-2, 4)$, we will use equation (1) to find a second point for that part of the graph where $x > -2$ and equation (2) to find a second point for that part of the graph where $x < -2$.

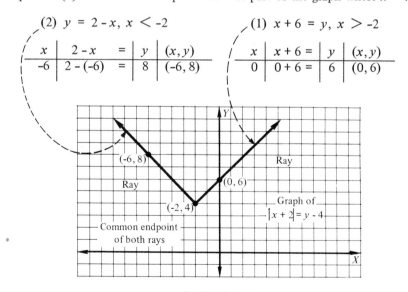

FIGURE 8-21

368 8/The Graph of an Open Sentence in Two Variables

In the plotting of earlier graphs we found four points although we knew that *only two are necessary*. At this time to shorten our work we are going to find but two points of each *ray*. One of these two points will be the common endpoint of both rays (Figure 8-21).

Example 1

Graph the following equation:

$$|2x - 8| - 3 = y.$$

Explanation The first step is to solve the equation for the absolute value of $2x - 8$. The work is then completed as in the explanation above.

Solution
$$|2x - 8| - 3 = y.$$
$$|2x - 8| = y + 3$$

(1)

$2x - 8 = +(y + 3)$	when $2x - 8 > 0$
$2x - 8 = y + 3$	when $2x > 8$
$2x - 8 - 3 = y$	when $x > 4$
$2x - 11 = y$	when $x > 4$

(2)

$2x - 8 = -(y + 3)$	when $2x - 8 < 0$
$2x - 8 = -y - 3$	when $2x < 8$
$2x - 8 + y = -3$	when $x < 4$
$y = -3 + 8 - 2x$	when $x < 4$
$y = 5 - 2x$	when $x < 4$

x	$2x - 11 =$	y	(x, y)
6	$2 \cdot 6 - 11 =$	1	$(6, 1)$

x	$5 - 2x =$	y	(x, y)
0	$5 - 2 \cdot 0 =$	5	$(0, 5)$

For $x = 4$, $|2 \cdot 4 - 8| - 3 = y$ (Original Equation)
$$0 - 3 = y$$
$$-3 = y$$

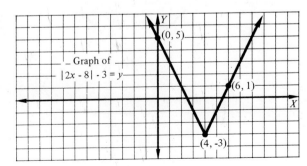

Graph of $|2x - 8| - 3 = y$

Example 2

Graph the following equation:

$$|y| - x = 5.$$

Explanation The first step is to solve the equation for the absolute value of y. The work is then completed as in Example 1, however now the variable between the absolute bars is y rather than x.

Solution

$$|y| - x = 5$$
$$|y| = 5 + x$$

(1)

$y = +(5 + x)$	when $y > 0$
$y = 5 + x$	when $y > 0$
$y - 5 = x$	when $y > 0$

(2)

$y = -(5 + x)$	when $y < 0$
$y = -5 - x$	when $y < 0$
$x = -5 - y$	when $y < 0$

y	$y - 5 =$	x	(x, y)
2	$2 - 5 =$	-3	$(-3, 2)$

y	$-5 - y$	$=$	x	(x, y)
-1	$-5 - (-1)$	$=$	-4	$(-4, -1)$

For $y = 0$, $|0| - x = 5$ (Original Equation)

$$-x = 5$$
$$-5 = x$$

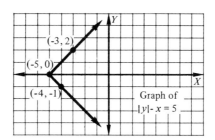

Graph of $|y| - x = 5$

Explanation continued Notice that equation (1) represents the original equation $|y| - x = 5$, only when replacements for y are numbers greater than 0. Hence, since we will be making a replacement for y and then finding the corresponding value of x, equation (1) is solved for x. A similar situation is true in equation (2) for that equation represents the equation $|y| - x = 5$ only when replacements for y are numbers less than 0.

EXERCISES 8.4

Graph each of the following equations.

1. $|x + 2| = y$
2. $|2x - 4| = y$
3. $|3x + 3| = y - 2$
4. $|x - 5| + y = 3$
5. $|y + 6| = |x - 3| + 4$
6. $|2x + 6| - y = 7$
7. $|x| = y$
8. $|y| = x$
9. $|y| - 6 = x$
10. $|2y - 8| = x + 3$

8.5 THE GRAPHS OF LINEAR EQUATIONS IN ONE VARIABLE

Thus far we have examined the method for determining the graph in the coordinate plane of a linear equation in *two* variables. At this time we want to turn our attention to discovering a method for determining the graph in the coordinate plane of a linear equation in *one* variable. In reality, we learned how to do this at the very outset of this chapter.

You may recall that the linear equation

$$x = 5$$

is actually a direction which states that

The point is 5 units to the right of the vertical axis.

We found that there are many points that are 5 units to the right of the vertical axis. These are the points of the line in Figure 8-22 that is parallel to the vertical axis. Each of the points of this line is 5 units from the vertical axis.

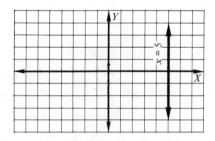

FIGURE 8-22

Similarly, if the linear equation involves the *y* variable only such as

$$y = -3.$$

this, too, represents a direction. In this situation the direction is

The point is 3 units below the horizontal axis.

8.5/The Graphs of Linear Equations in One Variable

Here again there are many points that are named by this direction. These are the points of the line in Figure 8-23 that is parallel to the horizontal axis. Each of the points of this line is 3 units below the horizontal axis.

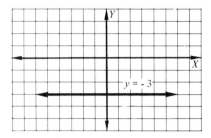

FIGURE 8-23

It would seem then that the graph of a linear equation in *one* variable will be a line that is parallel to either the X-axis or the Y-axis.

The linear equation may appear in a form somewhat more general than the two we have just examined. This is the case in the equation below:

$$3x + 14 = 2.$$

In this event, we need simply solve the equation for the variable to make it resemble the form with which we are familiar. Thus,

$$3x + 14 = 2$$
$$3x = 2 - 14$$
$$3x = -12$$
$$x = -4$$

At this point we recognize the equation $x = -4$ as the direction:

The point is 4 units to the left of the Y-axis.

Therefore the graph of the equation $3x + 14 = 2$ is the set of points in Figure 8-24.

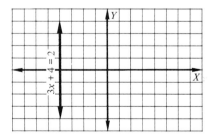

FIGURE 8-24

Example 1

Graph the equation $6x - 1 = 11$.

Solution

$$6x - 1 = 11$$
$$6x = 11 + 1$$
$$6x = 12$$
$$x = 2$$

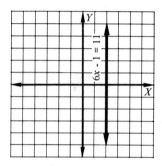

Example 2

Graph the equation $12 - 3y = 2y - 3$.

Solution

$$12 - 3y = 2y - 3$$
$$12 - 3y - 2y = -3$$
$$-3y - 2y = -3 - 12$$
$$-5y = -15$$
$$y = 3$$

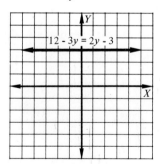

EXERCISES 8.5

Graph each of the following equations.

1. $x = 5$
2. $x = -6$
3. $y = 2$
4. $y = -5$

5. $2x + 1 = 7$
6. $5 - x = 9$
7. $5 + 2y = 11$
8. $8 - 5y = -7$
9. $3x + 4 = 2x - 1$
10. $2(x + 3) + 1 = 9$
11. $7 - (y - 3) = 6$
12. $3(2y - 1) - 5 = 4y - 12$
13. $x = 0$
14. $2y + 5 = 5$

8.6 THE GRAPH OF THE LINEAR INEQUALITY

PART 1

It is just a short step from interpreting the equation

$$x = 4$$

as a direction to interpreting the inequality

$$x > 4$$

as a direction. If $x = 4$ is understood to be

The point is 4 units to the right of the vertical axis

then $x > 4$ will imply

The point is *more* than 4 units to the right of the vertical axis.

In view of this, the points that fulfill this condition will be the points in the shaded region of the coordinate plane shown in Figure 8-25. Each of these points is more than 4 units to the right of the vertical axis.

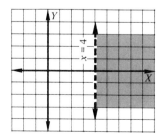

FIGURE 8-25

The set of points in a plane that include all the points on one side of a line is called a **half-plane**. Thus, the graph of the inequality

$$x > 4$$

is the half-plane to the right of the line $x = 4$. Notice that the line $x = 4$ is drawn as a dotted line. This is done to indicate the fact that the points of this line are *not* included in the graph.

A sentence such as

$$x \geqslant 4$$

is read as

x is either greater than 4 or x is equal to 4.

Hence, if we interpret this sentence as a direction, it becomes

---

The set of points that are either more than 4 units to the right of the vertical axis or that are exactly 4 units to the right of the vertical axis.

The graph of the sentence

$$x \geqslant 4$$

will be identical to the graph in Figure 8-26 except that the line $x = 4$ will be a solid line rather than a dotted line.

Example 1

Graph the following sentence:

$$3x - 5 \leqslant 10.$$

Explanation As in Section 8.5, the first step is to solve this sentence for x. The final sentence is then interpreted as a direction.

Solution
$$3x - 5 \leqslant 10.$$
$$3x \leqslant 10 + 5$$
$$3x \leqslant 15$$
$$x \leqslant 5$$

Direction: The set of points that are either less than 5 units to the right of the Y-axis or that are exactly 5 units to the right of the Y-axis.

8.6/The Graph of the Linear Inequality

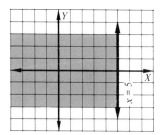

Rather than express the direction as we have in the solution above, we find it somewhat simpler to express the direction for the sentence

$$x \leqslant 5$$

as

The set of points in the half-plane to the left of the line $x = 5$ and also the set of points that include the line $x = 5$.

Example 2

State the directions given by the sentence

$$x \geqslant -3.$$

Solution The set of points in the half-plane to the right of the line $x = -3$ and also the set of points that include the line $x = -3$.

Example 3

State the directions given by the sentence

$$y < 4.$$

Solution The set of points in the half-plane *below* the line $y = 4$.

Example 4

Graph the following sentence.

$$9 - 2y \leqslant 13$$

Solution

$9 - 2y \leqslant 13$	(1)
$-2y \leqslant 13 - 9$	(2)
$-2y \leqslant 4$	(3)
$y \geqslant -2$	(4)

Direction: The set of points in the half-plane *above* the line $y = -2$ and also the set of points that include the line $y = -2$.

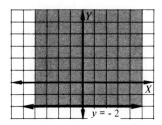

Explanation Notice that in going from step (3) to step (4) both sides of sentence (3) were multiplied by a *negative* $\frac{1}{2}$. Hence it was necessary to change the order of inequality from $<$ to $>$.

EXERCISES 8.6 (Part 1)

A

State the directions given by each of the sentences below. Use the method shown in Examples 2 and 3.

1. $x > 7$
2. $x > 1$
3. $x < 9$
4. $x < 4$
5. $y > 6$
6. $y > 8$
7. $y < 2$
8. $y < 5$
9. $x > -6$
10. $x > -3$
11. $x < -4$
12. $y > -8$
13. $y > -1$
14. $y < -5$
15. $x > 0$
16. $y < 0$
17. $x \geq 10$
18. $y \leq 6$
19. $x \leq -2$
20. $y \geq -3$

B

Graph each of the following sentences.

1. $x > 3$
2. $x < -1$
3. $y > 5$
4. $y < 6$
5. $x > -4$
6. $y \leq -2$
7. $x \geq -5$
8. $x \leq 8$
9. $y \geq 0$
10. $x \leq 0$

8.6 / The Graph of the Linear Inequality

C

Graph each of the following sentences.

1. $2x - 5 > 3$
2. $3x + 8 < 17$
3. $4y - 1 > 15$
4. $2 < 14 - 3y$
5. $2x > 12 - 2x$
6. $7y \geq y + 18$
7. $x \leq 5 - 4x$
8. $2y \geq 5y - 18$
9. $x \leq 10 + 6x$
10. $4x - 3 \geq 2x - 3$

PART 2

The method used in Part 1 for graphing a linear inequality in *one* variable can be applied equally well for graphing a linear inequality in two variables. Previously, the sentence

$$x > 5$$

implied

> The set of points in the half-plane to the *right* of the line $x = 5$.

Now, the sentence

$$x > 2y + 1$$

implies

> The set of points in the half-plane to the *right* of the line $x = 2y + 1$.

In view of this, graphing a sentence such as

$$x > 2y + 1$$

resolves itself to first graphing the equation of the line $x = 2y + 1$. This is then followed by shading the half-plane to the right of that line. The shaded region represents the set of points of the graph of $x > 2y + 1$. This procedure is shown below.

$$x = 2y + 1$$

y	$2y + 1$	$=$	x	(x, y)
0	$2 \cdot 0 + 1$	$=$	1	$(1, 0)$
-3	$2(-3) + 1$	$=$	-5	$(-5, -3)$

378   8/The Graph of an Open Sentence in Two Variables

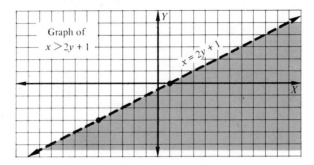

Notice that the line $x = 2y + 1$ is drawn above as a dotted line. As before, the purpose of this is to indicate that the points of this line are not included in the graph of the sentence

$$x > 2y + 1.$$

Had the sentence been

$$x \geqslant 2y + 1$$

then the line $x = 2y + 1$ would have been drawn as a solid line to indicate that these points are to be included in the graph of this sentence.

Example 1

Describe the half-plane that represents the graph of the sentence $5x + 2y < 6$.

Explanation We must first solve this inequality for either x or y.

Solution
$$5x + 2y < 6$$
$$2y < 6 - 5x$$
$$y < 3 - \frac{5x}{2}$$

Since we find that y is *less* than $3 - 5x/2$, the graph will be the half-plane *below* the line $y = 3 - 5x/2$.

Example 2

Graph the sentence $2x + 3y < 6$.

Solution
$$2x + 3y < 6$$
$$2x < 6 - 3y$$
$$x < 3 - \frac{3y}{2}$$

8.6/The Graph of the Linear Inequality

y	$3 - \frac{3y}{2}$	$=$	x	(x,y)
0	$3 - \frac{3 \cdot 0}{2}$	$=$	3	$(3, 0)$
4	$3 - \frac{3 \cdot 4}{2}$	$=$	-3	$(-3, 4)$

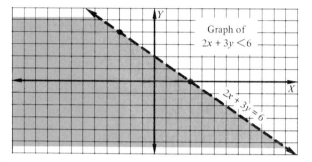

Explanation The inequality $2x + 3y < 6$ is solved for x. The sentence $x < 3 - 3y/2$ implies that the points of the graph will be in the half-plane to the *left* of the line $x = 3 - 3y/2$. After graphing this line, we shade the region of the plane to its left.

When examining Example 2 you may have wondered what would have happened had the inequality been solved for y rather than x. That is, would the graph have turned out to be the same half-plane as just found? The answer to this question is "Yes." Were this not so, then much of what we had learned in this course would have been in error. If we adhere to the principles we have established, then any path we take will lead to the same answer as any other path.

Example 3

Graph the sentence $5x \leq 3y + 15$.

Solution

$$5x \leq 3y + 15$$
$$5x - 3y \leq 15$$
$$-3y \leq 15 - 5x$$
$$y \geq -5 + \frac{5x}{3}$$

x	$-5 + \frac{5x}{3}$	$=$	y	(x,y)
0	$-5 + \frac{5 \cdot 0}{3}$	$=$	-5	$(0, -5)$
3	$-5 + \frac{5 \cdot 3}{3}$	$=$	0	$(3, 0)$

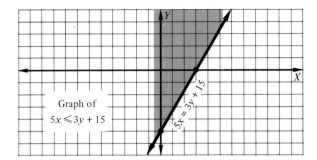

Explanation Notice that the graph of the equation $5x = 3y + 15$ is drawn as a solid line to indicate that the points of this line are to be included in the graph.

EXERCISES 8.6 (Part 2)

A

Describe the half-plane that represents the graph of each of the following sentences. (See the solution in Example 1.)

1. $x > 5y - 3$
2. $x < 2 + 3y$
3. $y > 4x - 7$
4. $y < 7x + 1$
5. $x + 2y < 5$
6. $3x + y > 4$
7. $5x + 2y < 4$
8. $4x - y > 9$
9. $3y - x < 6$
10. $7x - 3y < 12$
11. $6x > 10 - 5y$
12. $9y > 21 + 7x$

B

Graph each of the following sentences.

1. $x > y + 3$
2. $x < y - 5$
3. $y > 2x + 1$
4. $y < 5 - 3x$
5. $x + y < 6$
6. $x - 3y > 5$
7. $2x + y < 7$
8. $x + 5y \leq 7$
9. $3x - 2y \leq 6$
10. $5y \leq 8 - 2x$
11. $7x \geq 9 + 3y$
12. $5y \geq x + 4$
13. $2x - 3y \geq 0$
14. $5x - 2y \geq 0$

CHAPTER REVIEW

1. Express each of the following ordered pairs of numbers as a pair of directions.

   a. $(+3, +7)$
   b. $(-4, 2)$
   c. $(5, -3)$
   d. $(0, 4)$

2. Write each of the following pairs of equations as an ordered pair of numbers.

   a. $x = 5, y = -3$
   b. $x = -2, y = 0$

3. Write each of the following points as an ordered pair of numbers.

   a. The abscissa is -4; the ordinate is $+3$.
   b. The ordinate is 5; the abscissa is -1.

4. What are the coordinates of each of the points below.

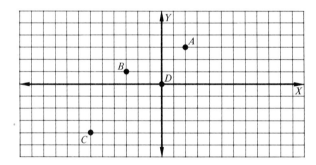

   a. A
   b. B
   c. C
   d. D

5. Answer each of the following questions.

   a. If the abscissa of a point is positive and the ordinate of the point is negative, in which quadrant does the point lie?
   b. If the abscissa and the ordinate are both negative, in which quadrant does the point lie?
   c. What conclusion can be drawn concerning the coordinates of a point that lies on the graph of an equation?
   d. What conclusion can be drawn concerning an ordered pair of numbers that does not satisfy an equation?

6. In each of the exercises below, what conclusion can be drawn concerning the coordinates of point P?

   a. 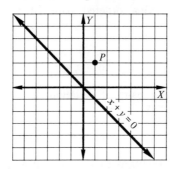   b.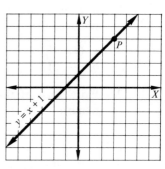

7. Solve each of the following equations for the variable indicated.
   a. $y + 2x = 5$ for y
   b. $2x + 5y = 6$ for x
   c. $3x - 2y = 7$ for y
   d. $7y - x = 3$ for x

8. Graph each of the following equations.
   a. $3x + y = 4$
   b. $x - 3y = 6$
   c. $5x - 3y = 8$
   d. $3x + 6 = x - 2$

9. In each of the following exercises the coordinates of two points of a line are given. Find the slope of each of these lines.
   a. $(4, 2)$ and $(7, 9)$
   b. $(-3, 8)$ and $(0, -2)$

10. Determine the slope of the line in each of the exercises below.

    a. 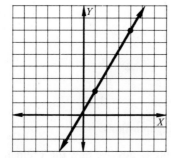   b.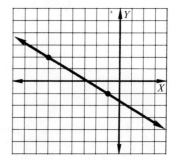

11. Determine the slope of each of the lines whose equations are given below.
    a. $y = \frac{2}{3}x + 6$
    b. $x = 2y - 3$

12. In each of the following exercises the slope of a line and the coordinates of a point on that line are given. Graph the line in each exercise.
    a. $(3, 1)$; slope $= 4/5$
    b. $(-4, 0)$; slope $= -3$

13. By using the point-slope method draw the graph of the equation

$$5x + 2y = 4.$$

14. In each of the following exercises the equations of two lines are given. You are to state whether the lines are "parallel" or "not parallel" and justify your answer.

  a. $2x + 3y = 5$ and $3x - 2y = 4$
  b. $12x - 8y = 5$ and $21x = 14y + 3$

15. (Optional) Graph the following equation:

$$|3x - 6| + 4 = y$$

16. Graph each of the following sentences.

  a. $x > 1$ c. $3y - 5 \geqslant 7$
  b. $y \leqslant 3$ d. $9 - 2x > 1$

17. Describe the half-plane that represents the graph of each of the following sentences.

  a. $x > 9 - 2y$ b. $y < 3x + 4$

18. Graph each of the following sentences.

  a. $x - 3y \leqslant 4$ b. $5x - 2y < 6$

Systems of Open Sentences in Two Variables

In the previous chapter we discovered that there are infinitely many pairs of values that will satisfy an open sentence in two variables. We shall now turn our attention to finding ways for determining pairs of values that will satisfy at the very same time two open sentences in two variables.

Generally speaking, several equations that impose conditions on a number of variables are called a **system of equations**. For instance, in the following equations,

1. $2x + y = 12$
2. $5x - y = 7$

the condition placed on the variables x and y in the first equation is:

1. If the variable x is multiplied by 2 and that product is added to y, the total will be 12.

Similar conditions are placed on these same variables by the second equation.

In situations of this nature, the objective is to try to determine values for the variables that will satisfy both equations at the same time. The *process* of finding the numbers that will satisfy the system of equations is called **solving the system of equations**.

9.1 SOLVING A LINEAR SYSTEM OF EQUATIONS BY GRAPHING

PART 1

We learned that one of the relationships that exists between an equation and the graph of that equation is that the coordinates of every point of the graph satisfy the equation. Hence, in Figure 9-1, the coordinates of a point, such as A will have to satisfy the equation $x + y = 6$, for point A is a point of the graph of that equation.

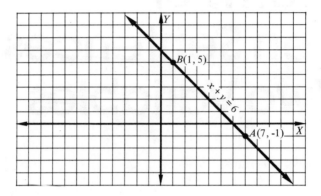

FIGURE 9-1

In Figure 9-2 the equation $2x - y = 6$ has been graphed on the same set of axes as the equation $x + y = 6$. From what was just stated, we know that the coordinates of every point of the line $x + y = 6$ will satisfy the equation $x + y = 6$. In the same way, the coordinates of every point of the line $2x - y = 6$ will satisfy the equation $2x - y = 6$. Since point P is a point of **both** lines, then the coordinates of point P must satisfy both equations.

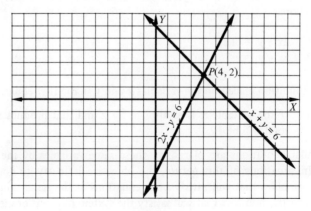

FIGURE 9-2

9.1/Solving a Linear System of Equations by Graphing

By counting, the coordinates of P are found to be (4, 2). Using this pair of numbers as replacements for x and y we can soon determine whether they actually do satisfy both equations.

$$2x - y = 6 \qquad x + y = 6$$
$$2 \cdot 4 - 2 \stackrel{?}{=} 6 \qquad 4 + 2 \stackrel{?}{=} 6$$
$$6 = 6 \qquad 6 = 6$$

Apparently, it would seem that the process of finding a pair of numbers that will satisfy two equations at the same time is simply one in which both equations are graphed on the same axes. The coordinates of the point that is common to both graphs will be the pair of numbers that satisfies both equations. This pair of numbers is called the **solution** of the two equations.

Our graphing of linear equations in two variables has taught us that the graphs of these equations are lines. Since two *distinct* lines can intersect in but one point at most, then the number of pairs of values that will satisfy two linear equations of this nature is just one. In the illustration above this pair of values was (4, 2).

There is a possibility though, that two lines may not have a point of intersection. This will happen when the two lines are parallel. And, as you recall, two lines are parallel when their slopes are equal.

Consider the equations whose graphs are shown in Figure 9-3. Since the slope of both lines is $\frac{-2}{1}$, the lines are parallel. In view of this, the two equations have no solution; that is, there is no pair of values that satisfies both equations simultaneously.

$$2x + y = -4 \qquad\qquad 2y = 12 - 4x$$
$$y = -4 - 2x \qquad\qquad y = 6 - 2x$$
$$\text{Slope} = \frac{-2}{1} \qquad\qquad \text{Slope} = \frac{-2}{1}$$

x	$-4 - 2x$	=	y
0	$-4 - 2 \cdot 0$	=	-4

x	$6 - 2x$	=	y
0	$6 - 2 \cdot 0$	=	6

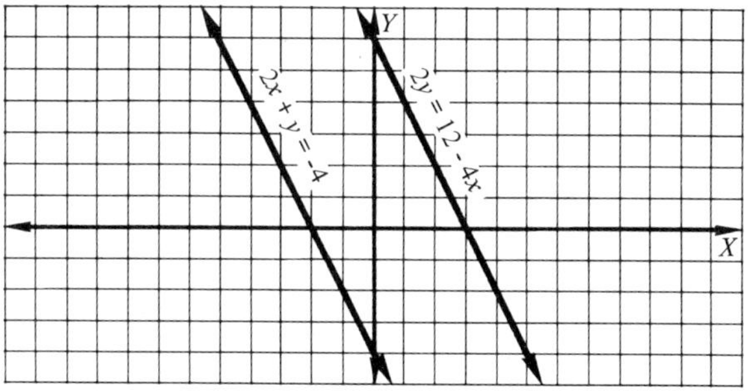

FIGURE 9-3

If a system of equations has no solution, then the equations of this system are said to be **inconsistent**. On the other hand, if the system does have a solution, then its equations are **consistent equations**.

There is still a third situation that might arise when we graph two linear equations in two variables. To discover what this is, let us take a look at the two equations here.

$$2x + 6y = 8 \quad \text{and} \quad 3x = 12 - 9y$$

In order to graph each of the equations by using the slope and a point of each line, it is necessary to solve each of them for y.

$$\begin{array}{ll} 2x + 6y = 8 & 3x = 12 - 9y \\ 6y = 8 - 2x & 9y = 12 - 3x \end{array}$$

(1) $\quad y = \dfrac{4}{3} - \dfrac{1}{3}x \qquad$ (2) $\quad y = \dfrac{4}{3} - \dfrac{1}{3}x$

Upon solving both for y we discover that not only are the slopes of the two lines the same but, also, every term of equation (1) is a term of equation (2). This implies that every point of the graph of

$$2x + 6y = 8$$

will be a point of the graph of

$$3x = 12 - 9y.$$

And this, in turn, implies that there are infinitely many pairs of values that satisfy both of these equations at the same time.

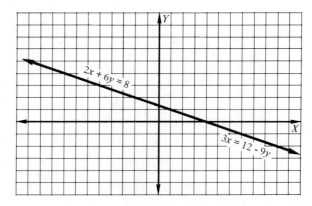

FIGURE 9-4

Equations that have identically the same graph are called **dependent equations**.

Example 1

Determine whether the following pair of equations are consistent, inconsistent, or dependent. Justify your answer.

$$3x - 5y = 4 \quad (1)$$
$$2x = 3y - 5 \quad (2)$$

Solution (1) $\quad 3x - 5y = 4 \qquad$ (2) $\quad 2x = 3y - 5$

$\qquad\qquad\quad 3x - 4 = 5y \qquad\qquad\quad 2x - 5 = 3y$

$\qquad\qquad\quad \frac{3}{5}x - \frac{4}{5} = y \qquad\qquad\quad \frac{2}{3}x - \frac{5}{3} = y$

$\qquad\qquad\qquad$ Slope $= \frac{3}{5} \qquad\qquad\qquad$ Slope $= \frac{2}{3}$

Since the slopes of the two equations are different, the graphs will intersect, and therefore, the equations are consistent.

Example 2

Determine whether the following pair of equations are consistent, inconsistent, or dependent. Justify your answer.

$$4x = 6y - 2 \quad (1)$$
$$6x - 9y = 12 \quad (2)$$

Solution (1) $\quad 4x = 6y - 2 \qquad$ (2) $\quad 6x - 9y = 12$

$\qquad\qquad\quad 4x + 2 = 6y \qquad\qquad\quad 6x - 12 = 9y$

$\qquad\qquad\quad \frac{2}{3}x + \frac{1}{3} = y \qquad\qquad\quad \frac{2}{3}x - \frac{4}{3} = y$

$\qquad\qquad\qquad$ Slope $= \frac{2}{3} \qquad\qquad\qquad$ Slope $= \frac{2}{3}$

Since the slopes of the two equations are equal but the equations are not identical *when solved for y*, then the lines are parallel. Hence, the equations are inconsistent.

Example 3

Determine whether the following pair of equations are consistent, inconsistent, or dependent. Justify your answer.

$$(1) \quad x + y = 3 \qquad (2) \quad 5x = 15 - 5y$$

Solution   (1) $x + y = 3$ (2) $5x = 15 - 5y$

$y = 3 - x$ $5y = 15 - 5x$

Slope $= \dfrac{-1}{1}$ $y = 3 - x$

Slope $= \dfrac{-1}{1}$

Since the slopes of the two equations are the same and the equations are identical *when solved for y*, then the graphs will be the same line. Hence, the equations are dependent.

EXERCISES 9.1 (Part 1)

Determine whether the equations in each of the following systems are consistent, inconsistent, or dependent. Justify each answer.

1. $y = 2x - 5$
   $y = 3x + 4$

2. $y = 7 - 5x$
   $y = 3 + 8x$

3. $y = \dfrac{2}{3}x + 5$
   $y = \dfrac{2}{3}x - 4$

4. $y = 6x + 10$
   $y = 9 + 6x$

5. $2x + 3y = 4$
   $3x + 2y = 7$

6. $4x - 3y = 9$
   $3y = 8 + 4x$

7. $6x - 2y = 10$
   $15x = 5y - 10$

8. $x - y = 2$
   $x + y = 2$

9. $9x = 12y + 6$
   $8y = 6x - 4$

10. $8x = 24 - 4y$
    $18 - 3y = 6x$

11. $3x - y = 0$
    $y = 2x + 3$

12. $6x = 8y$
    $21x - 5 = 28y$

PART 2

Knowing whether a system of linear equations is consistent or not is in itself interesting, however it is more important that we be able to determine the common solution of that system if the equations are consistent. The method for finding the common solution by graphing was suggested in Part 1. The following example will show just how this is done.

In order to make certain that the equations are consistent, it would be best that they be graphed by the point-slope method. Should we discover that the slopes are equal, we will then have to determine whether the equations are inconsistent and have no solution or are dependent and have infinitely many solutions.

9.1/Solving a Linear System of Equations by Graphing

Example 1

Determine the common solution for the following system of equations.

$$y = 3x + 4 \quad (1)$$
$$y = -x + 4 \quad (2)$$

Solution (1) $y = 3x - 4$ (2) $y = -x + 4$

Slope $= \dfrac{3}{1}$ Slope $= \dfrac{-1}{1}$

x	$3x - 4$	$=$	y
0	$3 \cdot 0 - 4$	$=$	-4

x	$-x + 4$	$=$	y
0	$0 + 4$	$=$	4

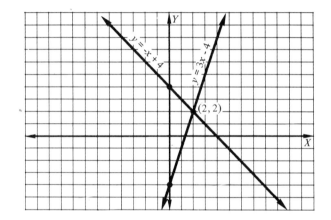

Common solution is (2, 2).

Check in equation (1):   Check in equation (2):

$y = 3x - 4$ $y = -x + 4$

$2 \stackrel{?}{=} 3(2) - 4$ $2 \stackrel{?}{=} -2 + 4$

$2 = 2$ $2 = 2$

Explanation   Once the coordinates of the point of intersection of the two lines are found, it is best to check this pair of values in both equations to make certain that no errors were made. This pair of values must satisfy both equations.

Example 2

Determine the common solution for the following system of equations.

$$2x + 3y = 0 \quad (1)$$
$$2x + 8 = y \quad (2)$$

Solution (1) $2x + 3y = 0$

$3y = -2x$

$y = \dfrac{-2}{3}x$

Slope $= \dfrac{-2}{3}$

(2) $2x + 8 = y$

Slope $= \dfrac{2}{1}$

x	$-\dfrac{2}{3}x =$	y
0	$-\dfrac{2}{3} \cdot 0 =$	0

x	$2x + 8 =$	y
0	$2 \cdot 0 + 8 =$	8

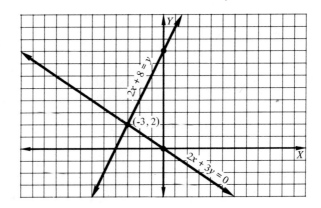

Common solution is $(-3, 2)$.

Check in equation (1):

$2x + 3y = 0$

$2(-3) + 3(2) \stackrel{?}{=} 0$

$-6 + 6 \stackrel{?}{=} 0$

$0 = 0$

Check in equation (2):

$2x + 8 = y$

$2(-3) + 8 \stackrel{?}{=} 2$

$-6 + 8 \stackrel{?}{=} 2$

$2 = 2$

Example 3

Determine the common solution for the following system of equations.

$9x - 3y = 14$ \hfill (1)

$15x = 8 + 5y$ \hfill (2)

9.1/Solving a Linear System of Equations by Graphing

Solution
(1) $9x - 3y = 14$
$9x - 14 = 3y$
$3x - \dfrac{14}{3} = y$
Slope $= \dfrac{3}{1}$

(2) $15x = 8 + 5y$
$15x - 8 = 5y$
$3x - \dfrac{8}{5} = y$
Slope $= \dfrac{3}{1}$

These equations are inconsistent. They have no common solution.

Explanation As soon as we discover that the slopes are equal we investigate to determine if the equations when solved for y are identical. Since they are not we conclude that the graphs are parallel lines and therefore have no point of intersection.

Example 4

Determine the common solution for the following system of equations.

$$4x - 32 = 16y \qquad (1)$$
$$7x = 56 + 28 \qquad (2)$$

Solution
(1) $4x - 32 = 16y$
$\dfrac{1}{4}x - 2 = y$
Slope $= \dfrac{1}{4}$

(2) $7x = 56 + 28y$
$7x - 56 = 28y$
$\dfrac{1}{4}x - 2 = y$
Slope $= \dfrac{1}{4}$

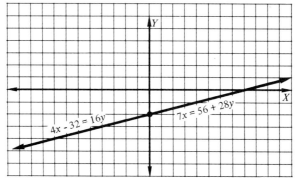

Explanation Upon discovering that the slopes are equal, we then examined both equations after they had been solved for y. These two equations are identical. This implied that the solutions of one are also the solutions of the other. The coordinates of every point of their common graph above will satisfy both equations.

EXERCISES 9.1 (Part 2)

A

The equations of each of the following systems are consistent. Determine the common solution of each system and check your answer.

1. $y = x + 1$
   $y = -x + 7$

2. $y = 2x + 1$
   $y = 7 - x$

3. $y = \frac{2}{3}x - 3$
   $y = \frac{-3}{2}x + 10$

4. $y = x + 6$
   $y = -\frac{3}{4}x - 1$

5. $y = 6 - x$
   $y = \frac{2}{3}x - 4$

6. $x + y = -5$
   $y - 2x = 10$

7. $3x + y = 1$
   $x - y = 7$

8. $2x + y = 2$
   $5x - y = -2$

9. $2x + y = -7$
   $-3x + y = 3$

10. $x - y = 0$
    $3x + y = 0$

11. $3x + y = 6$
    $-5x + 3y = 4$

12. $x + y = -4$
    $2x - 3y = 12$

13. $2x - y = 13$
    $x + 2y = -6$

14. $2x = 16 - y$
    $4x = 5y - 10$

15. $3y = 3 - 2x$
    $3x = 24 + 2y$

16. $x + 2y + 12 = 0$
    $5x + 4 = 4y$

17. $y = 6$
    $x + y = 9$

18. $2y + 8 = 0$
    $3x + y = -1$

19. $x = 4$
    $3x - y = 10$

20. $2x - 6 = 0$
    $3y - 12 = 0$

B

Determine the common solution of each of the following systems of equations. If the equations are inconsistent, express your answer as in Example 3. If the equations are dependent, draw the graph and express your answer as in Example 4.

1. $y = 2x + 5$
   $y = 2x - 5$

2. $3x - 2y = 6$
   $15x = 10y + 30$

3. $x = y + 1$
   $5x + 4 = 2y$

4. $3y + 8 = 5x$
   $20x - 9 = 12y$

5. $6x - 9y = 15$
   $6y = 4x - 10$

6. $8x = 28 - 12y$
   $18 - 9y = 6x$

7. $x - y = 4$
   $x = -3y$

8. $10x = 8y + 15$
   $25x - 15 = 20y$

9. $2x = 3y$
   $3x - 2y = 0$

10. $3y - 12 = 0$
    $6 + 2y = 0$

9.2 SOLVING A LINEAR SYSTEM OF EQUATIONS BY SUBSTITUTION

PART 1

By this point it should be quite evident that finding the common solution to a system of linear equations by graphing has a number of drawbacks. Not the least of these is the fact that this method is far too time consuming. In addition to this there are two other major weaknesses with this approach. Should the common solution turn out to be the point (1,978, -2,001), we would need a rather large piece of graph paper in order to eventually locate the point of intersection of the two graphs. Even more important than this is the fact that if the common solution is (3.059, 2.468), the graphic method cannot attain this accuracy. Because of this, in both this section and the next, we shall turn to other ways for finding the common solution of a linear system.

Very early in this course we were asked to find the value of an expression, such as

$$2x + 3,$$

where we were told that

$$x = 5.$$

It was possible for us to replace x with 5 in the expression $2x + 3$ since both x and 5 are but two different names for the same number.

This very same approach can be used when finding the common solution to certain systems of linear equations. Consider, as an example, the system

$$x = 6 \qquad (1)$$

$$2x + y = 8. \qquad (2)$$

As we know, there are infinitely many ordered pairs of numbers that satisfy each of these equations. We are searching for *the* one pair of numbers that will satisfy both of the equations at the same time. In view of this, we will now think of the x in both equations as being the very same number. We will also consider the y in both equations to be the very same number. This pair of numbers is the coordinates of the point of intersection of their graphs.

Equation (1) above implies that x and 6 name the very same number. In view of this, it is possible to replace x with 6 in equation (2), thus obtaining

$$2(6) + y = 8. \qquad (2)$$

Equation (2) can now be recognized as a linear equation in one variable that can readily be solved for y.

$$\begin{aligned} 2(6) + y &= 8 \qquad (2) \\ 12 + y &= 8 \\ y &= 8 - 12 \\ y &= -4 \end{aligned}$$

The numbers just found for x and y are the common solution of this system. The check below shows that this is so.

The common solution is $(6, -4)$.

Check for equation (1):

$$x = 6$$
$$6 = 6$$

Check for equation (2):

$$2x + y = 8$$
$$2(6) + (-4) \stackrel{?}{=} 8$$
$$12 - 4 \stackrel{?}{=} 8$$
$$8 = 8$$

Example 1

Solve the following system of equations and check the answer.

$$\begin{aligned} y &= 8 \\ 3x + 2y &= 25 \end{aligned}$$

Solution (1) $\quad y = 8$
(2) $\quad 3x + 2y = 25$

(2) $\quad \begin{aligned} 3x + 2(8) &= 25 \qquad (y \text{ is replaced with } 8) \\ 3x + 16 &= 25 \\ 3x &= 9 \\ x &= 3 \end{aligned}$

The common solution is $(3, 8)$.

9.2/Solving a Linear System of Equations by Substitution

Check in equation (1):
$$y = 8$$
$$8 = 8$$

Check in equation (2):
$$3x + 2y = 25$$
$$3(3) + 2(8) \stackrel{?}{=} 25$$
$$9 + 16 \stackrel{?}{=} 25$$
$$25 = 25$$

Example 2

Solve the following system of equations and check the answer.

$$x = -4y$$
$$3x - 2y = 28$$

Solution (1) $\quad x = -4y$
(2) $\quad 3x - 2y = 28$

(2) $\quad 3(-4y) - 2y = 28 \quad$ (x is replaced with -4y)
$$-12y - 2y = 28$$
$$-14y = 28$$
$$y = -2$$
$$x = -4y \qquad\qquad (1)$$
$$x = -4(-2) \quad\quad (y \text{ is replaced with } -2)$$
$$x = 8$$

The common solution is (8, -2).

Check in equation (1):
$$x = -4y$$
$$8 \stackrel{?}{=} -4(-2)$$
$$8 = 8$$

Check in equation (2):
$$3x - 2y = 28$$
$$3(8) - 2(-2) \stackrel{?}{=} 28$$
$$24 + 4 \stackrel{?}{=} 28$$
$$28 = 28$$

Explanation In the solution above, the information in equation (1) is that x and -4y name the same number. Hence, it is possible to replace x in equation (2) with -4y. By doing this, the only variable that will now appear in equation (2) is y. Solving this equation gives us -2 as the value of y at the common point. To determine the corresponding value of x at the common point, we replace y with -2 in equation (1). The value of x at the common point is found to be 8. The pair of numbers (8, -2) is then tested in both equations.

Example 3

Solve the following system of equations and check the answer.

$$b = 1 - 2a$$
$$5a + 2b = -2$$

Solution (1) $\quad b = 1 - 2a$
(2) $\quad 5a + 2b = -2$

(2) $\quad 5a + 2(1 - 2a) = -2 \quad$ (b is replaced with $1 - 2a$)
$\quad\quad\quad 5a + 2 - 4a = -2$
$\quad\quad\quad 5a - 4a = -2 - 2$
$\quad\quad\quad\quad\quad a = -4$

$\quad\quad\quad\quad b = 1 - 2a \quad\quad\quad\quad\quad\quad\quad\quad\quad\quad\quad (1)$

$\quad\quad\quad\quad b = 1 - 2(-4) \quad$ (a is replaced with -4)
$\quad\quad\quad\quad b = 1 + 8$
$\quad\quad\quad\quad b = 9$

The common solution is $(-4, 9)$.

Check in equation (1):
$b = 1 - 2a$
$9 \stackrel{?}{=} 1 - 2(-4)$
$9 \stackrel{?}{=} 1 + 8$
$9 = 9$

Check in equation (2):
$5a + 2b = -2$
$5(-4) + 2(9) \stackrel{?}{=} -2$
$-20 + 18 \stackrel{?}{=} -2$
$-2 = -2$

Explanation Notice that when the common solution is written as an ordered pair of numbers, the numbers are arranged in alphabetical order; that is, the of a is written first and the value of b is written second, $(-4, 9)$.

Since this method for solving a system of equations involves *replacing* one quantity with another, it is called the replacement or **substitition method.**

EXERCISES 9.2 (Part 1)

Solve each of the following systems of equations by the substitution method. Check your answers.

1. $y = 3$
   $2x + 5y = 17$

2. $x = 5$
   $4x - 3y = 11$

9.2/Solving a Linear System of Equations by Substitition

3. $3x + 4y = 4$
   $x = 4$

4. $y = 4x$
   $5x + y = 18$

5. $y = -2x$
   $3x + 2y = 3$

6. $9x + 2y = 4$
   $y = -4x$

7. $a = 5b$
   $a + 6b = 22$

8. $a = 3b$
   $2a - 5b = -4$

9. $a + 8b = 2$
   $a = -7b$

10. $b = -5a$
    $5a - b = 20$

11. $5c - d = 27$
    $d = -4c$

12. $d = 5c + 1$
    $3c + 2d = 15$

13. $w = 2z + 3$
    $2w + 5z = 33$

14. $m = 3p + 4$
    $4p - 3m = -12$

15. $m = 3p - 1$
    $m + 5p = -9$

16. $t = 2s + 13$
    $3s - t = -18$

17. $v - w = -25$
    $w = 2v + 13$

18. $3t - s = -19$
    $s = -2t - 1$

19. $v = 10w + 1$
    $v - 8w = 2$

20. $8p - 5q = -13$
    $q = 4p + 5$

PART 2

Each of the systems in Part 1 was designed so that one of the equations had already been solved for one of the two variables. This enabled us to make an immediate replacement for that variable in the other equation. In the event that neither equation is solved for one variable, our first task is to do so.

The question now arises as to which equation to select and which variable to solve for. To make our work as simple as possible, we look for an equation in which one of the variables has a coefficient of either +1 or -1. We then solve that equation for that particular variable. If more than one variable has a coefficient of +1 or -1, then we can choose whichever one we desire.

Example 1

Solve the following system of equations and check the answer.

$$3x + y = 2$$
$$5x + 2y = 1$$

Solution   (1) $3x + y = 2$

(2) $5x + 2y = 1$

(1) $\quad y = 2 - 3x \quad$ (Equation (1) is solved for y)
(2) $\quad 5x + 2(2 - 3x) = 1 \quad$ (y is replaced with $2 - 3x$)
$$5x + 4 - 6x = 1$$
$$5x - 6x = 1 - 4$$
$$-1x = -3$$
$$x = 3$$
(1) $\quad y = 2 - 3x$
$\quad y = 2 - 3(3) \quad$ (x is replaced with 3)
$\quad y = 2 - 9$
$\quad y = -7$

The common solution is $(3, -7)$.

Check for equation (1):

$$3x + y = 2$$
$$3(3) + (-7) \stackrel{?}{=} 2$$
$$9 - 7 \stackrel{?}{=} 2$$
$$2 = 2$$

Check for equation (2):

$$5x + 2y = 1$$
$$5(3) + 2(-7) \stackrel{?}{=} 1$$
$$15 - 14 \stackrel{?}{=} 1$$
$$1 = 1$$

Explanation The coefficient of y in equation (1) is the only coefficient that fulfills the requirement of being either +1 or -1. Hence, equation (1) is solved for y. From that point on the approach is the same as that used in Part 1. Notice that when checking our answers, we are concerned with knowing whether that pair of numbers satisfies the **original equations**.

Example 2

Solve the following system of equations and check the answer.

$$5a - 4b = 10$$
$$2a - b = 1$$

Solution
(1) $\quad 5a - 4b = 10$
(2) $\quad 2a - b = 1$
_____
(2) $\quad 2a = 1 + b$
(2) $\quad 2a - 1 = b \quad$ (Equation (2) is solved for b)

(1) $5a - 4(2a - 1) = 10$ (b is replaced with $2a - 1$)
$5a - 8a + 4 = 10$
$5a - 8a = 10 - 4$
$-3a = 6$
$a = -2$

(2) $2a - 1 = b$
$2(-2) - 1 = b$ (a is replaced with -2)
$-4 - 1 = b$
$-5 = b$

The common solution is $(-2, -5)$

Check for equation (1):
$5a - 4b = 10$
$5(-2) - 4(-5) \stackrel{?}{=} 10$
$-10 + 20 \stackrel{?}{=} 10$
$10 = 10$

Check for equation (2):
$2a - b = 1$
$2(-2) - (-5) \stackrel{?}{=} 1$
$-4 + 5 \stackrel{?}{=} 1$
$1 = 1$

EXERCISES 9.2 (Part 2)

Solve each of the following systems of equations by the substitution method. Check your answers.

1. $x + 2y = 8$
   $2x + 3y = 13$

2. $2x + y = 15$
   $3x - 2y = -2$

3. $4x - 5y = 9$
   $x - 3y = 4$

4. $3a + 5b = 11$
   $2a + b = -2$

5. $5a + 4b = 10$
   $-a + 2b = 12$

6. $2c - d = -3$
   $4c - 5d = -3$

7. $x - y = 2$
   $2y - x = 6$

8. $x + 1 = 8y$
   $2x - 7y = 7$

9. $2x = y - 12$
   $2x - 12 = 3y$

10. $3a = 2 - 5b$
    $a + 2b = 0$

11. $d - 6 = 3c$
    $3d + c + 22 = 0$

12. $2x + y - 7 = 0$
    $6x - y - 5 = 0$

13. $6t = 9 - 2s$
    $s - 11 = 10t$

14. $10p = 7 - 8q$
    $q - p = 2$

9.3 SOLVING A LINEAR SYSTEM OF EQUATIONS BY ADDITION OR SUBTRACTION

PART 1

The systems of equations we have examined thus far have all been designed so that at least one of the coefficients is either a +1 or a -1. If this is not the situation, the system can still be solved using the substitition method, however the computation becomes rather involved. For instance, consider the system

$$2x + 3y = 13 \qquad (1)$$
$$5x - 3y = 22. \qquad (2)$$

Were we to solve equation (1) for x we would find its value to be:

$$x = \frac{13 - 3y}{2}.$$

Replacing x in equation (2) with $(13 - 3y)/2$ leads to the cumbersome equation below.

$$5 \cdot \frac{13 - 3y}{2} - 3y = 22. \qquad (2)$$

In order to avoid difficult substititions involving fractions a third method has been developed for solving a system of linear equations. This method makes use of the addition principle of equality. Using specific numbers, we will take a moment to recall this principle.

$$\text{If } 14 = 14$$
$$\text{and } 5 = 5$$

then we can conclude that

$$14 + 5 = 14 + 5.$$

This very same process can be applied to the system of equations we examined earlier.

$$2x + 3y = 13 \qquad (1)$$
$$5x - 3y = 22 \qquad (2)$$

Were we to add the *left* member of (1) to the *left* member of (2), their sum will be exactly the same as the sum obtained by adding the *right* member of (1) to the *right* member of (2).

$$(2x + 3y) + (5x - 3y) = 13 + 22.$$

9.3/Solving a Linear System of Equations by Addition or Subtraction

At this point it is possible to remove the parentheses obtaining

$$2x + 3y + 5x - 3y = 13 + 22. \qquad (3)$$

When we collect like terms on both sides of equation (3), we discover that we now have an equation in one variable.

$$7x = 35 \qquad (4)$$

By solving this equation for x we obtain

$$x = 5.$$

The corresponding value for y at the common solution of (1) and (2) can be found as usual by replacing x with 5 in equation (1).

$$\begin{align} 2x + 3y &= 13 \qquad (1)\\ 2 \cdot 5 + 3y &= 13 \\ 10 + 3y &= 13 \\ 3y &= 13 - 10 \\ 3y &= 3 \\ y &= 1 \end{align}$$

Hence, the common solution is (5, 1).

The important feature about the example above is that the y term in equation (1) is the additive inverse of the y term in equation (2). Hence, when the left member of the first equation is added to the left member of the second equation the y terms are eliminated, thus leaving us with an equation that has but one variable.

Apparently, then, the addition property of equality can be a great advantage to us *if the first equation has a variable term that is the additive inverse of a variable term in the second equation.*

Example 1

Solve and check the following system of equations.

$$\begin{align} 5x - 2y &= 16 \\ 7x + 2y &= 8 \end{align}$$

Explanation Rather than add left member to left member and indicate that sum by writing it horizontally as

$$(5x - 2y) + (7x + 2y),$$

404   9/Systems of Open Sentences in Two Variables

we simply add the left member to the left member vertically as they appear.

$$\begin{array}{r} 5x - \cancel{2y} \\ 7x + \cancel{2y} \\ \hline 12x \end{array}$$

The same is done when finding the sum of the right members of the two equations:

$$\begin{array}{r} 16 \\ 8 \\ \hline 24 \end{array}$$

Solution (1) $5x - \cancel{2y} = 16$
(2) $7x + \cancel{2y} = 8$
(3) $12x = 24$
$x = 2$

(1) $5x - 2y = 16$
$\downarrow$
$5(2) - 2y = 16$ (x is replaced with 2 in equation (1))
$10 - 2y = 16$
$-2y = 16 - 10$
$-2y = 6$
$y = -3$

The common solution is (2, -3).

Check in equation (1):   Check in equation (2):

$5x - 2y = 16$ $7x + 2y = 8$
$5(2) - 2(-3) \stackrel{?}{=} 16$ $7(2) + 2(-3) \stackrel{?}{=} 8$
$10 + 6 \stackrel{?}{=} 16$ $14 - 6 \stackrel{?}{=} 8$
$16 = 16$ $8 = 8$

Although we had no need for the subtraction principle of equality prior to now, it will be an aid in solving some systems of equations. If a variable term in the first equation is *exactly the same* as a variable term in the second equation, then the addition principle of equality will not eliminate that variable as it had in Example 1. When the two terms are identical we can eliminate these terms by subtracting the left member from the left member rather than adding—and the same, of course, for the right members.

9.3/Solving a Linear System of Equations by Addition or Subtraction

Subtraction Principle of Equality

If $\quad a = b$

and $\quad c = d$

then $a - c = b - d$.

Example 2

Solve and check the following system of equations.

$$3x - 8y = 1$$
$$3x + 5y = -25$$

Explanation Since the x term in the first equation is identical to the x term in the second equation we can eliminate this variable by using the subtraction principle of equality. Recall that the very first thing we must do when subtracting one polynomial from another is to mentally replace each of the terms of the subtrahend with its additive inverse.

Solution
(1) $\quad 3x - 8y = 1$
(2) $\quad 3x + 5y = -25$
(3) $\quad\quad\;\; -13y = +26 \quad$ (Subtraction)
$\quad\quad\quad\quad\quad\; y = -2$

(1) $\quad 3x - 8y = 1$
$\quad\quad 3x - 8(-2) = 1$
$\quad\quad\quad 3x + 16 = 1$
$\quad\quad\quad\quad\;\; 3x = 1 - 16$
$\quad\quad\quad\quad\;\; 3x = -15$
$\quad\quad\quad\quad\quad x = -5$

The common solution is $(-5, -2)$.

Check for equation (1):

$3x - 8y = 1$

$3(-5) - 8(-2) \stackrel{?}{=} 1$

$-15 + 16 \stackrel{?}{=} 1$

$1 = 1$

Check for equation (2):

$3x + 5y = -25$

$3(-5) + 5(-2) \stackrel{?}{=} -25$

$-15 - 10 \stackrel{?}{=} -25$

$-25 = -25$

Considering the fact that finding the common solutions in Example 1 and 2 involved the application of the addition or subtraction principles of equality it seems only natural that this method be called the **addition or subtraction method** for solving a linear system.

Example 3

Solve and check the following system of equations.

$$3a + 2b = 4$$
$$-5a + 2b = 36$$

Explanation Since the b term in the second equation is identical to the b term in the first equation, the subtraction principle is applied.

Solution
(1) $\quad 3a + 2b = 4$
(2) $\quad -5a + 2b = 36$
(3) $\quad\quad 8a = -32 \quad$ (Subtraction)
$\quad\quad\quad a = -4$

(1) $\quad 3a + 2b = 4$
$\quad\quad 3(-4) + 2b = 4$
$\quad\quad -12 + 2b = 4$
$\quad\quad\quad 2b = 4 + 12$
$\quad\quad\quad 2b = 16$
$\quad\quad\quad b = 8$

The common solution is $(-4, 8)$.

Check in equation (1):
$$3a + 2b = 4$$
$$3(-4) + 2(8) \stackrel{?}{=} 4$$
$$-12 + 16 \stackrel{?}{=} 4$$
$$4 = 4$$

Check in equation (2):
$$-5a + 2b = 36$$
$$-5(-4) + 2(8) \stackrel{?}{=} 36$$
$$20 + 16 \stackrel{?}{=} 36$$
$$36 = 36$$

EXERCISES 9.3 (Part 1)

Solve each of the following systems of equations by the addition or subtraction method. Check your answers.

1. $\quad x + 2y = 13$
$\quad\; 3x - 2y = 23$

2. $\quad x + 2y = 25$
$\quad -x + 3y = 35$

9.3/Solving a Linear System of Equations by Addition or Subtraction

3. $3a + 4b = 19$
   $5a - 4b = -43$

4. $2a + 5b = 2$
   $a + 5b = -4$

5. $3a + b = 15$
   $5a + b = 29$

6. $4c - d = -7$
   $4c + 3d = -43$

7. $5w - 3z = 33$
   $2w - 3z = -3$

8. $7v - 2w = -64$
   $-4v - 2w = 2$

9. $-4c + 3d = -2$
   $-4c - 5d = 22$

10. $-6b + 7c = 6$
    $6b + 5c = 30$

11. $-3x - 4y = 10$
    $-3x + 4y = 20$

12. $8x + 4y = 7$
    $-8x + 12y = 5$

13. $12s - 16t = 15$
    $20s - 16t = 17$

14. $6m - 8n = 13$
    $6m + 12n = -57$

PART 2

As we solved the systems of equations in Part 1, it may have occurred to us that each pair of equations had a term in one that was either identical to a term of the other or the additive inverse of the other. This leads to the question, "How can the addition or subtraction method be applied if this is not the case?"

Consider the system of equations

$$4x + 3y = -4 \qquad (1)$$

$$7x + 6y = -13. \qquad (2)$$

When examining the y terms of these two equations, we notice that the coefficient of y in the second equation is a multiple of the coefficient of y in the first equation. This suggests the possibility of applying the multiplication principle of equality, whereby we multiply both members of equation (1) by the same number. The number selected is one that will make the coefficient of y in equation (1) the same as the coefficient of y in equation (2).

Quite apparently, both members of equation (1) will have to be multiplied by 2 in order to change the coefficient 3 into a 6. This process is shown below:

$$2 \cdot (1) \quad 8x + 6y = -8$$

$$(2) \quad 7x + 6y = -13$$

Now that the y term in equation (1) is identical to the y term in equation (2), the solution of the two equations is found in exactly the same manner as in Part 1.

Example 1

Solve and check the following system of equations.

$$7x + 12y = 11$$
$$5x + 4y = 17$$

Explanation The coefficient of the y term in the first equation is a multiple of the coefficient of the y term in the second equation. Hence, if both members of the second equation are multiplied by 3, the y terms in both equations will be identical.

Solution

$$\begin{array}{rl} (1) & 7x + 12y = 11 \\ (2) & 5x + 4y = 17 \\ \hline (1) & 7x + 12y = 11 \\ 3 \cdot (2) & 15x + 12y = 51 \\ \hline (3) & -8x = -40 \quad \text{(Subtraction)} \\ & x = 5 \end{array}$$

$$\begin{array}{rl} (1) & 7x + 12y = 11 \\ & 7(5) + 12y = 11 \\ & 35 + 12y = 11 \\ & 12y = 11 - 35 \\ & 12y = -24 \\ & y = -2 \end{array}$$

The common solution is $(5, -2)$.

Check in equation (1):

$$7x + 12y = 11$$
$$7(5) + 12(-2) \stackrel{?}{=} 11$$
$$35 - 24 \stackrel{?}{=} 11$$
$$11 = 11$$

Check in equation (2):

$$5x + 4y = 17$$
$$5(5) + 4(-2) \stackrel{?}{=} 17$$
$$25 - 8 \stackrel{?}{=} 17$$
$$17 = 17$$

Explanation Notice that in the fourth step of the solution the number 3 is written to the left of equation (2). This is done to remind us that both members of that equation are multiplied by 3.

Example 2

Solve and check the following system of equations.

$$-c + 4b = -6$$
$$4c - 3b = -15$$

Explanation In this situation if both members of equation (1) are multiplied by 4, the c term of the first equation will be the additive inverse of the c term of the second equation.

Solution
(1) $-c + 4b = -6$
(2) $4c - 3b = -15$
$4 \cdot (1)$ $-4c + 16b = -24$
(2) $4c - 3b = -15$
$13b = -39$ (Addition)
$b = -3$
(1) $-c + 4(-3) = -6$
$-c - 12 = -6$
$-12 = -6 + c$
$-12 + 6 = c$
$-6 = c$

The common solution is $(-3, -6)$.

Check in equation (1):

$-c + 4b = -6$
$-(-6) + 4(-3) \stackrel{?}{=} -6$
$6 - 12 \stackrel{?}{=} -6$
$-6 = -6$

Check in equation (2):

$4c - 3b = -15$
$4(-6) - 3(-3) \stackrel{?}{=} -15$
$-24 + 9 \stackrel{?}{=} -15$
$-15 = -15$

EXERCISES 9.3 (Part 2)

Solve and check each of the following systems of equations.

1. $3x + 4y = 11$
   $5x + 8y = 21$

2. $5a + 2b = 29$
   $10a - 3b = 9$

3. $4c + 3d = 7$
   $5c - 6d = -40$

4. $10w + 7z = -3$
   $2w + 5z = 3$

5. $-2v + 15w = -35$
   $3v - 5w = 0$

6. $s - 2t = 10$
   $3s + 5t = -47$

7. $4x + 5y = -12$
   $3x - y = -9$

8. $3m + 2n = 10$
   $9m - 5n = -14$

9. $3b - 8c = -20$
   $5b + 16c = 18$

10. $3p + q = 11$
    $5p + 3q = -3$

PART 3

The approach to solving a linear system by addition or subtraction has thus far followed the same basic pattern we used when adding or subtracting fractions. Initially we encountered fractions whose denominators were the same. This was followed by the situation where the denominator of one of the fractions was a multiple of the denominator of the other fraction. And finally we were asked to add fractions where neither denominator was a multiple of the other. In this last stage, as you recall, it was necessary to find the lowest common multiple of both denominators.

We are at this last stage now in solving a linear system. In the two earlier stages we examined those situations where two coefficients were the same:

$$5x + 4y = 13$$

$$3x + 4y = 11.$$

We also examined those situations where one coefficient was a multiple of the other:

$$4x + 3y = 10$$

$$7x + 9y = 25.$$

Hence, we are now at the point where neither coefficient is a multiple of the other:

$$5x + 6y = 17$$

$$7x + 4y = 15.$$

When this occurs we follow exactly the same course we took when adding fractions. Thus, to eliminate the y variable we find the lowest common multiple of the 6 and 4. That number is 12. Hence, to rewrite the coefficient of y in the first equation as 12 we must multiply both members of that equation by 2. Similarly, to rewrite the coefficient of y in the second equation as 12 we

9.3/Solving a Linear System of Equations by Addition or Subtraction

must multiply both members of that equation by 3. These steps are shown below.

$$
\begin{align*}
(1) \quad & 5x + 6y = 17 \quad \text{L.C.M.} = 12 \\
(2) \quad & 7x + 4y = 15 \\
2 \cdot (1) \quad & 10x + 12y = 34 \\
3 \cdot (2) \quad & 21x + 12y = 45
\end{align*}
$$

The two equations are now of the same form as those we solved in Part 1. Hence, the solution is completed in the same manner as earlier. Notice that we wrote three things above to remind us where we were going. To the right of the system we wrote the L.C.M. of the coefficients of the y variable in each equation. To the left of each equation we showed the number by which each member of that equation is to be multiplied.

Would it have been possible to have found the L.C.M. of the x variable rather than the y variable? Yes; whether we eliminate one variable or the other makes no difference for the common solution of the system must be the same no matter how the system is solved.

Example 1

Solve and check the following system of equations.

$$
\begin{align*}
6x - 5y &= 45 \\
8x + 9y &= 13
\end{align*}
$$

Solution

$$
\begin{align*}
(1) \quad & 6x - 5y = 45 \quad \text{L.C.M.} = 24 \\
(2) \quad & \underline{8x + 9y = 13} \\
4 \cdot (1) \quad & 24x - 20y = 180 \\
3 \cdot (2) \quad & \underline{24x + 27y = 39} \\
(3) \quad & -47y = 141 \quad \text{(Subtraction)} \\
& y = -3 \\
(1) \quad & 6x - 5y = 45 \\
& 6x - 5(-3) = 45 \\
& 6x + 15 = 45 \\
& 6x = 45 - 15 \\
& 6x = 30 \\
& x = 5
\end{align*}
$$

The common solution is $(5, -3)$.

Check for equation (1):

$$6x - 5y = 45$$
$$6(5) - 5(-3) \stackrel{?}{=} 45$$
$$30 + 15 \stackrel{?}{=} 45$$
$$45 = 45$$

Check for equation (2):

$$8x + 9y = 13$$
$$8(5) + 9(-3) \stackrel{?}{=} 13$$
$$40 - 27 \stackrel{?}{=} 13$$
$$13 = 13$$

EXERCISES 9.3 (Part 3)

Solve and check each of the following systems of equations.

1. $2x + 3y = 31$
   $3x + 2y = 29$

2. $5x - 4y = 12$
   $3x + 5y = 22$

3. $4a - 3b = 27$
   $6a + 5b = -7$

4. $7a - 8b = 33$
   $5a + 6b = 47$

5. $2c + 9d = -1$
   $3c + 5d = -10$

6. $3c + 5d = -7$
   $4c - 9d = 22$

7. $5r + 7s = -15$
   $3r - 5s = 37$

8. $5r - 6s = -7$
   $4r - 9s = 7$

9. $10m + 3n = 42$
   $4m + 7n = 40$

10. $4m + 12n = 25$
    $-6m + 8n = -5$

PART 4

In each of the systems of linear equations that we have solved by the addition or subtraction method, the equations were always arranged in **standard form**; that is, the x term appeared first, followed by the y term, followed by the equality sign and finally, followed by the constant term. This is shown in the equation below:

$$5x - 4y = 7.$$

If it so happens that either one or both of the equations are not written in standard form, we must rewrite the equations in standard form. Thus, in the equation

$$4x = 5 - 3(x - 2y),$$

we will first find the product of -3 and $x - 2y$:

$$4x = 5 - 3x + 6y.$$

9.3/Solving a Linear System of Equations by Addition or Subtraction

At this point we will rewrite the equation so that all variable terms appear on the left side of the equation and all constant terms appear on the right side.

$$4x + 3x - 6y = 5$$

When we collect like terms, the equation will be in standard form.

$$7x - 6y = 5$$

Should the equation contain fractions, then our first objective will be to multiply both members of the equation by the lowest common multiple of the denominators. As an example, consider the equation

(1) $\quad \dfrac{x}{2} - \dfrac{y+3}{5} = \dfrac{2}{5}.$

The L.C.M. is 10. Hence,

(2) $\quad 10\left(\dfrac{x}{2} - \dfrac{y+3}{5}\right) = 10 \cdot \dfrac{2}{5}$

(3) $\quad \cancel{10}^{5} \cdot \dfrac{x}{\cancel{2}} - \cancel{10}^{2} \cdot \dfrac{y+3}{\cancel{5}} = \cancel{10}^{2} \cdot \dfrac{2}{\cancel{5}}$

(4) $5x - 2(y + 3) = 4$

(5) $\quad 5x - 2y - 6 = 4$

(6) $\quad\quad 5x - 2y = 4 + 6$

(7) $\quad\quad 5x - 2y = 10$

After eliminating fractions as is the case in step (4) we continued as we had in the previous example. Equation (7) is equivalent to equation (1), however equation (7) is expressed in standard form.

Example 1

Solve the following system of equations.

$$5x = 2y - 1$$

$$6x = 5y - 3(4 - x)$$

Solution (1) $\quad 5x = 2y - 1$

(1) $\quad\underline{5x - 2y = -1}\quad$ Standard form for equation (1)

(2) $\quad 6x = 5y - 3(4 - x)$

$\quad 6x = 5y - 12 + 3x$

$\quad 6x - 3x - 5y = -12$

(2) $\quad\underline{3x - 5y = -12}\quad$ Standard form for equation (2)

Having expressed each equation in standard form we now write (2) below the (1):

(1) $\quad 5x - 2y = -1 \quad$ L.C.M. = 15

(2) $\quad\underline{3x - 5y = -12}$

$3 \cdot (1) \quad 15x - 6y = -3$

$5 \cdot (2) \quad \underline{15x - 25y = -60}$

(3) $\quad 19y = 57 \quad$ (Subtraction)

$\quad y = 3$

(1) $\quad 5x - 2y = -1$

$\quad 5x - 2(3) = -1$

$\quad 5x - 6 = -1$

$\quad 5x = -1 + 6$

$\quad 5x = 5$

$\quad x = 5$

The common solution is $(1, 3)$.

Example 2

Solve the following system of equations.

$$\frac{a}{6} + \frac{b}{4} = 3$$

$$\frac{a+3}{2} = \frac{a+b}{4} + 1$$

9.3/Solving a Linear System of Equations by Addition or Subtraction

Solution

(1) $\quad \dfrac{a}{6} + \dfrac{b}{4} = 3$

$\qquad 12\left(\dfrac{a}{6} + \dfrac{b}{4}\right) = 12 \cdot 3$

$\qquad \overset{2}{\cancel{12}} \cdot \dfrac{a}{\cancel{6}} + \overset{3}{\cancel{12}} \cdot \dfrac{b}{\cancel{4}} = 12 \cdot 3$

(1) $\qquad 2a + 3b = 36 \qquad$ Standard form for equation (1)

(2) $\quad \dfrac{a+3}{2} = \dfrac{a+b}{4} + 1$

$\qquad 4\left(\dfrac{a+3}{2}\right) = 4\left(\dfrac{a+b}{4} + 1\right)$

$\qquad \overset{2}{\cancel{4}} \cdot \dfrac{a+3}{\cancel{2}} = \overset{1}{\cancel{4}} \cdot \dfrac{a+b}{\cancel{4}} + 4 \cdot 1$

$\qquad 2a + 6 = a + b + 4$

$\qquad 2a - a - b = 4 - 6$

(2) $\qquad a - b = -2 \qquad$ Standard form for equation (2)

Having expressed each equation in standard form, we now write (2) below (1):

(1) $\quad 2a + 3b = 36 \qquad$ L.C.M. = 2

(2) $\quad \underline{a - b = -2}$

$1 \cdot (1) \quad 2a + 3b = 36$

$2 \cdot (2) \quad \underline{2a - 2b = -4}$

(3) $\qquad 5b = 40 \qquad$ (Subtraction)

$\qquad\qquad b = 8$

(1) $\quad 2a + 3b = 36$

$\qquad 2a + 24 = 36$

$\qquad\qquad 2a = 36 - 24$

$\qquad\qquad 2a = 12$

$\qquad\qquad a = 6$

The common solution is (6, 8).

EXERCISES 9.3 (Part 4)

A

Determine the standard form of each of the following equations.

1. $5x = 6y - 3$
2. $3y = 4 - 7x$
3. $3a + 4 = 5b + 6$
4. $7a - 2 = a - 3b$
5. $6c - 2d + 5 = 0$
6. $5c - d = 3c + 4 - 2d$
7. $2(b - 3c) = 5$
8. $4b - 3 = 3(2b + 5c)$
9. $\frac{a}{2} + \frac{b}{3} = 5$
10. $\frac{3a}{4} - \frac{5}{4} = \frac{b}{2}$
11. $\frac{5w - 3z}{2} = \frac{2}{3}$
12. $7w = \frac{3z - 2}{5}$
13. $\frac{2x + 3y}{6} = \frac{5x - 4}{3}$
14. $\frac{7x}{4} + \frac{x + y}{5} = 2$
15. $\frac{5y - 2}{6} - \frac{5z}{4} = 3$
16. $\frac{2y - 3z}{9} - \frac{5z - 4}{6} = \frac{7y}{3}$

B

Solve each of the following systems of equations.

1. $3x = 14 - 2y$
   $5x - 6 = y$

2. $3b = 4a - 9$
   $3a = 11 - 2b$

3. $5a - 2b = b - 1$
   $4a - 5b = 3 - 2a$

4. $5(x + y) = 14 - x$
   $3(x + y) = 12 - x$

5. $x = 3(y + 4)$
   $4(x + 2y) = 10 + y$

6. $2(2x + 3) = -1(3y + 2)$
   $3(x + y) = 3 - 2(y - 1)$

7. $y + 4x = 36$
   $\frac{x}{2} + \frac{y}{5} = \frac{18}{5}$

8. $6(a - b) = 7 - b$
   $\frac{a}{3} - \frac{b}{4} = \frac{1}{4}$

9. $\frac{3a}{4} + \frac{b}{2} = \frac{-11}{4}$
   $\frac{3a}{2} = 7 + \frac{b}{4}$

10. $\frac{y - 10}{4} = \frac{1}{6} + \frac{x + 2}{3}$
    $\frac{x + y}{3} - 1 = \frac{1 - x}{15}$

9.4 SOLVING A LINEAR SYSTEM OF INEQUALITIES BY GRAPHING

Earlier in this chapter we found that the common solutions of two open sentences were the pairs of values that satisfy both of the open sentences at the same time. These pairs of values name points that are common to the graphs of the open sentences.

When the two open sentences are linear equations, their graphs are straight lines. Hence, in general, the number of points the two lines have in common is only one and, therefore, there is only one pair of values that satisfies both equations at the same time.

However, the graph of an inequality consists of a half-plane. In view of this, the common solutions of two inequalities will consist of the region of the plane where two half-planes intersect. That region will have infinitely many points, which in turn implies that there will be infinitely many pairs of values that will satisfy the two linear inequalities at the same time. Since there are far more common solutions to two inequalities than can possibly be listed, the only manner in which these solutions can be expressed is by pointing to the region common to their graphs. The way in which this is done will be shown in the following examples.

Example 1

Determine the common solutions for the following system of inequalities.

$$y > 3$$
$$x < -2$$

Explanation Each of these inequalities is graphed on the same set of axes. In the following graph the points in the half-plane "above" the line $y = 3$ is the graph of $y > 3$. Similarly, the points to the "left" of the line $x = -2$ represent the half-plane that is the graph of $x < -2$. The darker shaded region is the intersection of the two half-planes. The coordinates of each of the points in the darker shaded region will satisfy both inequalities.

Solution

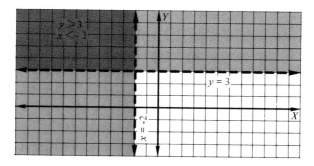

Example 2

Determine the common solutions for the following system of inequalities.

$$y < 2x - 3$$
$$x > 1 - 2y$$

Explanation Our initial step is to determine the graph of each of these inequalities. The points of the half-plane below the line $y = 2x - 3$ is the graph of $y < 2x - 3$. The points of the half-plane to the right of the line $x = 1 - 2y$ is the graph of $x > 1 - 2y$. The darker shaded region is the intersection of the two half-planes. This region contains the points whose coordinates satisfy both inequalities.

Solution $y = 2x - 3$ 　　　　　　　　 $x = 1 - 2y$

x	$2x - 3 =$	y
0	$2 \cdot 0 - 3 =$	-3
2	$2 \cdot 2 - 3 =$	1

y	$1 - 2y =$	x
0	$1 - 2 \cdot 0 =$	1
3	$1 - 2 \cdot 3 =$	-5

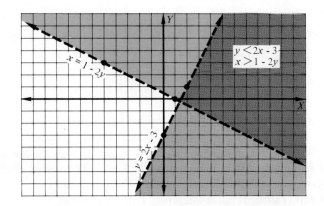

Although it is possible to solve both inequalities for the same variable, it is best to solve the first for one of the variables and the second for the other. By doing this, the graph of the intersection of the two half-planes shows up more clearly than if both inequalities are solved for the same variable.

Example 3

Determine the common solutions for the following system of inequalities.

$$3y + x > -5$$
$$3x - 2y > 6$$

9.4/Solving a Linear System of Inequalities by Graphing

Solution
(1) $3y + x > -5$
$x > -5 - 3y$

y	$-5 - 3y$	=	x
0	$-5 - 3 \cdot 0$	=	-5
-2	$-5 - 3(-2)$	=	1

(2) $3x - 2y < 6$
$-2y < 6 - 3x$
$y > -3 + \dfrac{3x}{2}$

x	$-3 + \dfrac{3x}{2}$	=	y
0	$-3 + \dfrac{3 \cdot 0}{2}$	=	-3
4	$-3 + \dfrac{3 \cdot 4}{2}$	=	3

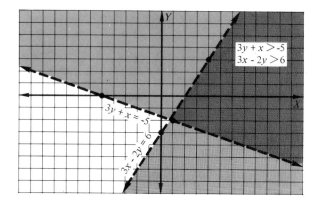

EXERCISES 9.4

Determine the common solutions for each of the following systems of inequalities.

1. $y < 5$
   $x > 2$

2. $y < -1$
   $x > -2$

3. $y > 0$
   $x > 0$

4. $x < 3y + 1$
   $y > 4 - x$

5. $x > 1 - 2y$
   $y > 2 + \dfrac{1}{3}x$

6. $x < \dfrac{1}{2}y$
   $y > 4 - \dfrac{1}{3}x$

7. $x + 3y < 5$
   $y - 2x < 1$

8. $x + 4y > 0$
   $y + 3 < 3x$

9. $4x - 8 > 0$
   $y + x < 1$

10. $x - 5y + 1 < 0$
    $3x - y > 0$

11. $2x + 3y < 6$
    $3x - 2y > -6$

12. $5x - y > 4$
    $5y + 2x < 8$

9.5 PROBLEM SOLVING INVOLVING TWO VARIABLES

PART 1

Our ability to solve systems of linear equations in two variables has given us a powerful tool in our investigation of narrative problems. In the past, each time we encountered a narrative in which we were asked to determine two different numbers, we began by representing one of these numbers by some variable. We then had to represent the other number in terms of this *same* variable. This will no longer be necessary for we can now represent each of the numbers by its own variable.

As we discovered earlier, every narrative problem that calls upon us to determine two numbers must have two pieces of information relating these two numbers. Each of these pieces of information will now lead us to a separate equation. Once we have established these two equations, then finding their common solution will involve nothing more than what we have been doing throughout this chapter.

A number of the illustrations and exercises that follow will be much the same as those investigated earlier. Now, though, we shall apply this new approach to determining the numbers in question.

Example 1

The sum of two numbers is 27. The sum of 3 times the first and 4 times the second is 96. What are the numbers?

Explanation The two pieces of information in this example seem quite apparent. They are:

1. The sum of the two numbers is 27.
2. The sum of 3 times the first and 4 times the second is 96.

Hence, should we represent the first number by x and the second by y, then the first piece of information can be translated into the mathematical sentence of

1. $x + y = 27$.

In addition, the translation of the second sentence is

2. $3x + 4y = 96$.

Our work is then completed by finding the common solution to these two equations.

Solution Let x represent the first number.
Let y represent the second number.

Therefore, (1) $\quad x + y = 27$
(2) $\quad 3x + 4y = 96$

9.5/Problem Solving Involving Two Variables

$3 \cdot (1)$ $3x + 3y = 81$
$1 \cdot (2)$ $3x + 4y = 96$
(3) $-1y = -15$
                $y = 15$ (Second number)
(1) $x + y = 27$
                $x + 15 = 27$
                $x = 27 - 15$
                $x = 12$ (First number)

Check:
1. The sum of the two numbers is $12 + 15 = 27$.
2. The sum of 3 times the first and 4 times the second is
   $3 \cdot 12 + 4 \cdot 15 = 36 + 60 = 96$.

Example 2

John's age is twice Mary's age. When 8 is added to John's age the sum will be 1 less than 3 times Mary's age. What are their ages?

Explanation The two pieces of information in this problem are:

1. John's age is twice Mary's age.
2. 8 added to John's age is 1 less than 3 times Mary's age.

Hence, by representing John's age by x and Mary's age by y, both of these sentences can be readily translated.

1.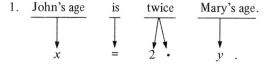

2.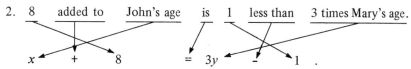

The common solution is found after arranging each of these equations in standard form.

Solution Let x represent John's age.
Let y represent Mary's age.

Therefore, (1) $x = 2y$
(2) $x + 8 = 3y - 1$
───────────────
(1) $x - 2y = 0$
(2) $x - 3y = -9$
───────────────
(3) $y = 9$ (Mary's age)
(1) $x = 2y$
        $x = 2 \cdot 9$
        $x = 18$ (John's age)

Check: 1. John's age is twice Mary's age. $18 = 2 \cdot 9$
        2. 8 added to John's age is 1 less than 3 times Mary's age.
        $18 + 8 \stackrel{?}{=} 3 \cdot 9 - 1$
        $26 \stackrel{?}{=} 27 - 1$
        $26 = 26$

Example 3

A person purchased one variety of apples at 32¢ per pound and another variety at 37¢ per pound. The number of pounds at 32¢ per pound is one more than twice the number at 37¢ per pound. How many pounds of each variety were purchased if the total cost was $5.37?

Explanation This very same problem was examined in Chapter 5. You may want to refer to it at this time. Basically the solution of problems of this nature depends upon realizing that the total cost of a number of items that are all priced at the same cost per item is found by multiplying

the cost per item   times   the number of items.

Thus, in this example if the number of pounds at 32¢ per pound is represented by x, the total value of these x pounds is $32x$. Similarly, if the number of pounds of apples at 37¢ per pound is represented by y, then the total cost of these apples is $37y$.

The two pieces of information in this problem are:

1. The number of pounds at 32¢ per pound is one more than twice the number at 37¢ per pound.
2. The total cost of the 32¢ apples and the 37¢ apples is $5.37.

Solution Let x represent the number of pounds at 32¢ per pound.
Let y represent the number of pounds at 37¢ per pound.

9.5/Problem Solving Involving Two Variables

Therefore, (1) $\quad x = 2y + 1$

(2) $\quad \underline{32x + 37y = 537}$

(2) $32(2y + 1) + 37y = 537$

$64y + 32 + 37y = 537$

$64y + 37y = 537 - 32$

$101y = 505$

$y = 5 \quad$ (Number of pounds @ 37¢)

(1) $\quad x = 2y + 1$

$x = 2 \cdot 5 + 1$

$x = 10 + 1$

$x = 11 \quad$ (Number of pounds @ 32¢)

Check: 1. The number of pounds at 32¢ per pound is 1 more than twice the number at 37¢ per pound:

$11 \stackrel{?}{=} 2 \cdot 5 + 1$

$11 \stackrel{?}{=} 10 + 1$

$11 = 11$

2. The total cost is $5.37:

Cost @ 32¢		Cost @ 37¢		Total Cost
11 × 32¢	+	5 × 37¢	$\stackrel{?}{=}$	537¢
352¢	+	185¢	$\stackrel{?}{=}$	537¢
		537¢	=	537¢

Explanation continued Notice that when writing $5.37 in equation (2) it appears as 537¢. Since the 32 and 37 on the left side of the equation are expressed in "cents," the right side has to be expressed in cents also.

EXERCISES 9.5 (Part 1)

Find the numbers requested in each of the following problems.

1. The sum of two numbers is 43. The difference of these numbers is 5. What are the numbers?
2. The sum of Tina's and Arthur's age is 59. Arthur is 3 years older than Tina. What are their ages?

3. The total cost of a house and the land on which it is built is $75,000. If the cost of the house is 4 times the cost of the land, what is the cost of each?

4. Jane purchased a coat and dress for a total cost of $240. The price of the coat was $45 more than twice the cost of the dress. How much did she pay for each of these items?

5. During the first two days of their trip the Petersons drove 670 miles. The number of miles they drove the first day was 20 less than the number they drove the second day. How many miles did they drive each day?

6. The varsity basketball team scored a total of 149 points during the first two games of the season. The score of the second game was 22 less than twice the score of the first game. How many points were scored in each game?

7. Victor's raise last year was $\frac{1}{4}$ of his weekly salary at that time. After receiving the raise he earned $275 per week. How much was his weekly salary and how much was the raise he received?

8. The larger of two numbers is 4 more than 3 times the smaller. If 5 is added to the larger, the sum will be 2 less than 4 times the smaller. What are the numbers?

9. The length of a rectangle is 3 times the width. The perimeter of the rectangle is 64 meters. What are the length and width of the rectangle?

10. Three times the larger of two numbers added to 5 times the smaller is 2. If 3 times the smaller is subtracted from twice the larger, the difference is 14. What are the two numbers?

11. If the larger of two numbers is divided by 2 and that quotient is added to the smaller number, the sum will be 3. If the smaller of the two numbers is divided by 5 and that quotient is added to the larger number, the sum will be 24. What are the numbers?

12. Sally purchased 50 stamps for which she paid $6.20. If she purchased only 11¢ and 13¢ stamps, how many of each did she buy?

13. When a coin bank was opened it was found that it contained $6.55 in dimes and quarters. The number of dimes was 6 more than the number of quarters. How many coins of each variety were there?

14. The receipts at the performance of a school play was $1,675. The students paid $2 per ticket while nonstudents paid $3 per ticket. The total number of people attending the performance was 700. How many student tickets were sold and how many nonstudent tickets were sold?

15. Nuts at 89¢ per pound are mixed with nuts at 95¢ per pound to make a blend whose total cost is $27.48. The number of pounds at 89¢ per pound is 4 more than the number at 95¢ per pound. How many pounds of each variety are in the blend?

16. If Betty purchases 3 large grapefruit and 5 small ones, it will cost her $1.60. However, if she purchases 1 large one and 7 small ones, her cost will be $1.28. What is the cost of each large grapefruit and each small one?

17. If 5 pounds of red grapes and 4 pounds of green grapes are purchased, the total price will be $3.13. If 4 pounds of red grapes and 5 pounds of green grapes are purchased, the total price will be $3.08. What is the price per pound of each variety of grapes?

PART 2

In order to understand the meaning of a number in the decimal or base-ten system with which we are familiar, that number is frequently expressed in *expanded notation form.* For instance, the number 37 when written in expanded notation form will appear as

$$37 = 3 \cdot 10 + 7 \cdot 1.$$

Occasionally, the expanded notation form of 37 will even be written as

$$37 = 3 \text{ tens} + 7 \text{ ones}.$$

or possibly

$$37 = 3 \text{ tens} + 7 \text{ units}.$$

In view of this, the digit 3 in the number 37 is called the *tens digit* while the digit 7 is called the *units digit*. Similarly, in the number 64, the 6 is the tens digit while the 4 is the units digit.

On the other hand, if we are told that the tens digit of a number is 8 and the units digit is 2, it is possible to determine the number itself by realizing that the number when written in expanded form will be

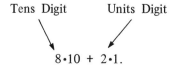

In the same way, if the tens digit happens to be represented by the variable t while the units digit is represented by the variable u, then the number itself when written in expanded notation appears as

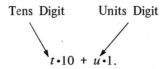

By this point, though, we are well aware of the fact that the numerical coefficient usually appears before the variable rather than after it. Hence, the quantity above is rewritten as

$$t \cdot 10 + u \cdot 1 = 10t + 1u$$
$$= 10t + u$$

Thus, if t represents the tens digit and u represents the units digit, then the number itself is represented by

$$10t + u.$$

Example 1

In a two-digit number the units digit is twice the tens digit. The number itself is 4 more than 5 times the units digit. What is the number?

Explanation The tens digit in the number will be represented by t while the units digit will be represented by u. The two pieces of information in the narrative can be translated into the mathematical sentences that appear below them.

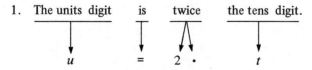

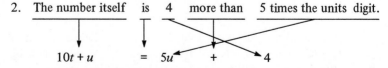

Solution Let t represent the tens digit.
Let u represent the units digit.

Therefore, (1) $\qquad u = 2t$

(2) $\qquad 10t + u = 5u + 4$

9.5/Problem Solving Involving Two Variables

$$
\begin{align}
(2) \quad & 10t + u - 5u = 4 \\
(2) \quad & 10t - 4u = 4 \\
(3) \quad & 10t - 4(2t) = 4 \quad (u \text{ is replaced with } 2t) \\
& 10t - 8t = 4 \\
& 2t = 4 \\
& t = 2 \quad \text{(tens digit)} \\
(1) \quad & u = 2t \\
& u = 2 \cdot 2 \\
& u = 4 \quad \text{(units digit)}
\end{align}
$$

The number is 24.

Example 2

In a two-digit number the tens digit is 4 more than the units digit. If the digits are reversed, then twice the new number will be 12 more than the original number. What was the original number?

Explanation If the digits of a number are reversed, then whereas t was the original tens digit and u the units digit, in the new number, u will be the tens digit and t the units digit.

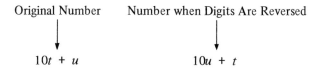

For instance, when the digits of the number 83 are reversed, the new number will be 38. Thus,

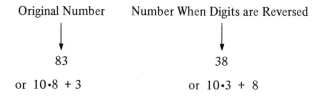

The two pieces of information in this example and their translations follow.

1.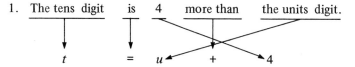

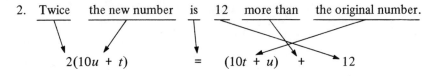

Solution Let t represent the tens digit.
Let u represent the units digit.

Therefore, (1) $\qquad t = u + 4$

(2) $\qquad 2(10u + t) = (10t + u) + 12$

(2) $\qquad 20u + 2t = 10t + u + 12$

$\qquad 20u + 2t - 10t - u = 12$

$\qquad 19u - 8t = 12$

(3) $\qquad 19u - 8(u + 4) = 12 \qquad (u + 4 \text{ replaces } t)$

$\qquad 19u - 8u - 32 = 12$

$\qquad 19u - 8u = 12 + 32$

$\qquad 11u = 44$

$\qquad u = 4 \qquad \text{(units digit)}$

(1) $\qquad t = u + 4$

$\qquad t = 4 + 4$

$\qquad t = 8 \qquad \text{(tens digit)}$

The number is 84.

EXERCISES 9.5 (Part 2)

Find the number requested in each of the following problems. Each of the problems involve two-digit numbers only.

1. The units digit is 2 more than the tens digit. The number itself is 5 more than 6 times the units digit. What is the number?

2. The tens digit is 1 less than the units digit. The number itself is 2 more than 9 times the units digit. What is the number?

3. The units digit is 4 times the tens digit. The number itself is 4 more than 3 times the units digit. What is the number?

4. Four times the units digit is 1 more than 3 times the tens digit. The number itself is 1 less than 11 times the tens digit. What is the number?

5. The units digit is 2 less than 3 times the tens digit. If 2 is subtracted from the number, the difference will be 5 times the units digit. What is the number?

6. The sum of the digits is 10. Seven times the tens digit added to 8 times the units digit is the number itself. What is the number?

7. The units digit is 1 more than the tens digit. When the digits are reversed, the new number is 7 times the sum of the digits. What was the original number?

8. If the tens digit is subtracted from the units digit, the difference is 1. If the digits are reversed, the new number will be 2 less than 14 times the original tens digit. What is the original number?

9. The units digit is 4 more than the tens digit. If the digits are reversed, the sum of the original number and the new number will be 154. What is the original number?

10. The units digit and the tens digit are two consecutive even numbers where the units digit is the larger. The number itself is 6 times the units digit. What is the number?

11. The units digit and the tens digit are two consecutive integers where the units digit is the larger. The number itself is 2 more than 8 times the units digit. What is the number?

12. The units digit and the tens digit are two consecutive odd integers where the tens digit is the smaller. If the digits are interchanged the sum of the new number and the original number will be 88. What is the original number?

PART 3

Narrative problems often examined at this point in the study of algebra are those that are known as "motion" problems. Under this heading we encounter situations where boats are rowed upstream to a certain landing at which time they are turned around and rowed downstream returning to their original dock. Similarly, we find narratives that involve people or fish swimming upstream followed by their return downstream. Also falling into this same group of problems is the airplane that first flies with the wind and then returns against the wind. In each of these problems we are called upon to determine certain speeds under the conditions presented in the narrative.

Before we can turn our attention to these problems we will first have to investigate the relation that exists between the distance an object will travel and the speed and time during which it is traveling. For instance, consider the situation where an automobile is traveling at 30 miles per hour. The term 30 miles per hour (mph) implies that during each hour the car will have moved a distance of 30 miles. Hence, in 1 hour it will have traveled a distance of 30 miles while in 2 hours the distance will be 60 miles.

The number 60 is determined by the fact that since the car travels 30 miles each hour then in 2 hours it will travel

$$2 \text{ times } 30 \text{ miles} \quad \text{or} \quad 60 \text{ miles.}$$

Similarly, in 5 hours it will travel

$$5 \text{ times } 30 \text{ miles} \quad \text{or} \quad 150 \text{ miles.}$$

In general, we can say that the *distance* a car travels can be found by multiplying the *number of hours* during which the car travels by the *number of miles per hour* at which it travels. The *number of hours* which the car has traveled is called the *time* it travels. The *number of miles per hour* at which it travels is referred to by several names such as

velocity, speed, or rate.

Using these new terms we can express the relation between the distance, rate, and time as

Time   times   Rate = Distance

or as a formula in the form of

$$TR = D.$$

It is important to realize that the "rate" and the "time" in the above formula must both be expressed in the same unit. For instance, if a runner is traveling at 8 feet per *second*, then the "time" stated in the formula must also be expressed in *seconds*. Similarly, if the speed of a plane is expressed in terms of 70 miles per *minute*, then so, too, must the "time" be stated in *minutes* before the formula can be applied.

Now let's examine this information in terms of the narratives that are of interest to us. Consider, as an example, a person who, when expending a certain amount of energy, is able to row a boat on a lake at a rate of 5 miles per hour. Lake water is usually considered as "still" water for it does not move. Now, if this same person should expend the same amount of energy and row his boat on a river, something quite different than before will occur. Should he be rowing downstream, then not only will the boat be moving at the 5 mph that it had before but, also, it will be moved along by the current of the river. Thus, if the current of the river is 2 mph, then the total speed at which the boat will be traveling is

$$5 \text{ mph} + 2 \text{ mph} \quad \text{or} \quad 7 \text{ mph.}$$

9.5/Problem Solving Involving Two Variables

On the other hand, should he be rowing upstream, that is, against the current, then the flow of the river moving in the direction *opposite* to which the boat is moving will slow down the boat. Hence, rather than moving at the rate of 5 mph, the speed will be

$$5 \text{ mph} - 2 \text{ mph} \quad \text{or} \quad 3 \text{ mph}.$$

In general, if the rate of a boat in *still* water is x miles per hour and if the rate of the current of the river is y miles per hour, then when the boat is traveling downstream (with the current), its rate will be

$$(x + y) \text{ miles per hour}.$$

On the other hand, when it is traveling upstream (against the current), its rate will be

$$(x - y) \text{ miles per hour}.$$

With this background we are prepared to examine the following examples.

Example 1

A motorboat takes 5 hours to travel a distance of 45 miles upstream on a river. The return trip takes only 3 hours. What is the rate at which the boat travels in still water and what is the rate of the current of the river?

Explanation The two pieces of information stated in the problem are:

1. It takes 5 hours to travel a *distance* of 45 miles upstream.
2. It takes 3 hours to travel a *distance* of 45 miles downstream.

Hence, if we represent the rate of the boat in still water as x and the rate of the current as y, then by applying the formula $TR = D$, each of the above pieces of information can be translated as

$$
\begin{array}{cccccccl}
1. & T & \cdot & R & = & D & & \\
 & \downarrow & & \downarrow & & \downarrow & & \\
 & 5 & \cdot & (x - y) & = & 45 & & \text{(Upstream)} \\
2. & T & \cdot & R & = & D & & \\
 & \downarrow & & \downarrow & & \downarrow & & \\
 & 3 & \cdot & (x + y) & = & 45 & & \text{(Downstream)}
\end{array}
$$

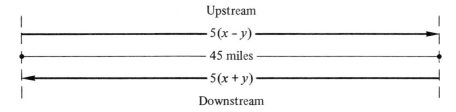

Solution Let x represent the rate of the boat in still water. Let y represent the rate of the current.

Therefore, $x - y$ represents rate upstream

and $x + y$ represents rate downstream.

Hence,
(1) $5(x - y) = 45$
(2) $3(x + y) = 45$
(1) $\overline{5x - 5y = 45}$
(2) $\overline{3x + 3y = 45}$
$3 \cdot (1)$ $\overline{15x - 15y = 135}$
$5 \cdot (2)$ $\overline{15x + 15y = 225}$
(3) $\overline{30x = 360}$
$x = 12$ mph   (Rate in still water)
(2) $3x + 3y = 45$
$3 \cdot 12 + 3y = 45$
$36 + 3y = 45$
$3y = 45 - 36$
$3y = 9$
$y = 3$ mph   (Rate of current)

Explanation Notice that since the distance is stated in "miles" and the time is stated in "hours," then the rate must be expressed in "miles per hour."

The solution of the two equations in Example 1 could have been found somewhat more easily than we had done above. Since this same situation will occur quite frequently in the exercises that follow it will be well to examine how our work can be simplified.

Consider equation (1) in Example 1:

$$5(x - y) = 45. \qquad (1)$$

Using the multiplication principle of equality, we can multiply both members of this equation by $\frac{1}{5}$. When this is done, the new equation will be far simpler than the one above:

$$\tfrac{1}{5} \cdot 5(x - y) = \tfrac{1}{5} \cdot 45$$
$$x - y = 9$$

However, $\frac{1}{5}$ times 45 is exactly the same as 45 divided by 5:

$$45 \div 5 = 9.$$

Similarly, $\frac{1}{5}$ times $5(x-y)$ is exactly the same as $5(x-y)$ **divided** by 5:

$$5(x-y) \div 5 = \frac{5(x-y)}{5} = x - y$$

In view of this, rather than considering both members of the equation as being multiplied by $\frac{1}{5}$, the mathematicians prefer to think in terms of both members being divided by 5.

Equation (2) of Example 1 can also be simplified by dividing both members by 3:

$$3(x+y) = 45 \qquad (2)$$

$$x + y = 15.$$

Quite apparently, in order to apply this method for simplifying an equation, it is necessary that both members of the equation have a *common factor*. For instance, the equation

$$3x + 5y = 7$$

cannot be simplified by dividing both members by a common factor for the left and right members have no common factor. However, the equation

$$8x + 20y = 12$$

can be simplified to

$$2x + 5y = 3.$$

Both members of $8x + 20y = 12$ were divided by their common factor 4 to arrive at the equation $2x + 5y = 3$.

Example 2

Simplify the following equation

$$150x + 50y = 200.$$

Explanation Since both members of this equation have the factor 50, it is possible to simplify this equation by dividing both members by 50.

Solution $\qquad\qquad 150x + 50y = 200$

$\qquad\qquad\qquad\quad 3x + 1y = 4 \qquad$ (Both members $\div$ 50)

In the following example you will find that the division principle can be applied although the divisor does *not* happen to be a *factor* of both members of the equation.

Example 3

Flying with the wind a plane traveled a distance of 1575 miles in 3 hours 30 minutes. Had the plane been traveling against the wind, the same trip would have taken 4 hours 30 minutes. What was the speed of the wind and what would the speed of the plane have been had no wind been blowing?

Explanation The only difference between this example and Example 1 is that in this situation the "time" is expressed in hours and minutes. Since the velocity of a plane is usually stated in terms of miles per *hour*, it is necessary to rename the "minutes" in terms of "hours." As there are 60 minutes in 1 hour, the 30 minutes of traveling time is represented as $\frac{30}{60}$ of 1 hour or $\frac{1}{2}$ hour. Hence, flying with the wind

the time is $3\frac{1}{2}$ hours or 3.5 hours.

Flying against the wind

the time is $4\frac{1}{2}$ hours or 4.5 hours.

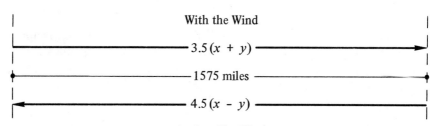

Against The Wind

Solution Let x represent the rate of the plane in still air.
Let y represent the rate of the wind.

Therefore, $x + y$ represents the rate with the wind
and $x - y$ represents the rate against the wind.
Hence,
(1) $3.5(x + y) = 1575$
(2) $4.5(x - y) = 1575$
_____
(1) $x + y = 450$ (Both members ÷ 3.5)
(2) $x - y = 350$ (Both members ÷ 4.5)
_____
(3) $2x = 800$
$x = 400$ mph (Rate in still air)

$$(1) \quad x + y = 450$$
$$400 + y = 450$$
$$y = 450 - 400$$
$$y = 50 \text{ mph} \quad \text{(Rate of wind)}$$

EXERCISES 9.5 (Part 3)

A

Simplify each of the following equations by dividing both members by their highest common factor.

1. $5(x + y) = 30$
2. $7(x - y) = 35$
3. $6(5x - y) = 42$
4. $3x + 3y = 3$
5. $4x - 6y = 4$
6. $9x - 6y = 12$
7. $25x + 25y = 100$
8. $40x - 8y = 24$
9. $60x + 48y = 24$
10. $70x - 90y = 120$

B

Find the numbers requested in each of the following problems.

1. Carl took a number of his friends for a ride in his motorboat. He spent 6 hours traveling a distance of 60 miles while going against the current of a river. The return trip down the river took only 5 hours. What was the rate of the current and what would the speed of the boat have been in still water?

2. Jim rowed his boat up the river a distance of 7.5 miles. The trip took him 2.5 hours. The trip down the river took him only 1.5 hours. What was the speed of the current and at what rate would Jim have been rowing had the boat been in still water?

3. During a trip of 1,500 miles a plane is constantly traveling against the wind. It took 6 hours to make this trip. Had the plane been traveling with the wind the trip would have taken only 5 hours. What was the velocity of the wind and at what rate would the plane have been traveling had it been flying in still air?

4. It takes a plane 5.5 hours to travel a distance of 1,320 miles when flying against the wind. On the return trip when flying with the wind of the same velocity, the trip took only 4 hours. What was the velocity of the wind, and what would the velocity of the plane have been had it been flying in still air?

5. In making a trip of 1,170 miles when flying against the wind the pilot will spend $6\frac{1}{2}$ hours in flight. With the same wind speed, when traveling with the wind the pilot will only need $4\frac{1}{2}$ hours to make this flight. What is the wind velocity and what is the speed of the plane in still air?

6. When making a *round-trip* of 7,888 miles, a plane takes 8 hours and 30 minutes when traveling in one direction and only 7 hours and 15 minutes when traveling in the other direction. Assuming that the wind velocity is always the same, what is that velocity and what would the velocity of the plane be if it were flying in still air?

7. Peggy took a 15 mile *round-trip* on a river in her motorboat. She spent 1 hour 15 minutes when going upstream and only 45 minutes when returning downstream. What is the rate of the current and how fast would the boat have been traveling had the trip been in still water?

8. A small private plane flies with the wind a distance of 96 miles in 32 minutes. It makes the trip back under the same wind conditions except that the return trip takes 48 minutes.

    a. What was the wind velocity in miles per *minute*?
    b. What was the wind velocity in miles per *hour*? (Multiply your answer to *a* by 60.)

9. A small plane flies a *round-trip* of 300 miles. It takes 1 hour 15 minutes going and only 1 hour returning. The wind velocity was the same in both directions.

    a. What was the wind velocity in miles per minute?
    b. What was the wind velocity in miles per hour?

10. A plane left on a trip at 7:00 A.M. and reached its destination at 11:00 A.M. It began its return trip at 1:30 P.M. and arrived at the original airport at 6:45 P.M. The round-trip distance is 1,008 miles while the wind velocity remained the same throughout the day. What was the wind velocity?

PART 4

Being able to solve linear systems of equations in two variables can be a valuable aid to people who invest their money under different circumstances. Before we can understand what is involved in the solution of their investment problems, we must first know what is meant by the statement that "money earns money."

Many of us keep our money in savings banks that presently pay an interest rate of approximately 5%. This implies that the bank will pay us

5% of the amount on deposit

if we keep our money in the bank for *one full year*.

9.5/Problem Solving Involving Two Variables

From our earlier work on percent we learned that the English word "of" can be translated into the mathematical operation of multiplication. Hence, the amount a deposit of $4,000 will earn in *one year* is

$$5\% \text{ of } \$4,000$$
$$\downarrow \quad \downarrow \quad \downarrow$$
$$\text{or } 5\% \cdot \$4,000.$$

By multiplying .05 by $4,000, the product of $200 is found. Thus,

$$\$200 = \$4,000 \times 5\%.$$

In the example above,

The $4,000 is called the *principal*.
The 5% is called the *interest rate*.
The $200 is called the *interest*.

Hence, the interest is found as follows,

$$\text{Interest} = \text{Principal} \times \text{Rate}$$

$$I = PR.$$

The example just examined referred to the fact that the money had been deposited in a bank. Actually, the same situation exists whether the money is invested in a business that pays 5% per year, or in bonds, or in stocks, or is on loan to someone.

Now let us examine a few examples where we apply the interest formula.

Example 1

Mary Paglia has $10,000 on deposit in two different accounts. One of the accounts pays an interest rate of 6%, and the other pays 5%. Her interest on both of these accounts amounts to $570 for the year. How much money does she have in each account?

The two pieces of information given in this problem are:

1. The total principal is $10,000.
2. The total interest is $570.

Explanation If we represent the principal in the 6% account as x dollars, then the interest received on this account will be 6%x for,

$$\text{Interest (first account)} = P \cdot R$$
$$\downarrow \quad \downarrow$$
$$= x \cdot 6\% \text{ or } 6\%x.$$

Similarly, if the amount on deposit in the 5% account is y dollars, then the interest received from this account will be,

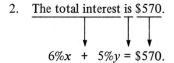

$$= y \cdot 5\% \text{ or } 5\%y.$$

Thus, the total interest on both accounts is

Interest on		Interest on
6% Account		5% Account
$6\%x$	$+$	$5\%y$.

Hence, the second piece of information can be translated as follows:

2.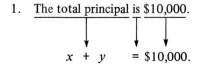

$$6\%x + 5\%y = \$570.$$

Also, since x dollars is the amount in one account and y dollars is the amount in the other account, then the first piece of information can be translated as

1. The total principal is $10,000.

$$x + y = \$10,000.$$

Solution Let x represent the number of dollars at the 6% rate.
Let y represent the number of dollars at the 5% rate.

$$
\begin{array}{rlrl}
\text{Therefore,} & (1) & x + y &= 10{,}000 \\
& (2) & 6\%x + 5\%y &= 570 \\
& (2) & .06x + .05y &= 570 \\
100 \cdot (2) & & 6x + 5y &= 57{,}000 \\
5 \cdot (1) & & 5x + 5y &= 50{,}000 \\
& & x &= \$7{,}000 & \text{(Amount at 6\% rate)} \\
& (1) & x + y &= 10{,}000 \\
& & 7{,}000 + y &= 10{,}000 \\
& & y &= 10{,}000 - 7{,}000 \\
& & y &= \$3{,}000 & \text{(Amount at 5\% rate)}
\end{array}
$$

9.5/Problem Solving Involving Two Variables

Equation (2) in the example above is simplified by multiplying both members of that equation by 100. This procedure is based on the fact that the fastest method for multiplying a number by 100 is to move the decimal point in that number 2 places to the right. Thus,

$$100 \times 5.264 = 5{\scriptstyle\curvearrowright}26.4 = 526.4$$

$$100 \times .3768 = {\scriptstyle\curvearrowright}37.68 = 37.68$$

$$100 \times .05 = {\scriptstyle\curvearrowright}05.0 = 5.$$

Similarly, multiplying a number by 10 will move the decimal point 1 place to the right, while multiplying a number by 1,000 will move the decimal point 3 places to the right in that number.

$$10 \times 5.264 = 5{\scriptstyle\curvearrowright}2.64 = 52.64$$

$$1,000 \times 4 = 4{\scriptstyle\curvearrowright}000. = 4,000.$$

In Equation (2) of Example 1 we are interested in expressing each coefficient as an integer. This implies the .06 and .05 has to be multiplied by 100. However, to do this we must multiply *both* members of that equation by 100.

$$(2) \quad .06x + .05y = 570$$
$$100 \cdot (2) \quad {\scriptstyle\curvearrowright}06.x + {\scriptstyle\curvearrowright}05.y = 570.00.$$
$$\text{or} \quad (2) \quad 6x + 5y = 57,000$$

Example 2

Simplify the following equation so that each of the coefficients is an integer.

$$.06x + .075y = 4,500$$

Explanation Were both members of this equation multiplied by 100, the coefficient .06 would become 6. However, the coefficient .075 would become 7.5 which is still a decimal. If both members are multiplied by 1,000, though, then .06 would become 60 and .075 would become 75. This will meet the conditions of the problem that each coefficient be an integer.

Solution $.06x + .075y = 4,500$

$${\scriptstyle\curvearrowright}060.x + {\scriptstyle\curvearrowright}075.y = 4500{\scriptstyle\curvearrowright}000. \quad \text{(Both members} \times 1,000\text{)}$$

$$\text{or} \quad 60x + 75y = 4,500,000$$

Example 3

A man invested a total of $20,000 in two different business enterprises that paid $7\frac{1}{2}\%$ and 8% each. The $7\frac{1}{2}\%$ investment earned him $260 more each year than the 8% investment. How much money did he invest in each business?

Explanation If we represent the number of dollars in the $7\frac{1}{2}\%$ investment by x and the number in the 8% investment by y, then each of the two pieces of information in the problem can be translated as follows:

1. The total investment is $20,000.

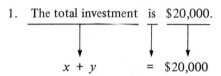

2. The interest on $7\frac{1}{2}\%$ is $260 more than the interest on 8%.

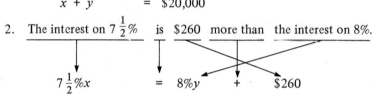

One more point of information—$7\frac{1}{2}\%$ is renamed as a decimal as follows:

$$7\frac{1}{2}\% = 7.5\% = .075$$

Solution Let x represent the number of dollars at the $7\frac{1}{2}\%$ rate.
Let y represent the number of dollars at the 8% rate.

$$
\begin{array}{rrrl}
\text{Therefore,} & (1) & x + y &= 20{,}000 \\
& (2) & 7\frac{1}{2}\%x &= 8\%y + 260 \\
\cline{2-4}
& (2) & .075x &= .08y + 260 \\
1{,}000 \cdot (2) & & 75x &= 80y + 260{,}000 \\
& (2) & 75x - 80y &= 260{,}000 \\
80 \cdot (1) & & 80x + 80y &= 1{,}600{,}000 \\
\cline{2-4}
& & 155x &= 1{,}860{,}000 \\
& & x &= \$12{,}000 \quad (\text{Amount at } 7\frac{1}{2}\% \text{ rate}) \\
(1) & & x + y &= 20{,}000 \\
& & 12{,}000 + y &= 20{,}000 \\
& & y &= 20{,}000 - 12{,}000 \\
& & y &= \$8{,}000 \quad (\text{Amount at } 8\% \text{ rate})
\end{array}
$$

9.5/Problem Solving Involving Two Variables

EXERCISES 9.5 (Part 4)

A

Simplify each of the following equations so that each of the coefficients is an integer.

1. $.05x + .06y = 2400$
2. $.07x - .04y = 4700$
3. $.065x - .075y = 684$
4. $.055x + .085y = 952$
5. $.03x + .045y = 5{,}600$
6. $.09x - .085y = 595$
7. $.075x + .04y = 12{,}500$
8. $.07x - .0525y = 746$

B

1. Raymond has two accounts at the Provident Savings Bank. On one of these he receives a 7% interest rate while on the other the rate is 5%. The total in both accounts is $8,000 and the total interest for the year that he expects to get from both is $520. How much money is deposited in each account?

2. Barbara opened two accounts at the Brookdale Trust Company. The total amount she deposited in both accounts was $12,000. One of the accounts pays a 5% interest rate and the other 6%. The total interest on both accounts at the end of the year will be $632. How much was deposited in each account?

3. The two accounts that Henry has at the Valley State Bank pay interest rates of 7% and 5.5%. The total deposited in both accounts is $14,000 and the total interest for the year was $882.50. How much is deposited in each account?

4. When Joe opened his two savings accounts at the United Savings Bank he was told that he would receive 7% on one account and 6% on the other. Both accounts had a total of $16,000. At the end of the year he found that the interest from the 7% account was $704 more than the interest from the 6% account. How much was deposited in each account?

5. Ethel invested $18,000 in bonds of two different companies. The bonds of one company paid an interest rate of 4.5% while those of the other paid 6.5%. She found that the annual interest on the 4.5% bonds was $40 more than on the 6.5% bonds. How much did she have invested in each of these companies?

6. Recently Myrtle purchased municipal bonds paying $8\frac{1}{2}$ %. A number of years back she had purchased bonds issued by this same city. She still owned these bonds which paid an annual interest rate of $4\frac{1}{2}$ %. Her new bonds paid her $120 less in interest each year than her old bonds. The total value of all her bonds issued by this city is $20,000. How much does she have invested at each rate?

7. Leonard has $3,000 more in one bank account than in the other. The first pays an interest rate of 6% while the second pays 4%. The annual interest he receives from both accounts is $1,380. How much is deposited in each account?

8. Martin has $6,000 more invested at $5\frac{1}{2}$ % than he has invested at 6%. His total annual interest from both investments is $1,710. How much does he have invested at each rate?

9. Stella invested $8,600 less in a 6% investment than she did in a 9% investment. The total annual interest she receives from both investments is $3,114. How much does she have invested at each rate?

10. Michael has $6,700 more invested at 7% than he has at 6%. His annual interest on the 7% investment is $617 more than on the 6% investment. How much does he have invested at each rate?

11. Roger has $4,110 more invested at 5% than he has at 8%. However, his annual return at 5% is $140 less than his return at 8%. How much is invested at each rate?

12. The Hallman Company invested $2,600 less in an enterprise that paid an interest rate of $6\frac{1}{2}$ % than in one that paid $8\frac{1}{2}$ %. If the $6\frac{1}{2}$ % investment returned $717 less each year than the $8\frac{1}{2}$ % investment, how much money was invested at each rate?

CHAPTER REVIEW

A

1. Determine whether the equations in each of the following systems are consistent, inconsistent, or dependent. Justify your answers.

   a. $y = 3x + 5$
      $y = 2x - 7$
   b. $y = 5x - 4$
      $y = 5x + 2$
   c. $2x + 3y = 7$
      $5x - 4y = 3$
   d. $4x - 6y = 12$
      $9y = 6x - 18$

2. Determine the common solution of each of the following systems by graphing.

   a. $2x + y = 11$
      $4x - y = 7$
   b. $x - 2y = 3$
      $3x + y = -5$

3. Solve each of the following systems of equations by the substitution method. Check your answers.

   a. $x = 4y$
      $2x - 5y = 6$
   b. $y = 1 - 2x$
      $3x + 4y = -6$
   c. $2x + y = 14$
      $3x + 2y = 23$
   d. $3x - 2y = 5$
      $2x - y = 1$

Chapter Review

4. Solve each of the following systems of equations by the addition or subtraction method. Check your answers.

    a. $3x + 2y = 23$
       $5x - 2y = 1$
    b. $2x + 3y = 13$
       $2x - 5y = 29$
    c. $3x + 2y = 6$
       $5x + 4y = 12$
    d. $4x + 5y = 28$
       $6x - 7y = -74$

5. Write each of the following equations in standard form.

    a. $3x = 5y + 4$
    b. $2a - 5 = 3b + 2$
    c. $\frac{x}{3} + \frac{y}{4} = 5$
    d. $\frac{2a - 3}{5} = \frac{4b + 1}{3}$

6. Solve each of the following systems of equations.

    a. $5x = 22 - 2y$
       $4y + 8 = 3x$
    b. $5x + 3y = 3x + 11$
       $3(x - y) = 2(12 - x)$
    c. $y + 2x = 20$
       $\frac{y}{5} + \frac{7}{5} = \frac{x}{2}$
    d. $5(x + y) = 2(x + 3) + 1$
       $\frac{x - y}{2} = \frac{x - y + 4}{3} + \frac{1 - x}{6}$

7. Determine the common solutions for each of the following systems of inequalities by graphing.

    a. $y > 2$
       $x < 6$
    b. $x > 2y + 3$
       $y > 5 - x$
    c. $x - 2y < 1$
       $2x + y > 3$
    d. $3x - y > 0$
       $x + 4y < -2$

B

Find the numbers requested in each of the following problems.

1. If the first of two numbers is added to twice the second, the sum is -13. However if twice the first is added to the second, the sum is 25. What are the numbers?

2. The length of a rectangle is 1 meter more than 3 times its width. If the perimeter of the rectangle is 50 meters, what are the length and width of the rectangle?

3. A coin bank contained 81 coins in nickles and dimes only. The total value of all the coins was $5.85. How many nickles and how many dimes were there?

4. The receipts from a basketball game were $2,810. Some tickets were sold at $2.50 each while others were sold at $1.50 each. How many were sold at each price if the total number sold was 1,376?

5. In a two-digit number the units digit is 2 more than the tens digit. The number itself is 4 more than 7 times the units digit. What is the number?

6. The sum of the digits of a two-digit number is 12. If the digits are reversed, the new number is 18 more than the original number. What is the original number?

7. It took a motorboat 8 hours to travel 24 miles upstream on a river. The return trip downstream took only 6 hours. What was the speed of the current and at what speed would the boat have traveled had it been in still water?

8. A small plane flying with the wind took 4 hours to travel a distance of 1,056 miles. Had it been flying against the wind, it would have taken 5 hours and 30 minutes to make the flight. What was the wind velocity and what would the velocity of the plane have been had it been flying in still air?

9. Joyce has a total of $18,000 in two savings accounts. One pays her an interest rate of 6% and the other 7%. The earnings from both accounts for one year amount to $1,150. How much money is there in each account?

10. Fred has $2,000 more invested at a 5% rate than at a 6.5% rate. The 6.5% investment pays him $62 more each year than the 5% investment. How much does he have invested at each rate?

Real Numbers Revisited

At an earlier point in our work we examined the solution of the quadratic equation by the factoring method. Unfortunately, all quadratic expressions are not factorable and hence, our objective now is to seek out methods by which we can solve those that fall into this group.

Although the equations in many of the examples and exercises on the next few pages can be solved by the factoring method, we plan to find their roots by another approach. The newer method is preferred at this time, for it will eventually lead us to a general solution that can be applied under conditions where the factoring method cannot.

10.1 THE SOLUTION OF THE PURE QUADRATIC EQUATION

You may recall that the **standard form** of the quadratic equation is

$$ax^2 + bx + c = 0 \qquad a \neq 0$$

where

a is the coefficient of the x^2 term.
b is the coefficient of the x term.
c is the constant term.

In the event that b is 0, then the quadratic equation becomes

$$ax^2 + 0x + c = 0$$

or simply,

$$ax^2 + c = 0.$$

A quadratic equation where the first power term is missing, such as the one above, is called a **pure quadratic equation.** An example of the simplest form of the pure quadratic is

$$x^2 - 25 = 0.$$

When this equation is rewritten as

$$x^2 = 25,$$

it can be translated into the English sentence

The number x when squared is equal to twenty-five.

Quite apparently, there are two numbers that when squared will equal 25. One of these is +5 while the other is −5.

$$(+5)^2 = 25 \qquad (-5)^2 = 25$$

Hence, the quadratic equation $x^2 = 25$ has two roots,

$$x = +5 \quad \text{and} \quad x = -5.$$

It would seem then that finding roots of a pure quadratic equation, such as $x^2 = 25$, involves finding the square root of the number that is the right member of the equation. This then leads to the question of where to turn in the event that the right member is not a perfect square? As an example, consider the pure quadratic

$$x^2 = 7.$$

Is there any way in which the roots of this equation can be expressed? Since there is no number with which we are familiar that when squared will equal 7, the mathematicians overcame this difficulty by inventing a number which they wrote as

$$\sqrt{7}.$$

This number is read as

The square root of 7.

10.1 / The Solution of the Pure Quadratic Equation

The symbol $\sqrt{}$ is called a **radical sign** while the quantity under the radical sign is called the **radicand**. In this situation the radicand is 7.

The $\sqrt{7}$ is the number such that when it is squared, the product will be 7.

$$(\sqrt{7})^2 = 7$$

In keeping with what we just learned, though, there must be *two* numbers whose squares will be 7. The other of these is written as $-\sqrt{7}$. This number, too, when squared will equal 7.

$$(-\sqrt{7})^2 = 7$$

Of the two square roots of 7, the positive one is called the **principal square root** of 7. That is,

The principal square root of 7 is $+\sqrt{7}$.

There is no special name to $-\sqrt{7}$.

In general, we will say that

The $\sqrt{b}$ represents a number that when squared will equal b.

Based on this interpretation, b can*not* be a negative number for there is *no* number with which we are familiar that when squared will be a negative number. A positive number when squared will yield a *positive* number and a negative number when squared will yield a *positive* number. Based on this, we will assume that in all our work **the radicand will always be a positive number**.

With this background, let us examine the solution of several pure quadratic equations.

Example 1

Find the roots of the equation

$$x^2 = 5.$$

Explanation The numbers whose squares are equal to 5 are represented as $+\sqrt{5}$ and $-\sqrt{5}$. These are read as, "The positive square root of 5" and "The negative square root of 5."

Solution
$$x^2 = 5$$
$$x = +\sqrt{5} \quad \text{and} \quad x = -\sqrt{5}$$

Example 2

Solve and check the following equation:

$$2x^2 - 5 = 13.$$

Explanation Since there is no first power term in x, we recognize this equation as a pure quadratic equation. Our initial step is to solve the equation for x^2. The solution can then be completed as in Example 1.

Solution
$$2x^2 - 5 = 13$$
$$2x^2 = 13 + 5$$
$$2x^2 = 18$$
$$x^2 = 9$$
$$x = +3 \quad \text{and} \quad x = -3$$

Check for $x = +3$:
$$2x^2 - 5 = 13$$
$$2(+3)^2 - 5 \stackrel{?}{=} 13$$
$$2 \cdot 9 - 5 \stackrel{?}{=} 13$$
$$18 - 5 \stackrel{?}{=} 13$$
$$13 = 13$$

Check for $x = -3$:
$$2x^2 - 5 = 13$$
$$2(-3)^2 - 5 \stackrel{?}{=} 13$$
$$2 \cdot 9 - 5 \stackrel{?}{=} 13$$
$$18 - 5 \stackrel{?}{=} 13$$
$$13 = 13$$

Example 3
Solve and check the following equation: $17 - 3x^2 = 2$.

Solution
$$17 - 3x^2 = 2$$
$$-3x^2 = 2 - 17$$
$$-3x^2 = -15$$
$$x^2 = 5$$
$$x = +\sqrt{5} \quad \text{and} \quad x = -\sqrt{5}$$

Check for $x = +\sqrt{5}$:
$$17 - 3x^2 = 2$$
$$17 - 3(+\sqrt{5})^2 \stackrel{?}{=} 2$$
$$17 - 3(5) \stackrel{?}{=} 2$$
$$17 - 15 \stackrel{?}{=} 2$$
$$2 = 2$$

Check for $x = -\sqrt{5}$:
$$17 - 3x^2 = 2$$
$$17 - 3(-\sqrt{5})^2 \stackrel{?}{=} 2$$
$$17 - 3(5) \stackrel{?}{=} 2$$
$$17 - 15 \stackrel{?}{=} 2$$
$$2 = 2$$

Example 4
Solve the following equation.

$$3 = 38 + 7x^2$$

Solution

$$3 = 38 + 7x^2$$
$$-7x^2 = 38 - 3$$
$$-7x^2 = 35$$
$$x^2 = -5$$

This equation has no roots.

Explanation When the equation was solved for x^2 it was found that x^2 was equal to -5. Since there is no number whose square is -5, there is no root to the original equation.

EXERCISES 10.1

Solve and check each of the following equations. If the equation has no roots, indicate this as is done in Example 4.

1. $x^2 = 16$
2. $2x^2 = 8$
3. $x^2 = 11$
4. $5x^2 = 15$
5. $x^2 = -3$
6. $2x^2 - 18 = 0$
7. $3x^2 - 48 = 0$
8. $2x^2 - 1 = 49$
9. $3x^2 - 21 = 0$
10. $2x^2 + 10 = 0$
11. $0 = 72 - 2x^2$
12. $75 = 3x^2$
13. $4x^2 - 21 = 23$
14. $3 = 101 - 2x^2$
15. $12 - 5x^2 = 2$
16. $25 = 1 - 6x^2$
17. $3(x^2 - 1) = 45$
18. $5(x^2 - 1) - 1 = 39$
19. $5(x^2 + 1) = 29 - x^2$
20. $7x^2 = 3(x^2 + 8)$

10.2 SOLUTION OF THE QUADRATIC EQUATION BY THE PURE QUADRATIC METHOD

Let's return for a moment to the pure quadratic equation

$$x^2 = 25.$$

This was translated into the English sentence,

"The number x when squared is equal to 25."

And as we discovered, x was equal to both +5 and -5.

A situation much the same as this exists in the equation

$$(y - 3)^2 = 25,$$

where $y - 3$ has replaced the variable x in the original equation. The translation of this new sentence is,

"The number $y - 3$ when squared is equal to 25."

Hence, as before, $y - 3$ will be equal to both +5 and −5.

(1) $y - 3 = +5$ and (2) $y - 3 = -5$

At this stage, both of these linear equations can be solved for y thus enabling us to obtain the roots of the equation $(y - 3)^2 = 25$.

(1) $y - 3 = +5$ (2) $y - 3 = -5$
$y = +5 + 3$ $y = -5 + 3$
$y = +8$ $y = -2$

It would appear then that the method we used for solving a pure quadratic equation can be used under the condition where the left member of the equation turns out to be the square of a binomial.

Example 1

Solve and check the following equation:

$$(2x + 3)^2 = 49.$$

Explanation In this situation, the binomial $2x + 3$ will be equal to both +7 and −7.

Solution $(2x + 3)^2 = 49$

$2x + 3 = +7$ and $2x + 3 = -7$
$2x = +7 - 3$ $2x = -7 - 3$
$2x = 4$ $2x = -10$
$x = 2$ $x = -5$

Check for $x = 2$: Check for $x = -5$:
$(2x + 3)^2 = 49$ $(2x + 3)^2 = 49$
$(2 \cdot 2 + 3)^2 \stackrel{?}{=} 49$ $(2 \cdot -5 + 3)^2 \stackrel{?}{=} 49$
$(4 + 3)^2 \stackrel{?}{=} 49$ $(-10 + 3)^2 \stackrel{?}{=} 49$
$7^2 \stackrel{?}{=} 49$ $(-7)^2 \stackrel{?}{=} 49$
$49 = 49$ $49 = 49$

Example 2

Solve the following equation:
$$(x - 4)^2 = 5$$

Solution
$$(x - 4)^2 = 5$$
$$x - 4 = +\sqrt{5} \quad \text{and} \quad x - 4 = -\sqrt{5}$$
$$x = 4 + \sqrt{5} \qquad\qquad x = 4 - \sqrt{5}$$

Explanation Quite often the roots of a quadratic equation are left in radical form as they are shown above. The method for approximating the roots will be presented in Section 10.4.

In both Examples 1 and 2 the left member of each equation was given as the *square of a binomial*. Unfortunately, this will not always be so. Quite often the left member of the equation is expressed as a *trinomial*. This places the responsibility upon us to factor that trinomial to show that it is the square of a binomial. The method for factoring a perfect square trinomial was examined in Section 6.4. You may want to review that topic.

Example 3

Solve and check the following equation:
$$x^2 - 12x + 36 = 81$$

Solution
$$x^2 - 12x + 36 = 81$$
$$(x - 6)^2 = 81$$

$x - 6 = +9$	$x - 6 = -9$
$x = +9 + 6$	$x = -9 + 6$
$x = +15$	$x = -3$

Check for $x = +15$:
$$x^2 - 12x + 36 = 81$$
$$(+15)^2 - 12(+15) + 36 \stackrel{?}{=} 81$$
$$225 - 180 + 36 \stackrel{?}{=} 81$$
$$81 = 81$$

Check for $x = -3$:
$$x^2 - 12x + 36 = 81$$
$$(-3)^2 - 12(-3) + 36 \stackrel{?}{=} 81$$
$$9 + 36 + 36 \stackrel{?}{=} 81$$
$$81 = 81$$

Explanation Upon factoring the trinomial $x^2 - 12x + 36$ we found that it can be expressed as the square of the binomial $x - 6$. After the equation $x^2 - 12x + 36 = 81$ is rewritten in the form $(x - 6)^2 = 81$, the solution follows the same pattern as used in Examples 1 and 2.

EXERCISES 10.2

A

Solve and check each of the following equations.

1. $(x + 1)^2 = 16$
2. $(x - 3)^2 = 9$
3. $(x + 5)^2 = 1$
4. $(x - 7)^2 = 25$
5. $(2x - 3)^2 = 25$
6. $(2x + 5)^2 = 81$
7. $(3x - 4)^2 = 64$
8. $(5x + 1)^2 = 121$
9. $x^2 + 6x + 9 = 4$
10. $x^2 + 8x + 16 = 9$
11. $x^2 - 12x + 36 = 16$
12. $x^2 - 10x + 25 = 64$
13. $x^2 + 2x + 1 = 1$
14. $x^2 - 2x + 1 = 100$
15. $x^2 - 20x + 100 = 1$
16. $x^2 + 18x + 81 = 144$

B

Solve each of the following equations.

1. $(x + 3)^2 = 5$
2. $(x - 5)^2 = 3$
3. $(x + 7)^2 = 10$
4. $(x - 6)^2 = 8$
5. $(2x + 3)^2 = 7$
6. $(x + 8)^2 = -5$

10.3 SOLUTION OF THE QUADRATIC EQUATION BY COMPLETING THE SQUARE

PART 1

Time and again in our study of algebra we have developed ways of examining situations, only to discover later that our methods applied only under rather limited conditions. Once again we face this same difficulty. In Section 10.2 we appeared to have found a way to solve any quadratic equation. Closer examination brings to light the fact that this method will work *only* if the left member of the equation is a *perfect square trinomial*.

What do we do if the left member is not a perfect square trinomial? Since our approach depends upon this being the case, we must take drastic steps and force the left member into the form we require, that is, into being the square of a binomial. Just how this is done can only be shown after we have carefully examined those features of a trinomial that make it the square of a binomial. The general perfect square trinomial and its factored form are shown below.

$$x^2 + 2bx + b^2 = (x + b)^2 \qquad (1)$$

10.3/Solution of the Quadratic Equation by Completing the Square

Notice that the coefficient of x in the middle term of the trinomial is twice the second term of the binomial.

$$\underset{\underbrace{\qquad\qquad\qquad}_{2b \text{ is twice } b}}{\_\_ + 2bx + \_\_ = (\_\_ + b)^2} \qquad (2)$$

In addition, notice that the third term of the trinomial is the square of the second term of the binomial.

$$\_\_ + \_\_ + b^2 = (\_\_ + b)^2 \qquad (3)$$

where b^2 is the square of b.

Hence, were we told that the first two terms of a perfect square trinomial were

$$x^2 + 10x + \_\_,$$

we would immediately know that the second term of the binomial must be 5. (See (2) above.)

$$x^2 + 10x + \_\_ = (x + 5)^2$$

where 10 is twice 5.

However, knowing that the second term of the binomial is 5, we would also know the third term of the trinomial is 25. (See (3) above.)

$$x^2 + 10x + 25 = (x + 5)^2$$

where 25 is the square of 5.

In view of this, finding the third term of a perfect square trinomial when we know the first two terms is a two step process:

1. Divide the coefficient of the middle term by 2.
2. The square of the number just found is the third term of the trinomial.

Example 1

The third term of the following perfect square trinomial is missing. What is that term?

$$x^2 + 16x + \_\_$$

Solution

Step 1: Divide the coefficient of the middle term by 2.

$$16 \div 2 = 8$$

Step 2: Square the number just found.

$$8^2 = 64 \quad \text{(Third term of trinomial)}$$

The trinomial is: $x^2 + 16x + 64$.

Example 2

If the following trinomial is to be a perfect square, what must the third term be?

$$y^2 - 11y + \underline{}$$

Solution

Step 1: Divide the coefficient of the middle term by 2.

$$(-11) \div 2 = \frac{-11}{2}$$

Step 2: Square the number just found.

$$\left(\frac{-11}{2}\right)^2 = \frac{121}{4} \quad \text{(Third term of trinomial)}$$

The trinomial is: $y^2 - 11y + 121/4$.

The process of finding the third term of a trinomial that will make that trinomial a perfect square is often referred to as *completing the square*.

Example 3

Complete the square of the following trinomial.

$$x^2 + x + \underline{}$$

Explanation Since no number appears as the coefficient of $+x$ we know that the coefficient is understood to be 1.

Solution

Step 1: Divide the coefficient of the middle term by 2.

$$1 \div 2 = \frac{1}{2}$$

Step 2: Square the number just found.

$$(\tfrac{1}{2})^2 = \tfrac{1}{4} \qquad \text{(Third term of trinomial)}$$

The trinomial is: $x^2 + x + \tfrac{1}{4}$.

Example 4

Complete the square of the following trinomial.

$$x^2 + \tfrac{6}{5}x + \underline{}$$

Explanation The only concern in this situation is in determining the quotient when $\tfrac{6}{5}$ is divided by 2. The division, as you recall, is done by multiplying the first fraction by the reciprocal of the second.

$$\tfrac{6}{5} \div 2 = \tfrac{6}{5} \div \tfrac{2}{1} = \tfrac{\cancel{6}^{3}}{5} \times \tfrac{1}{\cancel{2}} = \tfrac{3}{5}$$

Solution

Step 1: Divide the coefficient of the middle term by 2.

$$\tfrac{6}{5} \div 2 = \tfrac{3}{5}$$

Step 2: Square the number just found.

$$(\tfrac{3}{5})^2 = \tfrac{9}{25} \qquad \text{(Third term of trinomial)}$$

The trinomial is: $x^2 + \tfrac{6}{5}x + \tfrac{9}{25}$.

EXERCISES 10.3 (Part 1)

Complete the square of each of the following trinomials.

1. $x^2 + 8x + \underline{}$
2. $x^2 + 12x + \underline{}$
3. $y^2 + 18y + \underline{}$
4. $w^2 - 10w + \underline{}$
5. $z^2 - 14z + \underline{}$
6. $m^2 - 24m + \underline{}$
7. $n^2 + 30n + \underline{}$
8. $s^2 + 3s + \underline{}$
9. $s^2 - 5s + \underline{}$
10. $v^2 + 11v + \underline{}$
11. $w^2 + 15w + \underline{}$
12. $p^2 + p + \underline{}$

13. $f^2 + \frac{2}{5}f + \underline{}$

14. $e^2 - \frac{2}{7}e + \underline{}$

15. $w^2 + \frac{2}{3}w + \underline{}$

16. $m^2 + \frac{8}{3}m + \underline{}$

17. $q^2 - \frac{4}{5}q + \underline{}$

18. $x^2 - \frac{10}{3}x + \underline{}$

19. $s^2 + \frac{5}{4}s + \underline{}$

20. $x^2 + \frac{x}{3} + \underline{}$

PART 2

Being able to complete the square of a trinomial has placed us in a position where we can determine the roots of any quadratic equation—assuming, of course, that the equation does have roots. The method we learned for solving a quadratic equation depends upon the fact that the left member of the equation be a perfect square trinomial. If it is not, our investigation in Part 1 gives us the tools to force the left member into that mold.

Consider the equation

$$x^2 + 6x - 15 = 40. \tag{1}$$

As it stands now, the trinomial $x^2 + 6x - 15$ is not a perfect square. However, if the third term -15 did not exist in the left member of this equation, it would be possible for us to supply our own third term that would make the left member of that equation a perfect square trinomial. Eliminating the -15 can easily be accomplished by using the addition principle of equality and adding a $+15$ to both members of the equation. Thus,

$$x^2 + 6x = 40 + 15. \tag{2}$$

Now, the path has been cleared for us to complete the square of the left member of the equation.

$$x^2 + 6x + \underline{} = 40 + 15 \tag{2}$$
$$x^2 + 6x + 9 = 40 + 15 + 9 \tag{3}$$

Notice that it was not enough to add 9 to the left member of the equation for by the addition principle we must add the same number to *both* members.

$$(x + 3)^2 = 64 \tag{4}$$

$x + 3 = +8$ and $x + 3 = -8$

$x = 8 - 3$ $\qquad\qquad x = -8 - 3$

$x = 5$ $\qquad\qquad\quad x = -11$

10.3/Solution of the Quadratic Equation by Completing the Square

Based on the solution above, the method for solving a quadratic equation involves three steps.

1. Eliminate the constant term from the left member of the equation.
2. Complete the square as we learned in Part 1.
3. Complete the solution as we learned in Section 10.2.

Example 1

Solve and check the following equation:

$$x^2 - 10x + 31 = 10.$$

Solution Step 1—Eliminate the constant term from the left member.

$$x^2 - 10x + 31 = 10$$

$$x^2 - 10x = 10 - 31$$

Step 2—Complete the square.

$$x^2 - 10x + 25 = 10 - 31 + 25$$

Step 3—Solve as in Section 2.

$$(x-5)^2 = 4$$

$$x - 5 = +2 \quad \text{and} \quad x - 5 = -2$$

$$x = 7 \qquad\qquad x = 3$$

Check for $x = 7$:

$$x^2 - 10x + 31 = 10$$

$$(7)^2 - 10(7) + 31 \stackrel{?}{=} 10$$

$$49 - 70 + 31 \stackrel{?}{=} 10$$

$$10 = 10$$

Check for $x = 3$

$$x^2 - 10x + 31 = 10$$

$$(3)^2 - 10(3) + 31 \stackrel{?}{=} 10$$

$$9 - 30 + 31 \stackrel{?}{=} 10$$

$$10 = 10$$

Example 2

Solve the following equation.

$$x^2 = 6 + 5x$$

Solution

$$x^2 = 6 + 5x$$
$$x^2 - 5x = 6$$
$$x^2 - 5x + \underline{} = 6$$
$$x^2 - 5x + \frac{25}{4} = 6 + \frac{25}{4}$$
$$(x - \frac{5}{2})^2 = \frac{49}{4}$$

$$x - \frac{5}{2} = \frac{7}{2} \quad \text{and} \quad x - \frac{5}{2} = -\frac{7}{2}$$
$$x = \frac{7}{2} + \frac{5}{2} \qquad\qquad x = -\frac{7}{2} + \frac{5}{2}$$
$$x = \frac{12}{2} \text{ or } 6 \qquad\qquad x = -\frac{2}{2} \text{ or } -1$$

Explanation The only step at which we may encounter a little difficulty is in adding $6 + \frac{25}{4}$. This can be done as follows:

$$6 + \frac{25}{4} = \frac{6}{1} + \frac{25}{4}$$
$$= \frac{24}{4} + \frac{25}{4}$$
$$= \frac{49}{4}$$

Example 3

Solve the following equation:

$$x^2 + 10 = 5 - 8x$$

Solution

$$x^2 + 10 = 5 - 8x$$
$$x^2 + 8x = 5 - 10$$
$$x^2 + 8x + 16 = 5 - 10 + 16$$
$$(x + 4)^2 = 11$$
$$x + 4 = +\sqrt{11} \quad \text{and} \quad x + 4 = -\sqrt{11}$$
$$x = -4 + \sqrt{11} \qquad\qquad x = -4 - \sqrt{11}$$

Example 4

Solve the following equation.

$$x^2 + 4x + 5 = 0$$

Solution
$$x^2 + 4x + 5 = 0$$
$$x^2 + 4x = -5$$
$$x^2 + 4x + 4 = -5 + 4$$
$$(x + 2)^2 = -1$$

This equation has no roots.

You may have noticed that the coefficient of the x^2 term in each of the four examples has been 1. These examples were designed in this manner because our method for completing the square applies only if that coefficient is 1.

In the event that the coefficient of the x^2 term is not 1 we can eliminate that difficulty by applying the division principle of equality. The example below shows how this is done.

Example 5

Solve the following equation.

$$2x^2 + 3x = 15 - 2x$$

Solution
$$2x^2 + 3x = 15 - 2x$$
$$2x^2 + 3x + 2x = 15$$
$$2x^2 + 5x = 15 \qquad \text{(Now divide both members by 2)}$$
$$x^2 + \frac{5}{2}x = \frac{15}{2} \qquad \text{(Division Principle)}$$
$$x^2 + \frac{5}{2}x + \frac{25}{16} = \frac{15}{2} + \frac{25}{16}$$
$$\left(x + \frac{5}{4}\right)^2 = \frac{120}{16} + \frac{25}{16}$$
$$\left(x + \frac{5}{4}\right)^2 = \frac{145}{16}$$

$$x + \frac{5}{4} = \frac{\sqrt{145}}{4} \quad \text{and} \quad x + \frac{5}{4} = -\frac{\sqrt{145}}{4} \qquad (8)$$

$$x = -\frac{5}{4} + \frac{\sqrt{145}}{4} \qquad x = -\frac{5}{4} - \frac{\sqrt{145}}{4}$$

$$x = \frac{-5 + \sqrt{145}}{4} \qquad x = \frac{-5 - \sqrt{145}}{4}$$

Explanation In step (8) we are faced with the problem of finding a fraction that when multiplied by itself (squared) will give $\frac{145}{16}$. The denominator of that fraction, of course, is the number 4. Since the numerator 145 is not a perfect square, then from what we have learned, we express its square root as $\sqrt{145}$. Hence, the fraction whose square is $\frac{145}{16}$ is $\sqrt{145}/4$.

EXERCISES 10.3 (Part 2)

A

Solve each of the following equations by the completing the square method. Check your answers.

1. $x^2 + 4x - 2 = 3$
2. $x^2 - 8x + 19 = 4$
3. $x^2 - 10x + 18 = -6$
4. $x^2 - 8x = -7$
5. $x^2 + 2x = 8$
6. $x^2 = 6x + 16$
7. $x^2 = 14x - 48$
8. $x^2 = 14x - 45$
9. $x^2 - 35 = 2x$
10. $x^2 + 27 = 12x$
11. $x^2 - 2x = 2x - 3$
12. $x^2 + x = 24 - x$
13. $x^2 - 9x = x - 9$
14. $x^2 - 30 = 8x + 3$
15. $x^2 - 30 = 2x + 33$
16. $2x = 48 - x^2$
17. $35 - 12x = -x^2$
18. $60 - 16x = 5 - x^2$
19. $x^2 = 3x + 28$
20. $x^2 - 8 = 7x$

B

Solve each of the following equations by the completing the square method. If the equation has no roots indicate this as is done in Example 4.

1. $x^2 - \frac{3}{2}x = \frac{5}{2}$
2. $x^2 - \frac{7}{2}x = \frac{15}{2}$
3. $x^2 - 5x + 7 = 0$
4. $x^2 - \frac{5}{3}x + \frac{2}{3} = 0$

5. $x^2 + \frac{4}{3} = \frac{13}{3}x$

6. $x^2 + \frac{8}{3} = \frac{-10}{3}x$

7. $x^2 + \frac{8x}{5} = \frac{4}{5}$

8. $x^2 = \frac{x}{5} + \frac{4}{5}$

C

Solve each of the following equations by the completing the square method.

1. $2x^2 - 4x - 16 = 0$ 7. $2x^2 + 5x = 3$
2. $2x^2 - 16x + 30 = 0$ 8. $2x^2 + 5x = 12$
3. $3x^2 + 6x = 12$ 9. $3x^2 + 2 = 7x$
4. $4x^2 = 20x - 8$ 10. $3x^2 - 1 = 2x$
5. $5x^2 - 25 = 30x$ 11. $5x^2 = 7x + 6$
6. $3x^2 + 15 = 12x$ 12. $5x^2 = 8 - 6x$

10.4 SOLUTION OF THE QUADRATIC EQUATION BY THE QUADRATIC FORMULA

PART 1

The process of solving the quadratic equation by the completing the square method seems to involve constant repetition. The steps we went through when solving the very first of these equations are the same steps we use over and over again for each new quadratic equation we encounter. To save time and a great deal of effort, the mathematicians decided to take the following approach, "Why not solve the *general* quadratic equation and then use these roots as a *formula* for finding the roots of any quadratic equation."

The *general* quadratic equation is the one that was expressed in **standard form** at the outset of this chapter.

$$ax^2 + bx + c = 0 \quad a \neq 0$$

We are going to find the roots of this equation by the completing the square method in exactly the same manner as we have been doing.

To guide you through each step we will also solve the equation $3x^2 + 11x + 4 = 0$. As you examine each step in the solution of this equation, glance immediately above where you will find the comparably numbered step in the solution of the general quadratic $ax^2 + bx + c = 0$.

(1) $\quad ax^2 + bx + c = 0$

(2) $\quad ax^2 + bx = -c$

(3) $\quad x^2 + \dfrac{b}{a} x = -\dfrac{c}{a}$

(4) $\quad x^2 + \dfrac{b}{a} x + \dfrac{b^2}{4a^2} = \dfrac{b^2}{4a^2} - \dfrac{c}{a}$

(5) $\quad \left(x + \dfrac{b}{2a}\right)^2 = \dfrac{b^2}{4a^2} - \dfrac{4ac}{4a^2}$

(6) $\quad \left(x + \dfrac{b}{2a}\right)^2 = \dfrac{b^2 - 4ac}{4a^2}$

(7) $\quad x + \dfrac{b}{2a} = +\dfrac{\sqrt{b^2 - 4ac}}{2a}\quad$ and $\quad x + \dfrac{b}{2a} = -\dfrac{\sqrt{b^2 - 4ac}}{2a}$

(8) $\quad x = -\dfrac{b}{2a} + \dfrac{\sqrt{b^2 - 4ac}}{2a}\quad$ and $\quad x = -\dfrac{b}{2a} - \dfrac{\sqrt{b^2 - 4ac}}{2a}$

(9) $\quad x = \dfrac{-b + \sqrt{b^2 - 4ac}}{2a}\quad$ and $\quad x = \dfrac{-b - \sqrt{b^2 - 4ac}}{2a}$

---

(1) $\quad 3x^2 + 11x + 4 = 0$

(2) $\quad 3x^2 + 11x = -4$

(3) $\quad x^2 + \dfrac{11}{3} x = -\dfrac{4}{3}$

(4) $\quad x^2 + \dfrac{11}{3} x + \dfrac{121}{36} = \dfrac{121}{36} - \dfrac{4}{3}$

(5) $\quad \left(x + \dfrac{11}{6}\right)^2 = \dfrac{121}{36} - \dfrac{48}{36}$

(6) $\quad \left(x + \dfrac{11}{6}\right)^2 = \dfrac{121 - 48}{36}$

(7) $\quad x + \dfrac{11}{6} = +\dfrac{\sqrt{121 - 48}}{6}\quad$ and $\quad x + \dfrac{11}{6} = -\dfrac{\sqrt{121 - 48}}{6}$

(8) $\quad x = -\dfrac{11}{6} + \dfrac{\sqrt{121 - 48}}{6}\quad$ and $\quad x = -\dfrac{11}{6} - \dfrac{\sqrt{121 - 48}}{6}$

(9) $\quad x = \dfrac{-11 + \sqrt{121 - 48}}{6}\quad$ and $\quad x = \dfrac{-11 - \sqrt{121 - 48}}{6}$

10.4/Solution of the Quadratic Equation by the Quadratic Formula

The two values found for x are:

$$x = \frac{-b + \sqrt{b^2 - 4ac}}{2a} \quad \text{and} \quad x = \frac{-b - \sqrt{b^2 - 4ac}}{2a}$$

These are the two roots of the general quadratic equation

$$ax^2 + bx + c = 0.$$

Actually, these values of x are merely *formulas* that will enable us to determine the roots of any quadratic equation. Notice that each root has been expressed in terms of a, b, and c. However, a and b are simply the coefficients of the x^2 and x terms respectively, while c is the constant term. For instance, in the equation

$$3x^2 - 5x - 6 = 0,$$

the values of a, b, and c are,

$$a = 3, \quad b = -5, \quad \text{and} \quad c = -6.$$

There is one very important point that we must recognize before we can apply these two formulas:

The quadratic equation *must* be arranged in **standard form**.

If the equation is not in that form, the first step will be to express it in that form.

Example 1

Express the following quadratic equation in standard form.

$$5x^2 - 2 = 3x$$

Explanation The standard form of the quadratic equation has as its left member the following three terms:

1. First term is the x^2 term.
2. Second term is the x term.
3. Third term is the constant term.

The right member is 0. Hence, if $-3x$ is added to both members, the right member will be 0.

Solution
$$5x^2 - 6 = 3x$$
$$5x^2 - 6 - 3x = 0$$
$$5x^2 - 3x - 6 = 0$$

Example 2

What are the values of $a, b,$ and c in the following quadratic equation?

$$3x = 4 - 2x^2$$

Solution

$$3x = 4 - 2x^2 \qquad (1)$$
$$2x^2 + 3x - 4 = 0 \qquad (2)$$
$$a = 2, \quad b = 3, \quad c = -4$$

Explanation Notice that in going from step (1) to step (2) not only is $+2x^2$ added to both members of the equation but so too is -4. In the future work we will add both of these terms to each member of the equation in the same step.

Example 3

What are the values of $a, b,$ and c in the following quadratic equation?

$$3x^2 = 5x$$

Solution

$$3x^2 = 5x \qquad (1)$$
$$3x^2 - 5x = 0 \qquad (2)$$
$$3x^2 - 5x + 0 = 0 \qquad (3)$$
$$a = 3, \quad b = -5, \quad c = 0$$

Explanation In order that the right member of the equation be 0, $-5x$ is added to both members of equation (1). Upon examining equation (2) we notice that there is *no* constant term. To obtain a constant term we have to add the same number to both members. However, the right member must remain 0. Hence, the only number we can add to 0 and have it remain 0 is 0 itself.

Example 4

What are the values of $a, b,$ and c in the following quadratic equation?

$$5x^2 = 2$$

Solution

$$5x^2 = 2 \qquad (1)$$
$$5x^2 - 2 = 0 \qquad (2)$$
$$5x^2 + 0x - 2 = 0 \qquad (3)$$
$$a = 5, \quad b = 0, \quad c = -2$$

10.4/Solution of the Quadratic Equation by the Quadratic Formula

Explanation By examining step (2) we discover that the x term is missing. As in Example 3, we add a 0 to both members of equation (2). However, the 0 we add to the left member is expressed as $0x$ in order to give us the x term. This is permissible since 0 times x is 0.

Rather than write the roots of the general quadratic equation as two separate formulas we usually combine them into the single formula

$$x = \frac{-b \pm \sqrt{b^2 - 4ac}}{2a}$$

We must not forget, though, that this single formula names *two* different values of x. This formula is called the **quadratic formula**.

Example 5

Solve the following quadratic equation:

$$6x^2 + 5 = 13x$$

Solution

$$6x^2 + 5 = 13x$$

$$6x^2 - 13x + 5 = 0$$

$$a = 6, \quad b = -13, \quad c = 5$$

$$x = \frac{-b \pm \sqrt{b^2 - 4ac}}{2a}$$

$$x = \frac{-(-13) \pm \sqrt{(-13)^2 - 4(6)(5)}}{2(6)}$$

$$x = \frac{+13 \pm \sqrt{169 - 120}}{12}$$

$$x = \frac{+13 \pm \sqrt{49}}{12}$$

$$x = \frac{13 + 7}{12} \quad \text{and} \quad x = \frac{13 - 7}{12}$$

$$x = \frac{20}{12} \text{ or } \frac{5}{3} \qquad x = \frac{6}{12} \text{ or } \frac{1}{2}$$

Explanation The first step in the solution is to express the equation in standard form. Doing this enables us to determine the values of a, b, and c. These numbers are then used as replacements for a, b, and c in the quadratic formula.

Example 6

Solve the following quadratic equation.

$$3x^2 = 7x$$

Solution

$$3x^2 = 7x$$
$$3x^2 - 7x = 0$$
$$3x^2 - 7x + 0 = 0$$
$$a = 3, \quad b = -7, \quad c = 0$$
$$x = \frac{-b \pm \sqrt{b^2 - 4ac}}{2a}$$
$$x = \frac{-(-7) \pm \sqrt{(-7)^2 - 4(3)(0)}}{2(3)}$$
$$x = \frac{+7 \pm \sqrt{49 - 0}}{6}$$
$$x = \frac{+7 \pm \sqrt{49}}{6}$$

$$x = \frac{+7 + 7}{6} \quad \text{and} \quad x = \frac{+7 - 7}{6}$$

$$x = \frac{14}{6} \text{ or } \frac{7}{3} \quad\quad\quad x = \frac{0}{6} \text{ or } 0$$

EXERCISES 10.4 (Part 1)

A

Write each of the following quadratic equations in standard form.

1. $3x^2 + 6 + 5x = 0$
2. $2 + 3x^2 + 4x = 0$
3. $3x - 5 + 6x^2 = 0$
4. $6 - 4x + 7x^2 = 0$
5. $6x^2 - 3x = 4$
6. $5x^2 - 4 = 3x$
7. $4x^2 = 2x + 3$
8. $3x^2 = 5 - 4x$
9. $2x = 4 - 3x^2$
10. $6 = 7x - 2x^2$
11. $5x^2 = 6x$
12. $4x^2 = 7$

10.4/Solution of the Quadratic Equation by the Quadratic Formula

B

Determine the values of a, b, and c in each of the following quadratic equations.

1. $3x^2 + 5x + 6 = 0$
2. $4x^2 - 2x + 1 = 0$
3. $2x^2 - 3x - 5 = 0$
4. $x^2 + 2x - 6 = 0$
5. $x^2 - 4x + 7 = 0$
6. $3x^2 + x - 3 = 0$
7. $5x^2 - x + 3 = 0$
8. $x^2 - x - 1 = 0$
9. $3x^2 + 5x = 6$
10. $2x^2 - 2x = 3$
11. $x^2 = 5x - 4$
12. $3x^2 = 5 - x$
13. $2x = 7 - 4x^2$
14. $3 - 2x = 5x^2$
15. $4x^2 = 9x$
16. $x^2 = -6x$
17. $2x^2 = 11$
18. $3x^2 = 5$
19. $x(2x - 3) = 4$
20. $3(x^2 - 5) = 7x$

C

Determine the roots of each of the following quadratic equations by using the quadratic formula.

1. $x^2 + 7x + 10 = 0$
2. $x^2 + x = 20$
3. $2x^2 + 3 = 5x$
4. $3x^2 = 7x - 2$
5. $2x^2 = 4 + 7x$
6. $6x^2 - 13x = -6$
7. $5x^2 - 8x = 0$
8. $2x^2 = 7x$
9. $2x^2 - 8 = 0$
10. $5x^2 = 5$
11. $12x^2 + 2 = 11x$
12. $12x^2 = 13x - 3$
13. $4x^2 = 25$
14. $6x^2 + 12 = 17x$
15. $6x^2 = 7x + 5$
16. $x(x + 2) = 15$
17. $3(x^2 + 1) + 10x = 0$
18. $x(x - 1) + 10(x + 2) = 0$
19. $3x(2x + 1) = 2x + 1$
20. $2x^2 = \dfrac{5x}{2} - \dfrac{1}{4}$

PART 2

Each of the quadratic equations in Part 1 was designed so that the quantity $b^2 - 4ac$ would turn out to be a perfect square. At this time we are going to examine the situation in which this is not so.

Example 1

Determine the roots of the following quadratic equation.

$$3x^2 - 5x - 1 = 0$$

10/Real Numbers Revisited

Solution

$$3x^2 - 5x - 1 = 0$$

$$a = 3, \quad b = -5, \quad c = -1$$

$$x = \frac{-b \pm \sqrt{b^2 - 4ac}}{2a}$$

$$x = \frac{-(-5) \pm \sqrt{(-5)^2 - 4(3)(-1)}}{2(3)}$$

$$x = \frac{+5 \pm \sqrt{25 + 12}}{6}$$

$$x = \frac{+5 \pm \sqrt{37}}{6}$$

Explanation Since 37 is not a perfect square, the two roots of the equation are left in the form as they are shown above.

As just stated, quite often when the radicand $b^2 - 4ac$ is not a perfect square the roots are expressed as they are shown in Example 1. There are times, though, when there is a need for determining the approximate values of the roots of the equation. On those occasions we use either the hand held calculator to find the approximate square root of the radicand or else find the approximate square root in a table similar to the one on page 470.

Example 2

Determine the approximate values of the roots of the following quadratic equation.

$$5x^2 + 7x = 3$$

Solution

$$5x^2 + 7x = 3$$

$$5x^2 + 7x - 3 = 0$$

$$a = 5, \quad b = 7, \quad c = -3$$

$$x = \frac{-b \pm \sqrt{b^2 - 4ac}}{2a}$$

$$x = \frac{-7 \pm \sqrt{(7)^2 - 4(5)(-3)}}{2(5)}$$

$$x = \frac{-7 \pm \sqrt{49 + 60}}{10}$$

$$x = \frac{-7 \pm \sqrt{109}}{10}$$

10.4/Solution of the Quadratic Equation by the Quadratic Formula

$$x = \frac{-7 + 10.440}{10} \quad \text{and} \quad x = \frac{-7 - 10.440}{10}$$

$$x = \frac{3.440}{10} \qquad\qquad\qquad x = \frac{-17.440}{10}$$

$$x = .3440 \qquad\qquad\qquad x = -1.7440$$

Solution The approximate value of the square root of 109 is found by locating the number 109 in one of the three columns headed by the word **No.** After placing the edge of a piece of paper along the *row* containing the number 109, we run our finger to the right until we reach the column headed by the words **Square Root.** The number 10.440 at which our finger is pointing is the approximate square root of 109.

Example 3

Determine the approximate values of the roots of the following quadratic equation.

$$2x^2 = 3x - 4$$

Solution

$$2x^2 = 3x - 4$$

$$2x^2 - 3x + 4 = 0$$

$$a = 2, \quad b = -3, \quad c = 4$$

$$x = \frac{-b \pm \sqrt{b^2 - 4ac}}{2a}$$

$$x = \frac{-(-3) \pm \sqrt{(-3)^2 - 4(2)(4)}}{2(2)}$$

$$x = \frac{+3 \pm \sqrt{9 - 32}}{4}$$

$$x = \frac{+3 \pm \sqrt{-23}}{4}$$

Since the radicand is the negative number −23, the quadratic equation has no roots in terms of the numbers we know.

TABLE OF SQUARES AND SQUARE ROOTS

No.	Square	Square Root	No.	Square	Square Root	No.	Square	Square Root
1	1	1.000	51	2,601	7.141	101	10,201	10.050
2	4	1.414	52	2,704	7.211	102	10,404	10.100
3	9	1.732	53	2,809	7.280	103	10,609	10.149
4	16	2.000	54	2,916	7.348	104	10,816	10.198
5	25	2.236	55	3,025	7.416	105	11,025	10.247
6	36	2.449	56	3,136	7.483	106	11,236	10.296
7	49	2.646	57	3,249	7.550	107	11,449	10.344
8	64	2.828	58	3,364	7.616	108	11,664	10.392
9	81	3.000	59	3,481	7.681	109	11,881	10.440
10	100	3.162	60	3,600	7.746	110	12,100	10.488
11	121	3.317	61	3,721	7.810	111	12,321	10.536
12	144	3.464	62	3,844	7.874	112	12,544	10.583
13	169	3.606	63	3,969	7.937	113	12,769	10.630
14	196	3.742	64	4,096	8.000	114	12,996	10.677
15	225	3.873	65	4,225	8.062	115	13,225	10.724
16	256	4.000	66	4,356	8.124	116	13,456	10.770
17	289	4.123	67	4,489	8.185	117	13,689	10.817
18	324	4.243	68	4,624	8.246	118	13,924	10.863
19	361	4.359	69	4,761	8.307	119	14,161	10.909
20	400	4.472	70	4,900	8.367	120	14,400	10.954
21	441	4.583	71	5,041	8.426	121	14,641	11.000
22	484	4.690	72	5,184	8.485	122	14,884	11.045
23	529	4.796	73	5,329	8.544	123	15,129	11.091
24	576	4.899	74	5,476	8.602	124	15,376	11.136
25	625	5.000	75	5,625	8.660	125	15,625	11.180
26	676	5.099	76	5,776	8.718	126	15,876	11.225
27	729	5.196	77	5,929	8.775	127	16,129	11.269
28	784	5.291	78	6,084	8.832	128	16,384	11.314
29	841	5.385	79	6,241	8.888	129	16,641	11.358
30	900	5.477	80	6,400	8.944	130	16,900	11.402
31	961	5.568	81	6,561	9.000	131	17,161	11.446
32	1,024	5.657	82	6,724	9.055	132	17,424	11.489
33	1,089	5.745	83	6,889	9.110	133	17,689	11.533
34	1,156	5.831	84	7,056	9.165	134	17,956	11.576
35	1,225	5.916	85	7,225	9.220	135	18,225	11.619
36	1,296	6.000	86	7,396	9.274	136	18,496	11.662
37	1,369	6.083	87	7,569	9.327	137	18,769	11.705
38	1,444	6.164	88	7,744	9.381	138	19,044	11.747
39	1,521	6.245	89	7,921	9.434	139	19,321	11.790
40	1,600	6.325	90	8,100	9.487	140	19,600	11.832
41	1,681	6.403	91	8,281	9.539	141	19,881	11.874
42	1,764	6.481	92	8,464	9.592	142	20,164	11.916
43	1,849	6.557	93	8,649	9.644	143	20,449	11.958
44	1,936	6.633	94	8,836	9.695	144	20,736	12.000
45	2,025	6.708	95	9,025	9.747	145	21,025	12.042
46	2,116	6.782	96	9,216	9.798	146	21,316	12.083
47	2,209	6.856	97	9,409	9.849	147	21,609	12.124
48	2,304	6.928	98	9,604	9.899	148	21,904	12.166
49	2,401	7.000	99	9,801	9.950	149	22,201	12.207
50	2,500	7.071	100	10,000	10.000	150	22,500	12.247

EXERCISES 10.4 (Part 2)

A

Use the table to find the approximate square root of each of the following numbers.

1. 10
2. 24
3. 56
4. 73
5. 87
6. 3
7. 95
8. 61
9. 107
10. 123
11. 19
12. 142

B

Find the roots of each of the following quadratic equations. Leave your answer in radical form as is shown in Example 1. If the equation has no roots indicate this as is done in Example 3.

1. $2x^2 + 5x - 4 = 0$
2. $5x^2 - 9x + 2 = 0$
3. $3x^2 - 2x = 4$
4. $5x^2 + 3 = 3x$
5. $6x^2 + 7x - 4 = 0$
6. $7x^2 - 2 + 4x = 0$
7. $5x - 6 = -2x^2$
8. $x^2 = 2x + 7$
9. $8x^2 - 3x = -2$
10. $9x = 3 - x^2$
11. $7x^2 = 5x + 6$
12. $3x^2 = 7$

C

Find the approximate values of the roots of each of the following quadratic equations.

1. $3x^2 + 2x - 2 = 0$
2. $5x^2 - 2x - 1 = 0$
3. $4x^2 + 5x - 3 = 0$
4. $3x^2 = 4x + 5$
5. $x^2 - 7x = -3$
6. $x^2 = 5x + 1$
7. $6x^2 - 2 = 3x$
8. $9x^2 = 2 - 6x$
9. $6x = 7x^2 - 3$
10. $12x + 5 = -3x^2$
11. $5x^2 = 3$
12. $x(3x - 4) = 6$
13. $3x(x - 2) = 5$
14. $x^2 = 9(x - 1)$

10.5 THE THEOREM OF PYTHAGORAS

One of the more interesting applications of the quadratic equation is in connection with the **right triangle**. A right triangle is a special triangle where one of the angles in that triangle is a **right angle**. The side of the triangle *opposite* the right angle is called the **hypotenuse** while the other two sides are called the **legs** (Figure 10-1).

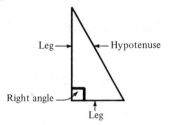

FIGURE 10-1

It so happens that in *every* right triangle,

If we square the length of one leg and add that number to the square of the length of the second leg, the sum turns out to be the same number as the square of the length of the hypotenuse.

As an example, in Figure 10-2 where,

a represents the length of one of the legs;
b represents the length of the other leg;
c represents the length of the hypotenuse;

then it will be true that

$$a^2 + b^2 = c^2$$

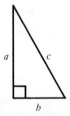

FIGURE 10-2

This relationship concerning the sides of certain *special* right triangles was known as far back as the time of the ancient Egyptians. For instance, they knew this relation was so for the special right triangle whose legs were 3 units and 4 units and whose hypotenuse was 5 units.

$$a^2 + b^2 = c^2$$
$$3^2 + 4^2 \stackrel{?}{=} 5^2$$
$$9 + 16 \stackrel{?}{=} 25$$
$$25 = 25$$

10.5/The Theorem of Pythagoras

However, it was not until sometime around 500 B.C. that the members of a secret Greek mathematical society called the Pythagorian Society discovered that this relationship held for *any* right triangle.

Example 1

The hypotenuse and one of the legs of a right triangle are 15 cm and 11 cm respectively. What is the length of the third side of the triangle?

Explanation Using a, b, and c as the lengths of the legs and the hypotenuse respectively of the right triangle, we can mark the right triangle as shown below.

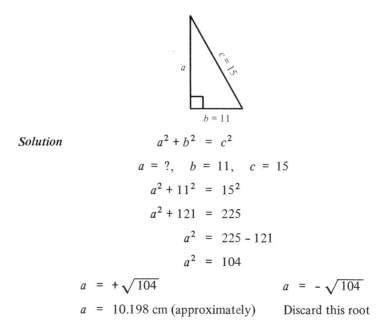

Solution

$$a^2 + b^2 = c^2$$
$$a = ?, \quad b = 11, \quad c = 15$$
$$a^2 + 11^2 = 15^2$$
$$a^2 + 121 = 225$$
$$a^2 = 225 - 121$$
$$a^2 = 104$$

$a = +\sqrt{104}$ $\qquad\qquad a = -\sqrt{104}$

$a = 10.198$ cm (approximately) $\qquad$ Discard this root

Explanation Since the equation $a^2 + 121 = 225$ turned out to be a pure quadratic equation, we solved it by using the pure quadratic method rather than the quadratic formula. Notice, also, that the root $-\sqrt{104}$ is discarded for although it is a root of the equation $a^2 + 121 = 225$, it has no meaning as the *length* of a leg of a right triangle.

Example 2

The values of a and b in a right triangle are given below. Find the value of c.

$$a = 7, \quad b = 9$$

Solution

$$a^2 + b^2 = c^2$$
$$a = 7, \quad b = 9, \quad c = ?$$

$$7^2 + 9^2 = c^2$$
$$49 + 81 = c^2$$
$$130 = c^2$$

$c = +\sqrt{130}$ $c = -\sqrt{130}$

$c = 11.402$ (approximately)   Discard this root.

Example 3

The side of a square is 6 cm. What is the length of the diagonal of the square?

Explanation The diagonal of a square, or, in fact, the diagonal of any rectangle, is the line segment whose endpoints are a pair of opposite vertices of the rectangle. In the figure below, we can see that the diagonal and the two of the sides of the square form a right triangle.

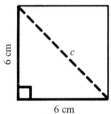

Solution
$$a^2 + b^2 = c^2$$
$$a = 6, \quad b = 6, \quad c = ?$$
$$6^2 + 6^2 = c^2$$
$$36 + 36 = c^2$$
$$72 = c^2$$

$c = +\sqrt{72}$ $c = -\sqrt{72}$

$c = 8.485$ cm (approximately)   Discard this root.

Example 4

One leg of a right triangle is 5 meters more than the other leg. If the hypotenuse is 9 meters in length, what are the lengths of the two legs?

Solution Let x represent the length of one leg.
Therefore, $x + 5$ represents the length of the other leg.

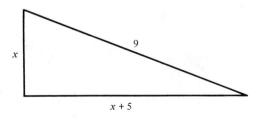

10.5/The Theorem of Pythagoras

$$a^2 + b^2 = c^2$$
$$a = x, \quad b = x + 5, \quad c = 9$$
$$x^2 + (x + 5)^2 = 9^2$$
$$x^2 + x^2 + 10x + 25 = 81$$
$$x^2 + x^2 + 10x + 25 - 81 = 0$$
$$2x^2 + 10x - 56 = 0$$
$$x^2 + 5x - 28 = 0 \qquad (6)$$

$$x = \frac{-5 \pm \sqrt{5^2 - 4(1)(-28)}}{2(1)}$$

$$x = \frac{-5 \pm \sqrt{25 + 112}}{2}$$

$$x = \frac{-5 \pm \sqrt{137}}{2}$$

$$x = \frac{-5 + 11.705}{2} \qquad\qquad x = \frac{-5 - 11.705}{2}$$

$$x = \frac{6.705}{2} \qquad\qquad x = \frac{-16.705}{2} \quad \text{(Discard)}$$

$$x = 3.353 \text{ m (approx.) one leg}$$
$$x + 5 = 3.353 + 5 = 8.353 \text{ m (approx.) other leg}$$

Explanation Notice that both members of the equation $2x^2 + 10x - 56 = 0$ in step (6) are divided by 2 so as to make the computation that follows less difficult.

EXERCISES 10.5

A

The values of a, b, and c below represent the lengths of the legs and the hypotenuse respectively of a right triangle. Determine the side that is not given in each exercise.

1. $a = 6, b = 8, c = ?$
2. $a = 12, b = ?, c = 15$
3. $a = ?, b = 16, c = 20$
4. $a = 2, b = 3, c = ?$
5. $a = 7, b = 10, c = ?$
6. $a = 5, b = ?, c = 13$
7. $a = ?, b = 6, c = 8$
8. $a = 16, b = ?, c = 19$
9. $a = ?, b = 24, c = 26$
10. $a = 40, b = ?, c = 41$
11. $a = 26, b = ?, c = 27$
12. $a = ?, b = 34, c = 35$
13. $a = x, b = x + 2, c = 6$
14. $a = x, b = x + 3, c = 7$
15. $a = x, b = x + 3, c = x + 5$
16. $a = x, b = 2x, c = x + 2$

B

Solve each of the following problems.

1. The side of a square is 8 centimeters long. What is the length of the diagonal of the square?

2. The side of a square is 7 meters long. How long is the diagonal of the square?

3. A diagonal of a square is 10 meters long. How long is the side of the square?

4. Two sides of a rectangle are 7 meters and 8 meters respectively. How long is the diagonal of the rectangle?

5. If the length and width of a rectangle are 9 inches and 2 inches respectively, how long is the diagonal of the rectangle?

6. One of the legs of a right triangle is 6 meters long. The hypotenuse is 2 meters longer than the other leg. What are the lengths of that leg and the hypotenuse?

7. One of the legs of a right triangle is 3 cm longer than the other leg. The hypotenuse is 5 cm long. What is the length of each of the legs?

8. The diagonal of a rectangle is 25 cm long. If one of the legs is 15 cm long, how long is the other leg?

9. Two cars start out from the same spot on two intersecting highways. The first car travels due north while the second travels due east. After a short period of time the first car has traveled 60 kilometers and the second has traveled 80 km. How far apart are the two cars at that moment?

10. Two airplanes start out from the same airfield at the same time. One travels due south at a speed of 150 km per hour while the other travels due east at 200 km per hour. How far apart will they be in 2 hours?

11. A ladder 27 feet long leans against a house and reaches a point on the house that is 25 feet above the ground. How far is the foot of the ladder from the side of the house?

12. A baseball diamond was built for the children of an elementary school, however it was made on a smaller scale than the normal baseball diamond. If the distance from home plate to first base is 60 feet, approximately how far is it from home plate to second base? The shape of a baseball diamond is a square. (Hint: In order to isolate the answer between two whole numbers look in the table at the column headed by the word "square.")

10.6 THE RATIONAL NUMBER

The study of the quadratic equation forced the mathematician into the need for inventing a new number. Before we can examine this new number, though, we must take a close look at those numbers with which we are already familiar. Our numbers, prior to now, could all be expressed as fractions. For instance, some of these numbers are

$$5, -6, 0, \frac{2}{3}, -\frac{7}{9}, 3\frac{1}{4}.$$

It is quite apparent that $\frac{2}{3}$ is a fraction for its numerator is 2 while its denominator is 3. Each of the others, however, are also fractions. The number 5 can be rewritten as

$$\frac{5}{1}.$$

Hence, in this fraction the numerator is 5 and the denominator is 1. Similarly, the number $3\frac{1}{4}$ can be rewritten as

$$\frac{13}{4}$$

where now we can see that 13 is the numerator and 4 is the denominator.

A number that can be written as a fraction where the numerator is an integer (whole number) and the denominator is an integer is called a **rational number**. The formal definition of a rational number usually appears as,

---
A rational number is a number that can be expressed in the form a/b where a and b are integers.

If we use $\frac{2}{3}$ as an example, then in this rational number,

$$a = 2 \quad \text{and} \quad b = 3.$$

Similarly, to show that 5 is a rational number we would first rewrite it as we had above and then point out that a and b are integers:

$$a = 5 \quad \text{and} \quad b = 1.$$

Example 1

Express $2\frac{1}{3}$ in the form a/b and state the values of a and b.

Explanation When we rename $2\frac{1}{3}$ as an improper fraction, we are really expressing it in the form of a/b.

478 10/Real Numbers Revisited

Solution $2\frac{1}{3} = \frac{7}{3}$

Therefore, $a = 7$ and $b = 3$.

It is interesting to note that should we try to rename any rational number as a *decimal*, one of two things will occur. Either

1. The division will turn out to be *exact*, or
2. The digits will repeat themselves either one at a time or two at a time or some fixed number of digits will repeat themselves over and over again.

When we say that the division is *exact* we mean that at some point the remainder will be 0. As an illustration, let's examine the rational number

$$\frac{19}{4}.$$

By dividing 4 into 19, we can find the decimal value of this fraction.

```
        4.75
    4 ) 19.00
        16
         3 0
         2 8
           20
           20
            0
```

Thus, the decimal value of $\frac{19}{4}$ is 4.75. The decimal 4.75 is called a **terminating decimal** for there are no digits other than 0 to the right of the 5. Hence, the rational number $\frac{19}{4}$ can be represented by the terminating decimal 4.75.

On the other hand, when we try to determine the decimal value of the rational number $\frac{125}{333}$ we find that it turns out to be .375375375...; that is, the digits 375 keep repeating themselves over and over. A decimal such as this is called a **repeating decimal**. The three dots, ..., called *ellipses* marks imply that the digits keep repeating themselves endlessly.

Every rational number can be expressed as either a terminating decimal or a repeating decimal. By simply dividing the denominator into the numerator we can soon discover whether the rational number names either the one or the other of them.

The reverse of this situation is a little more interesting; that is, how can we find the fraction named by either a termininating or repeating decimal? Before

10.6/The Rational Number

we attempt to show this, we must recall the simple method for multiplying a number by a power of 10. Consider the following:

$$34.26735 \times 10 = 342.6735$$

$$34.26735 \times 100 = 3426.735$$

$$34.26735 \times 1000 = 34267.35$$

$$34.26735 \times 10,000 = 342673.5$$

Notice that in each product *the decimal point is moved exactly the same number of places to the right as there are zeros following the number 1.*

Example 2

Determine the rational number a/b that is named by the terminating decimal .548.

Explanation Actually, we employ a little trickery to find the fraction a/b named by .548. Our first step is to allow N to be the fraction whose decimal is .548.

$$N = .548$$

Now by multiplying both members of this equation by 1,000, we convert the right member of the equation into an integer.

$$1,000\,N = 548.$$

Solving this equation we obtain

$$N = \frac{548}{1,000} = \frac{279}{500}$$

Thus, we have found that the decimal .548 can be expressed as the fraction $\frac{279}{500}$ where a is 279 and b is 500.

Solution Let N represent the rational number a/b.

Then $\qquad N = .548$

$\qquad 1,000\,N = 548.$

$$N = \frac{548}{1,000} = \frac{279}{500} \qquad (a = 279, \qquad b = 500)$$

480 10/Real Numbers Revisited

Example 3

Determine the rational number a/b that is named by the terminating decimal 37.61.

Solution Let N represent the rational number a/b.

Then
$$N = 37.61$$
$$100N = 3761.$$
$$N = \frac{3761}{100} \qquad (a = 3761, \qquad b = 100)$$

Explanation To change the right member of the equation $N = 37.61$ to an integer it is necessary to multiply both members of that equation by 100. The number 100 is selected as the multiplier in order to move the decimal point *two places* to the right in 37.61.

Example 4

Determine the rational number a/b that is named by the repeating decimal .434343...

Explanation Our approach in this situation is much the same as before. We begin by representing the rational number a/b by N. Thus,

$$N = .434343\ldots \qquad (1)$$

Since the digits repeat themselves two at a time, we multiply both members of the equation by 100 in order to move the decimal point two places to the right.

$$100N = 43.434343\ldots \qquad (2)$$

At this point we discover that that part of the number to the right of the decimal point in 43.434343... is exactly the same as that part of the number to the right of the decimal point in the original number .434343... Hence, if we subtract the right member of (1) from the right member of (2), all the digits to the right of the decimal point will be eliminated.

$$\begin{array}{ll} \text{Right member of (2):} & 43.434343\ldots \\ \text{Right member of (1):} & \underline{.434343\ldots} \\ \text{Difference:} & 43. \end{array}$$

However, if we subtract the right member of (1) from the right member of (2), we must do the same with the left members.

$$\begin{array}{ll} \text{Left member of (2):} & 100N \\ \text{Left member of (1):} & \underline{N} \\ \text{Difference:} & 99N \end{array}$$

10.6/The Rational Number

In view of this we obtain the equation

$$99N = 43.$$

Solving this equation we find N to be

$$N = \frac{43}{99}$$

Solution Let N represent the rational number a/b.

Then (1) $N = .434343\ldots$
(2) $100N = 43.434343\ldots$
(2) - (1) $99N = 43$

$$N = \frac{43}{99} \quad (a = 43, \quad b = 99)$$

Example 5

Determine the rational number a/b that is named by the repeating decimal $5.671671\ldots$

Solution Let N represent the rational number a/b.

Then (1) $N = 5.671671\ldots$
(2) $1000N = 5{,}671.671671\ldots$
(2) - (1) $999N = 5{,}666$

$$N = \frac{5{,}666}{999} \quad (a = 5{,}666, \quad b = 999)$$

Explanation Both members of the equation $N = 5.671671\ldots$ are multiplied by 1,000 because the digits are repeating *three* at a time.

EXERCISES 10.6

A

Using a/b as the general form of the rational number where a and b are integers, state the values of a and b for each of the following numbers.

1. 2/3
2. 4/5
3. 7/10
4. -5/8
5. 3/(-7)
6. 1/8
7. 6
8. -7
9. 0
10. -3/5

482 10/Real Numbers Revisited

B

Express each of the following terminating decimals as the rational number a/b where a and b are integers.

1. .43
2. .691
3. .8
4. .027
5. .007
6. .403
7. 52.6
8. 1.41
9. 3.07
10. 45.1
11. 9.371
12. 584.3
13. 32.492
14. 204.38
15. 5.6794
16. 300.05

C

Express each of the following repeating decimals as the rational number a/b where a and b are integers.

1. .272727...
2. .383838...
3. .476476...
4. .931931...
5. .740740...
6. .2222...
7. .5555...
8. .34613461...
9. 6.5353...
10. 12.4848...
11. 25.613613...
12. 17.045045...
13. 195.777...
14. .5343434...
15. .67531531...
16. 2.49717171...

10.7 THE IRRATIONAL NUMBER

We began Section 10.6 by saying that the study of the quadratic equation forced the mathematician into the need for inventing a new number. Each of the numbers we previously examined were rational numbers. However, many of the roots of quadratic equations turn out to be numbers such as

$$\sqrt{2}, \sqrt{3}, \sqrt{8}, \sqrt{257}, \sqrt{1{,}273}$$

and these numbers cannot be expressed in the form a/b where a and b are integers. Hence, none of these numbers is a rational number.

10.7/The Irrational Number

In general,

> A number that cannot be expressed in the form of a/b where a and b are integers is called an irrational number.

Numbers that are the square roots of nonperfect squares such as those above, $\sqrt{2}$, $\sqrt{3}$, comprise only a small fraction of the irrational numbers. The only irrational numbers we are concerned with, however, will be those that we encounter in the solution of the quadratic equation. Hence, our attention will focus on only those irrational numbers that are the square roots of nonperfect squares. The study of other irrational numbers will wait for a more advanced course in mathematics.

Our objective in both this section and the next few will be to examine how we might perform the fundamental operations on these new numbers. In order to keep our work as simple as possible at this stage, we must reemphasize the fact that the square root of a negative number will be excluded from our work for these numbers are meaningless to us at this time. Thus, should we encounter an expression such as

$$\sqrt{x},$$

it will be understood that x **must be a positive number or zero**. Numbers such as

$$\sqrt{-3}, \sqrt{-4}, \sqrt{-6}, \sqrt{-9}$$

do not exist for us.

There are a few general agreements that mathematicians have made concerning the square root of a number that are rather important. For instance, if the question is asked,

"What is the square root of 25?"

the answer is that there are two numbers that are the square roots of 25. The first of these is +5 for the product of +5 and +5 is 25. The second number is −5 for the product of −5 and −5 is 25. However, if the question is worded as,

"What is the value of $+\sqrt{25}$?"

then the answer is

$$+5.$$

Notice that there is *only one* number for the answer and that the *sign of that number is the same as the sign of the radical*.

Similarly, the question

"What is the value of $-\sqrt{25}$?"

has only one answer and that is

$$-5.$$

Here, too, *the sign of the answer is the same as the sign of the radical.*

Example 1
Simplify the following quantity.

$$-\sqrt{49}$$

Explanation The term "simplify" when used in a situation such as this implies that if the radicand is a perfect square, we find its square root.

Solution $\quad\quad\quad -\sqrt{49} = -7$

We run into a rather interesting situation in an expression such as

$$\sqrt{(5)^2} \ .$$

Here we are being asked to determine the number that when squared will be equal to 5 squared. To determine the number we proceed as follows:

$$\sqrt{(5)^2} \ = \ \sqrt{25} \ = \ 5.$$

Thus, we discover that the answer is exactly the same number as the quantity in the parentheses.

This number, 5, is the same as *this number*, 5.

Now let us consider the situation where the number in the parentheses is a negative number such as

$$\sqrt{(-5)^2} \ .$$

Proceeding as before, we find the following:

$$\sqrt{(-5)^2} \ = \ \sqrt{25} \ = \ 5.$$

Here we see that the answer is *not* the same as the quantity in the parentheses; the answer is the absolute value of that quantity.

This number, 5, is *not* the same as *this number*, -5.

In order to avoid this difficulty, we will assume that whenever we encounter an expression such as

$$\sqrt{x^2},$$

the replacements for x will be positive numbers.

EXERCISES 10.7

A

What are the square roots of each of the following numbers?

1. 16
2. 49
3. 1
4. 81
5. 144
6. 225

B

Simplify each of the following expressions.

1. $\sqrt{64}$
2. $\sqrt{81}$
3. $-\sqrt{4}$
4. $\sqrt{a^2}$
5. $-\sqrt{100}$
6. $-\sqrt{x^2}$
7. $\sqrt{y^6}$
8. $-\sqrt{121}$
9. $-\sqrt{1}$
10. $\sqrt{x^{16}}$

10.8 THE PRODUCT OF THE SQUARE ROOTS OF TWO QUANTITIES

PART 1

We initially ran across the need for the radical sign when trying to solve the equation

$$x^2 = 7.$$

Since we knew of no numbers that when squared would equal 7, we invented a number that would have this property. That number was expressed as

$\sqrt{7}$. (Using only the principal square root.)

Hence, this number is such that when it is squared, its product will be 7. Thus,

$$(\sqrt{7})^2 = 7 \tag{1}$$

However, another way of writing $(\sqrt{7})^2$ is as

$$\sqrt{7} \cdot \sqrt{7}.$$

In view of this, we can replace $(\sqrt{7})^2$ in (1) with $\sqrt{7} \cdot \sqrt{7}$ and write that expression as,

$$\sqrt{7} \cdot \sqrt{7} = 7. \tag{2}$$

Also, based on our investigation in Section 10.7 we know that

$$\sqrt{49} = 7. \tag{3}$$

By examining steps (2) and (3) we see that $\sqrt{7} \cdot \sqrt{7}$ is 7 and also that the $\sqrt{49}$ is 7. Hence, we can say,

$$\sqrt{7} \cdot \sqrt{7} = \sqrt{49}. \tag{4}$$

Inspection of step (4) gives us the clue we need for finding the product of the square roots of two numbers. This is done merely by multiplying the two radicands and then writing that product beneath the radical sign. In general this is expressed as

---

The Principle of Multiplication of Radicals

$$\sqrt{a} \cdot \sqrt{b} = \sqrt{ab}$$

Example 1
Find the product of the following radicals.

$$\sqrt{6} \cdot \sqrt{5}$$

Solution $\quad \sqrt{6} \cdot \sqrt{5} = \sqrt{6 \cdot 5}$

Example 2
Find the product of the following radicals.

$$(-\sqrt{2})(-\sqrt{5})$$

Explanation In finding the product of $-\sqrt{2}$ and $-\sqrt{5}$, we treat these two numbers as any two negative numbers. Hence, their product will be a positive number.

10.8/The Product of the Square Roots of Two Quantities

Solution
$$(-\sqrt{2})(-\sqrt{5}) = +\sqrt{2 \cdot 5}$$
$$= +\sqrt{10}$$

Example 3
Find the product of the following radicals.
$$\sqrt{-2} \cdot \sqrt{5}$$

Explanation As we learned earlier, the square root of -2 does not exist among the numbers with which we are familiar. Hence, it is not possible to determine the product of $\sqrt{-2}$ with $\sqrt{5}$.

Solution $\sqrt{-2} \cdot \sqrt{5}$ (This product does not exist.)

The principle of multiplication of two radicals is as valuable to us when used in the reverse direction as it is for finding the product of two radicals. From that principle we know that
$$\sqrt{a} \cdot \sqrt{b} = \sqrt{ab}.$$

By interchanging the left and right members we find that
$$\sqrt{ab} = \sqrt{a} \cdot \sqrt{b}.$$

This piece of information helps us to rename radicals in which the radicand has a perfect square as one of its factors. For instance, the $\sqrt{12}$ has the perfect square 4 as one of the factors of 12. Hence, the $\sqrt{12}$ can be rewritten as,
$$\sqrt{12} = \sqrt{4 \cdot 3}.$$

Now, by using the reverse of the principle of multiplication, we obtain,
$$\sqrt{4 \cdot 3} = \sqrt{4} \cdot \sqrt{3}.$$

However, the $\sqrt{4}$ is 2 which implies that
$$\sqrt{4} \cdot \sqrt{3} = 2\sqrt{3}.$$

Thus, we have shown that,

the $\sqrt{12}$ can be renamed as $2\sqrt{3}$.

Expressing the radicand as the product of a pair of its factors where one of the factors is a perfect square and then finding the square root of that factor is referred to as *simplifying a radical*. Quite apparently if the radicand does not have a perfect square as a factor, then it cannot be simplified.

Example 4

Simplify the following radical.

$$\sqrt{48}$$

Explanation 48 can be renamed as either 4·12 or 16·3 for both 4 and 16 are perfect squares. However, the term "simplify," when used with reference to a radical, implies that the radicand must not contain a perfect square as a factor when the answer is written in final form. In view of this, the 48 is renamed as 16·3 for 16 is the *largest* perfect square factor of 48.

Solution
$$\sqrt{48} = \sqrt{16 \cdot 3}$$
$$= \sqrt{16} \cdot \sqrt{3}$$
$$= 4\sqrt{3}$$

Example 5

Simplify the following radical.

$$\sqrt{a^7}$$

Solution
$$\sqrt{a^7} = \sqrt{a^6 \cdot a}$$
$$= \sqrt{a^6} \cdot \sqrt{a}$$
$$= a^3 \sqrt{a}$$

Explanation At the time we examined the method for factoring the difference of two squares we found that finding the square root of a power of a variable requires that the exponent by divided by 2. Thus,

"The square root of x^{12} is x^6." [12 ÷ 2 = 6]

Hence, in order to be divisible by 2 the exponent must be an even number. In the situation above, the exponent for a is *not* an even number. However, a^7 can be rewritten as

$$a^7 = a^6 \cdot a^1.$$

The factor a^6 is the *largest* perfect square factor of a^7.

Example 6

Simplify the following radical.

$$\sqrt{18a^6 b^9}$$

10.8 / The Product of the Square Roots of Two Quantities

Solution

$$\sqrt{18a^6b^9} = \sqrt{9a^6b^8 \cdot 2b}$$
$$= \sqrt{9a^6b^8} \cdot \sqrt{2b}$$
$$= 3a^3b^4 \sqrt{2b}$$

EXERCISES 10.8 (Part 1)

A

Find the product of the following radicals.

1. $\sqrt{7} \cdot \sqrt{5}$
2. $\sqrt{3} \cdot \sqrt{11}$
3. $-\sqrt{5} \, (\sqrt{2})$
4. $(-\sqrt{2})(-\sqrt{3})$
5. $(\sqrt{6})(-\sqrt{7})$
6. $\sqrt{-3} \cdot \sqrt{5}$
7. $\sqrt{2a} \cdot \sqrt{3b}$
8. $-\sqrt{5x} \cdot \sqrt{3y}$
9. $\sqrt{-2} \cdot \sqrt{-2}$
10. $3a \, (-\sqrt{5b})(-\sqrt{2c})$

B

Simplify each of the following radicals.

1. $\sqrt{8}$
2. $\sqrt{18}$
3. $\sqrt{20}$
4. $\sqrt{a^3}$
5. $\sqrt{b^5}$
6. $\sqrt{c^9}$
7. $\sqrt{a^2b}$
8. $\sqrt{ac^2}$
9. $\sqrt{4a}$
10. $\sqrt{3a^2}$
11. $\sqrt{12a}$
12. $\sqrt{9a^3}$
13. $\sqrt{a^3b^2}$
14. $\sqrt{a^4b^7}$
15. $\sqrt{9a^2b^5}$
16. $\sqrt{18a^4b^4}$
17. $\sqrt{25x^7y^5}$
18. $\sqrt{50x^2y^9}$
19. $\sqrt{16a^3b^2c}$
20. $\sqrt{32a^4b^3c^2}$

PART 2

Our objective now is to use the principle of multiplication of two radicals twice in the very same exercise. At first we will find the product of two radicals and then we will examine that product to see whether it can be simplified.

Example 1

Find the product of the radicals below.

$$\sqrt{6xy} \cdot \sqrt{2x^8y}$$

Solution
$$\sqrt{6xy} \cdot \sqrt{2x^8y} = \sqrt{6xy \cdot 2x^8y}$$
$$= \sqrt{12x^9y^2} \quad (2)$$
$$= \sqrt{4x^8y^2 \cdot 3x}$$
$$= \sqrt{4x^8y^2} \cdot \sqrt{3x}$$
$$= 2x^4y\sqrt{3x}$$

Explanation After having determined the product in (2), we find that the radicand $12x^9y^2$ has a number of perfect squares as its factors. Hence, the remainder of the solution is spent in simplifying that radical.

Example 2

Find the product below.

$$5w\sqrt{a} \cdot 2y\sqrt{b}$$

Explanation Based on our understanding of the commutative principle of multiplication, it is possible to rearrange the quantities above in any manner we desire. By arranging those quantities as shown below, we can find the product as we had in the past.

$$5w \cdot 2y \cdot \sqrt{a} \cdot \sqrt{b}$$

Solution
$$5w\sqrt{a} \cdot 2y\sqrt{b} = 5w \cdot 2y \cdot \sqrt{a} \cdot \sqrt{b}$$
$$= 10wy\sqrt{ab}$$

Explanation continued Henceforth, we will rearrange the quantities *mentally* and move directly from $5w\sqrt{a} \cdot 2y\sqrt{b}$ to $10wy\sqrt{ab}$.

Example 3

Find the product below.

$$2ab\sqrt{a^3b^5} \cdot a^2\sqrt{b^3}$$

Solution
$$2ab\sqrt{a^3b^5} \cdot a^2\sqrt{b^3} = 2a^3b\sqrt{a^3b^8}$$
$$= 2a^3b\sqrt{a^2b^8 \cdot a}$$
$$= 2a^3b\sqrt{a^2b^8}\sqrt{a}$$
$$= 2a^3b \cdot ab^4\sqrt{a}$$
$$= 2a^4b^5\sqrt{a}$$

10.9/The Quotient of the Square Roots of Two Quantities

EXERCISES 10.8 (Part 2)

A

Find the product in each of the exercises below. If necessary, simplify each answer.

1. $\sqrt{x^3} \sqrt{x^5}$
2. $\sqrt{a} \sqrt{a^3}$
3. $\sqrt{xy} \sqrt{x}$
4. $\sqrt{ab^3} \cdot \sqrt{b}$
5. $\sqrt{a^3b^3} \cdot \sqrt{ab^3}$
6. $\sqrt{a^2b^5} \cdot \sqrt{a^3b}$
7. $\sqrt{8x} \cdot \sqrt{2y}$
8. $\sqrt{5x^3} \cdot \sqrt{2x}$
9. $\sqrt{6ab} \sqrt{3b}$
10. $\sqrt{3x^2y} \cdot \sqrt{8y^3}$
11. $\sqrt{abc^3} \cdot \sqrt{a^3b}$
12. $\sqrt{10a^5} \cdot \sqrt{2b^3}$
13. $3\sqrt{x} \cdot 5\sqrt{y}$
14. $5a\sqrt{b} \cdot \sqrt{3c}$
15. $2x\sqrt{y} \cdot 5\sqrt{y}$
16. $3\sqrt{y^3} \cdot a\sqrt{xy}$
17. $2x\sqrt{xy} \cdot 5y\sqrt{x}$
18. $3xy\sqrt{xy} \cdot 2\sqrt{x^3}$
19. $2a\sqrt{6a} \cdot 3\sqrt{2}$
20. $3c\sqrt{12c^3} \cdot c\sqrt{3c^4}$
21. $bc\sqrt{bc} \cdot c^2\sqrt{5c^3}$
22. $x^2\sqrt{10x^3y} \cdot 3y\sqrt{5y^3}$
23. $\sqrt{x} \cdot \sqrt{xy} \cdot \sqrt{x^3y}$
24. $3a\sqrt{ab} \cdot 2b\sqrt{a} \cdot b\sqrt{b}$

B

Using the method for finding the product of a monomial with a binomial, find the product in each of the following exercises.

1. $\sqrt{2} (\sqrt{3} + \sqrt{5})$
2. $\sqrt{3} (\sqrt{3} + \sqrt{7})$
3. $\sqrt{a} (\sqrt{b} - \sqrt{c})$
4. $\sqrt{a} (\sqrt{a} + 2)$
5. $2\sqrt{a} (3\sqrt{a} + 5)$
6. $\sqrt{3a} (\sqrt{6a^3} - \sqrt{a})$
7. $\sqrt{5x} (y - x\sqrt{x})$
8. $2\sqrt{x} (3\sqrt{x} + 1)$
9. $\sqrt{xy} (\sqrt{x} - \sqrt{xy} + \sqrt{y})$
10. $x\sqrt{y} (\sqrt{x} - \sqrt{y})$

10.9 THE QUOTIENT OF THE SQUARE ROOTS OF TWO QUANTITIES

In the preceding section we developed the relation whereby,

$$\sqrt{7} \cdot \sqrt{7} = 7.$$

What we showed at that time for the number 7 will hold true for any number. Hence, in general we can say that

$$\sqrt{x} \cdot \sqrt{x} = x.$$

This piece of information will enable us to determine a method for dividing the square root of one quantity by the square root of another quantity.

The multiplication principle we learned enables us to write the product of two *separate* radicals as a *single* radical:

Separate Radicals   Single Radical

$$\sqrt{x} \cdot \sqrt{y} = \sqrt{xy}.$$

Our objective now is to show that the same situation will hold true in the operation of division: that is, that

Separate Radicals $\dfrac{\sqrt{x}}{\sqrt{y}} = \sqrt{\dfrac{x}{y}}$ Single Radical

In order to do this, we are going to show that the two fractions below name the same fraction.

$$(1)\ \dfrac{\sqrt{a^2}}{\sqrt{b^2}} \quad \text{and} \quad \sqrt{\dfrac{a^2}{b^2}}\ (2)$$

If this is so, then these two fractions will be equal to each other.

(1) $\quad \dfrac{\sqrt{a^2}}{\sqrt{b^2}} = \dfrac{\sqrt{a \cdot a}}{\sqrt{b \cdot b}} = \dfrac{\sqrt{a} \cdot \sqrt{a}}{\sqrt{b} \cdot \sqrt{b}} = \dfrac{a}{b}$

(2) $\quad \sqrt{\dfrac{a^2}{b^2}} = \sqrt{\dfrac{a \cdot a}{b \cdot b}} = \sqrt{\dfrac{a}{b}} \cdot \sqrt{\dfrac{a}{b}} = \dfrac{a}{b}$

These fractions are the same.

Since both

$$\dfrac{\sqrt{a^2}}{\sqrt{b^2}} \quad \text{and} \quad \sqrt{\dfrac{a^2}{b^2}} \quad \text{name the fraction } \dfrac{a}{b}$$

10.9/The Quotient of the Square Roots of Two Quantities

then we can conclude that

$$\frac{\sqrt{a^2}}{\sqrt{b^2}} = \sqrt{\frac{a^2}{b^2}}.$$

What we have shown for these two fractions we will accept in general for any two fractions involving radicals.

The Principle of Division of Radicals

$$\frac{\sqrt{c}}{\sqrt{d}} = \sqrt{\frac{c}{d}}$$

Example 1
Find the quotient in the expression below.

$$\frac{\sqrt{18x^3}}{\sqrt{2x}}$$

Solution

$$\frac{\sqrt{18x^3}}{\sqrt{2x}} = \sqrt{\frac{18x^3}{2x}}$$

$$= \sqrt{9x^2}$$

$$= 3x$$

In exactly the same manner as we used the reverse of the principle of multiplication of radicals to simplify certain radicals so, too, can we use the reverse of the principle of division of radicals to simplify certain radicals.

Example 2
Simplify the following radical.

$$\sqrt{\frac{5}{4y^2}}$$

Solution

$$\sqrt{\frac{5}{4y^2}} = \frac{\sqrt{5}}{\sqrt{4y^2}}$$

$$= \frac{\sqrt{5}}{2y}$$

Example 3
Simplify the following radical.

$$\sqrt{\frac{25x^3}{9y^4}}$$

Solution

$$\sqrt{\frac{25x^3}{9y^4}} = \frac{\sqrt{25x^3}}{\sqrt{9y^4}}$$

$$= \frac{\sqrt{25x^2}\sqrt{x}}{\sqrt{9y^4}}$$

$$= \frac{5x\sqrt{x}}{3y^2}$$

There are also some situations that arise where we must first divide one radical by another as we did in Example 1 and then simplify that answer as was necessary in Examples 2 and 3. This is shown below.

Example 4
Find the quotient in the expression below.

$$\frac{\sqrt{24x^3y^5}}{\sqrt{2x^5}}$$

Solution

$$\frac{\sqrt{24x^3y^5}}{\sqrt{2x^5}} = \sqrt{\frac{24x^3y^5}{2x^5}}$$

$$= \sqrt{\frac{12y^5}{x^2}}$$

$$= \frac{\sqrt{12y^5}}{\sqrt{x^2}}$$

$$= \frac{\sqrt{4y^4}\sqrt{3y}}{\sqrt{x^2}}$$

$$= \frac{2y^2\sqrt{3y}}{x}$$

10.9/The Quotient of the Square Roots of Two Quantities

EXERCISES 10.9

A

Find the quotient in each of the exercises below.

1. $\dfrac{\sqrt{45}}{\sqrt{5}}$

2. $\dfrac{\sqrt{27}}{\sqrt{3}}$

3. $\dfrac{\sqrt{20}}{\sqrt{5}}$

4. $\dfrac{\sqrt{50}}{\sqrt{2}}$

5. $\dfrac{\sqrt{48}}{\sqrt{3}}$

6. $\dfrac{\sqrt{100}}{\sqrt{5}}$

7. $\dfrac{\sqrt{x^3}}{\sqrt{x}}$

8. $\dfrac{\sqrt{ab^2}}{\sqrt{a}}$

9. $\dfrac{\sqrt{x^3 y^5}}{\sqrt{xy}}$

10. $\dfrac{\sqrt{a^7 b^3}}{\sqrt{ab^3}}$

11. $\dfrac{\sqrt{12a^3}}{\sqrt{3a}}$

12. $\dfrac{\sqrt{21xy}}{\sqrt{7x}}$

13. $\dfrac{\sqrt{33ab}}{\sqrt{3ab}}$

14. $\dfrac{\sqrt{98x^5 y}}{\sqrt{2xy}}$

15. $\dfrac{\sqrt{192a^3 bc^7}}{\sqrt{3abc^3}}$

16. $\dfrac{\sqrt{7a^3 b^5}}{\sqrt{3a^3 b^4 c}}$

B

Simplify each of the following radicals.

1. $\sqrt{\dfrac{4}{9}}$

2. $\sqrt{\dfrac{x^2}{y^2}}$

3. $\sqrt{\dfrac{x^6}{9y^2}}$

4. $\sqrt{\dfrac{7}{4a^2}}$

5. $\sqrt{\dfrac{3a^8}{16}}$

6. $\sqrt{\dfrac{5b^6}{c^4}}$

7. $\sqrt{\dfrac{x^2 y}{z^4}}$

8. $\sqrt{\dfrac{a^6}{b^4 c}}$

9. $\sqrt{\dfrac{9a}{25b^2}}$

10. $\sqrt{\dfrac{18a^2}{b^4}}$

11. $\sqrt{\dfrac{8b}{c^2}}$

12. $\sqrt{\dfrac{12x^2}{25a^2}}$

13. $\sqrt{\dfrac{27x^3y}{a^2}}$

14. $\sqrt{\dfrac{16a^5}{49b^8}}$

15. $\sqrt{\dfrac{8ab^3}{9c^6}}$

16. $\sqrt{\dfrac{x^5y^7}{81z^{18}}}$

C

Find the quotient in each of the exercises below. If possible, simplify your answer.

1. $\dfrac{\sqrt{60}}{\sqrt{5}}$

2. $\dfrac{\sqrt{x^7}}{\sqrt{x^4}}$

3. $\dfrac{\sqrt{x}}{\sqrt{x^5}}$

4. $\dfrac{\sqrt{2}}{\sqrt{18}}$

5. $\dfrac{\sqrt{24a}}{\sqrt{3a^3}}$

6. $\dfrac{\sqrt{60x^5}}{\sqrt{5xy^6}}$

7. $\dfrac{\sqrt{3xy^2}}{\sqrt{27xw^6}}$

8. $\dfrac{\sqrt{40a^5}}{\sqrt{18ab^6}}$

9. $\dfrac{\sqrt{12x^3yz}}{\sqrt{27y^5z}}$

10. $\dfrac{\sqrt{48a}}{\sqrt{100ab^2}}$

10.10 THE SUM OF THE SQUARE ROOTS OF QUANTITIES

Determining the sum of a number of radicals is exactly the same as determining the sum of a number of terms. If the terms are like terms, it is possible for us to "collect" them by adding the coefficients. Thus,

$$5a + 2a = 7a.$$

Similarly, if the radicands are the same, it is possible to "collect" the two radicals by adding the coefficients.

$$5\sqrt{11} + 2\sqrt{11} = 7\sqrt{11}.$$

In this example, the number 5 in the first term implies that there are *five* square roots of 11; to this must be added *two* square roots of 11, hence the sum is *seven* square roots of 11.

10.10/The Sum of the Square Roots of Quantities

In the event that the terms are unlike terms you may recall that we merely *expressed* the sum between the two terms. For instance, when determining the sum of $6x$ and $7y$ we simply showed that sum by writing

$$6x + 7y.$$

Similarly, the sum of $6\sqrt{5}$ and $7\sqrt{3}$ is expressed as

$$6\sqrt{5} + 7\sqrt{3}.$$

Example 1

Find the sum of the following radicals.

$$7\sqrt{3} + \sqrt{3} - 2\sqrt{3}$$

Explanation One of the points we had to be cautious of earlier in our work was to realize that whenever a coefficient is not specifically written in a term it is understood to be the number 1. The same situation exists in this example where we must remember to write the number 1 as the coefficient of $\sqrt{3}$ in the second term.

Solution $\quad 7\sqrt{3} + \sqrt{3} - 2\sqrt{3} = 7\sqrt{3} + 1\sqrt{3} - 2\sqrt{3}$

$$= 6\sqrt{3}$$

Example 2

Collect like terms in the following expression.

$$2\sqrt{7} - 4\sqrt{5} - 6\sqrt{7} + \sqrt{5}$$

Solution $2\sqrt{7} - 4\sqrt{5} - 6\sqrt{7} + \sqrt{5} \quad = 2\sqrt{7} - 4\sqrt{5} + 6\sqrt{7} + 1\sqrt{5}$

$$= 8\sqrt{7} - 3\sqrt{5}$$

At times it is necessary to simplify some of the radicals before it is possible to collect like terms. Thus, in a situation such as the one below by simplifying the second term we can then add the two terms.

$$5\sqrt{2} + 3\sqrt{8}$$

$$= 5\sqrt{2} + 3\sqrt{4 \cdot 2}$$

$$= 5\sqrt{2} + 3\sqrt{4} \cdot \sqrt{2}$$

$$= 5\sqrt{2} + 3 \cdot 2 \cdot \sqrt{2}$$

$$= 5\sqrt{2} + 6\sqrt{2}$$

$$= 11\sqrt{2}$$

In the event the radicand involves variables, the procedure is exactly the same as above. The examples that follow point out how this is done.

Example 3

Collect like terms in the following expression.

$$\sqrt{9x} + 3\sqrt{4x} - 5\sqrt{16x}$$

Solution
$$\begin{aligned}
\sqrt{9x} + 3\sqrt{4x} - 5\sqrt{16x} &= \sqrt{9}\sqrt{x} + 3\sqrt{4}\sqrt{x} - 5\sqrt{16}\sqrt{x} \\
&= 3\sqrt{x} + 3 \cdot 2\sqrt{x} - 5 \cdot 4\sqrt{x} \\
&= 3\sqrt{x} + 6\sqrt{x} - 20\sqrt{x} \\
&= -11\sqrt{x}
\end{aligned}$$

Example 4

Simplify and collect like terms in the following expression.

$$2\sqrt{4x^3} + 6\sqrt{9x^3} - x\sqrt{25x}$$

Solution

$$\begin{aligned}
2\sqrt{4x^3} - 6\sqrt{9x^3} - x\sqrt{25x} &= 2\sqrt{4x^2} \cdot \sqrt{x} - 6\sqrt{9x^2} \cdot \sqrt{x} - x\sqrt{25} \cdot \sqrt{x} \\
&= 2\sqrt{4x^2}\sqrt{x} - 6\sqrt{9x^2}\sqrt{x} - x\sqrt{25}\sqrt{x} \\
&= 2 \cdot 2x\sqrt{x} - 6 \cdot 3x\sqrt{x} - x \cdot 5\sqrt{x} \\
&= 4x\sqrt{x} - 18x\sqrt{x} - 5x\sqrt{x} \\
&= -19x\sqrt{x}
\end{aligned}$$

Explanation The only feature of the solution above that may seem strange to us is the step where we must add the terms in the expression

$$4x\sqrt{x} - 18x\sqrt{x} - 5x\sqrt{x}.$$

We treat the quantity $x\sqrt{x}$ in the same manner as we would treat the quantity ab when adding the following terms:

$$4ab - 18ab - 5ab.$$

The sum here is $-19ab$ and, similarly, the sum in the example 4 is $-19x\sqrt{x}$.

10.11/Rationalizing the Denominator of a Fraction

EXERCISES 10.10

A

Collect like terms in each of the following expressions.

1. $2\sqrt{5} + 7\sqrt{5}$
2. $5\sqrt{3} - 2\sqrt{3}$
3. $4\sqrt{7} + \sqrt{7}$
4. $\sqrt{5} + \sqrt{5}$
5. $-\sqrt{6} + 3\sqrt{6}$
6. $2\sqrt{7} - 2\sqrt{7}$
7. $2\sqrt{10} + 3\sqrt{10} - 4\sqrt{10}$
8. $\sqrt{7} - 2\sqrt{7} + \sqrt{7}$
9. $4\sqrt{11} - 7\sqrt{11} + \sqrt{11}$
10. $-5\sqrt{2} - \sqrt{2} + \sqrt{2}$
11. $3\sqrt{2} + 5\sqrt{2} - 4\sqrt{3}$
12. $2\sqrt{7} + 4\sqrt{5} + 3\sqrt{7}$
13. $\sqrt{5} + 2\sqrt{3} + 3\sqrt{5} + 5\sqrt{3}$
14. $2\sqrt{10} - \sqrt{5} + \sqrt{10} + 4\sqrt{5}$
15. $3\sqrt{a} + 5\sqrt{a} - 2\sqrt{b}$
16. $2\sqrt{a} + \sqrt{b} + \sqrt{a} + 5\sqrt{b}$
17. $5\sqrt{ab} - 6\sqrt{ab} + \sqrt{ab}$
18. $2\sqrt{3x} + 4\sqrt{3x} - 5\sqrt{3x}$
19. $2x\sqrt{y} + 3x\sqrt{y}$
20. $5a\sqrt{a} - 2a\sqrt{a} - a\sqrt{a}$

B

Collect like terms in each of the following expressions.

1. $\sqrt{4a} + \sqrt{9a}$
2. $\sqrt{25x} + \sqrt{16x}$
3. $\sqrt{49x} - \sqrt{64x}$
4. $\sqrt{2} + \sqrt{8}$
5. $\sqrt{3} - \sqrt{12}$
6. $\sqrt{12} + \sqrt{27}$
7. $2\sqrt{4x} + 3\sqrt{25x}$
8. $3\sqrt{25a} - 5\sqrt{9a}$
9. $\sqrt{18} + 3\sqrt{8}$
10. $\sqrt{75} - 2\sqrt{27}$
11. $\sqrt{8} + \sqrt{12} + \sqrt{18} + \sqrt{27}$
12. $2\sqrt{a^3} + 3a\sqrt{a}$
13. $5x\sqrt{x} - 6\sqrt{x^3}$
14. $\sqrt{4x^3} + x\sqrt{9x}$
15. $2x\sqrt{y} + x\sqrt{y} - y\sqrt{x}$
16. $\sqrt{x^2y} + \sqrt{9x^2y}$
17. $3\sqrt{9a^3} + 5\sqrt{16a^3} + a\sqrt{4a}$
18. $2a\sqrt{b} + \sqrt{4a^2b} - \sqrt{25a^2b}$
19. $3\sqrt{x^3y} - 4\sqrt{xy^3} + x\sqrt{9xy}$
20. $\sqrt{a^3} - \sqrt{b^3} - a\sqrt{25a} + b\sqrt{b}$

10.11 RATIONALIZING THE DENOMINATOR OF A FRACTION

In most situations involving radicals the mathematician prefers to leave his answer in radical form. There are times though when he needs an approximate value as an answer to a problem he is investigating. For instance, let us assume that as part of the computation he is doing he arrives at the quantity

$$\frac{3}{\sqrt{2}}$$

To find the approximate value of this fraction he will probably use a hand held calculator.

In the event he has no calculator, though, he might very well reach for the square root table where he will find the approximate value of the $\sqrt{2}$ as 1.414. He then divides 3 by 1.414.

$$1.414 \overline{) 3.000}^{\,2.121+}$$

This process itself is not difficult, however it is very tedious. The time spent in arriving at the answer is very lengthy for it requires division by a 4-digit number. And if the mathematician wants his answer even more accurate, he will have had to divide by an approximation of $\sqrt{2}$ that has more than four digits.

In order to avoid the need for having to divide by a number having many digits, the mathematician applies his background pertaining to fractions in the following way. His objective is to rename the fraction so that the

$$\frac{3}{\sqrt{2}}$$

denominator will be a *rational* number rather than the irrational number $\sqrt{2}$. He knows that a fraction can be renamed by multiplying it by some form of the number 1. The form he chooses is $\sqrt{2}/\sqrt{2}$ for this fraction when multiplied by $3/\sqrt{2}$ will give a product whose denominator is $\sqrt{4}$. However, $\sqrt{4}$ is 2 and 2 is a rational number.

This entire procedure is shown below.

$$\frac{3}{\sqrt{2}} = \frac{3}{\sqrt{2}} \cdot 1$$

$$= \frac{3}{\sqrt{2}} \cdot \frac{\sqrt{2}}{\sqrt{2}}$$

$$= \frac{3\sqrt{2}}{\sqrt{4}}$$

$$= \frac{3\sqrt{2}}{2}$$

Now, to find the approximate value of

$$\frac{3\sqrt{2}}{2}$$

10.11/Rationalizing the Denominator of a Fraction

we simply multiply 3 by 1.414 and divide that answer by 2. All of which can be done far more easily than the division by 1.414.

$$\frac{3\sqrt{2}}{2} = \frac{3(1.414)}{2}$$

$$= \frac{4.242}{2}$$

$$= 2.121$$

The process of changing the denominator of a fraction from an irrational number to a *rational* number is called **rationalizing the denominator of a fraction**.

Example 1

Rationalize the denominator of the following fraction.

$$\frac{5}{\sqrt{12}}$$

Explanation In order to make the work as simple as possible, we seek out the *smallest* number by which to multiply the denominator so that the *radicand* will be a *perfect square*. That number is $\sqrt{3}$. Hence, the form of 1 by which we must multiply the fraction $5/\sqrt{12}$ is $\sqrt{3}/\sqrt{3}$.

Solution

$$\frac{5}{\sqrt{12}} = \frac{5}{\sqrt{12}} \cdot \frac{\sqrt{3}}{\sqrt{3}}$$

$$= \frac{5\sqrt{3}}{\sqrt{36}}$$

$$= \frac{5\sqrt{3}}{6}$$

Example 2

Rationalize the denominator of the following fraction.

$$\frac{2\sqrt{3a}}{\sqrt{6a^3}}$$

Explanation The smallest quantity by which we can multiply $\sqrt{6a^3}$ so that the radicand will be a perfect square is $\sqrt{6a}$. Hence, the form of the number 1 in this situation is $\sqrt{6a}/\sqrt{6a}$.

Solution

$$\frac{2\sqrt{3a}}{\sqrt{6a^3}} = \frac{2\sqrt{3a}}{\sqrt{6a^3}} \cdot \frac{\sqrt{6a}}{\sqrt{6a}}$$

$$= \frac{2\sqrt{18a^2}}{\sqrt{36a^4}}$$

$$= \frac{2\sqrt{9a^2 \cdot 2}}{\sqrt{36a^4}}$$

$$= \frac{2 \cdot 3a\sqrt{2}}{6a^2}$$

$$= \frac{6a\sqrt{2}}{6a^2}$$

$$= \frac{\sqrt{2}}{a}$$

Example 3

Rationalize the denominator of the fraction below and then determine the approximate decimal value of the fraction.

$$\frac{7}{\sqrt{3}}$$

Solution

$$\frac{7}{\sqrt{3}} = \frac{7\sqrt{3}}{\sqrt{3}\sqrt{3}}$$

$$= \frac{7\sqrt{3}}{\sqrt{9}}$$

$$= \frac{7(1.732)}{3}$$

$$= \frac{12.124}{3}$$

$$= 4.041$$

EXERCISES 10.11

A

Rationalize the denominator of each of the fractions below.

1. $\dfrac{5}{\sqrt{2}}$

2. $\dfrac{3}{\sqrt{5}}$

10.11/Rationalizing the Denominator of a Fraction

3. $\dfrac{7}{\sqrt{8}}$

4. $\dfrac{\sqrt{2}}{\sqrt{7}}$

5. $\dfrac{\sqrt{5}}{\sqrt{12}}$

6. $\dfrac{a}{\sqrt{b}}$

7. $\dfrac{b}{\sqrt{b}}$

8. $\dfrac{1}{\sqrt{x}}$

9. $\dfrac{\sqrt{2}}{\sqrt{3x}}$

10. $\dfrac{\sqrt{x}}{\sqrt{xy}}$

11. $\dfrac{\sqrt{a}}{\sqrt{x^3}}$

12. $\dfrac{\sqrt{y}}{\sqrt{2x^3}}$

13. $\dfrac{3\sqrt{a}}{\sqrt{2b}}$

14. $\dfrac{\sqrt{5a}}{\sqrt{6a}}$

15. $\dfrac{\sqrt{2a}}{\sqrt{6ab}}$

16. $\dfrac{5\sqrt{x}}{\sqrt{5x}}$

17. $\dfrac{4\sqrt{a}}{\sqrt{8a^3}}$

18. $\dfrac{\sqrt{6b}}{4\sqrt{2a}}$

19. $\sqrt{\dfrac{a}{b}}$

20. $\sqrt{\dfrac{3a}{5b}}$

B

Rationalize the denominator of each of the following fractions and then determine the approximate value of the fraction. Use the square root table on page 470.

1. $\dfrac{1}{\sqrt{2}}$

2. $\dfrac{1}{\sqrt{3}}$

3. $\dfrac{5}{\sqrt{5}}$

4. $\dfrac{2}{\sqrt{6}}$

5. $\dfrac{\sqrt{2}}{\sqrt{3}}$

6. $\dfrac{\sqrt{5}}{\sqrt{8}}$

7. $\dfrac{\sqrt{7}}{\sqrt{12}}$

8. $\sqrt{\dfrac{4}{5}}$

9. $\sqrt{\dfrac{5}{7}}$

10. $\dfrac{6}{\sqrt{18}}$

10.12 RADICAL EQUATIONS

Before leaving the topic of radicals we want to take a look at a method by which we can find the roots of an equation where a *variable appears in a radicand*. An equation of this nature is called a **radical equation**. An example of a radical equation is the one below.

$$\sqrt{x-4} + 5 = 11$$

In this situation the variable x appears as one of the terms of the radicand $x - 4$ thus making this equation a radical equation.

Let's begin our examination of radical equations with the relatively simple one

$$\sqrt{x} = 5.$$

This mathematical sentence can be translated into the English sentence

The principal square root of what number is 5?

Notice that the translation called for the "principal" square root of the number. We can tell that this must be so for we had agreed at the outset of this chapter that the expression

$$\sqrt{x}$$

calls for the *positive* square root of the number x.

We know that the principal square root of 25 is 5. Hence, the value of x is 25. Thus, we can say,

$$\text{if} \quad \sqrt{x} = 5 \tag{1}$$

$$\text{then} \quad x = 25. \tag{2}$$

By examining these two equations we notice that the left member of equation (2) is the square of the left member of equation (1)

$$(\sqrt{x})^2 = x.$$

Also, the right member of equation (2) is the square of the right member of equation (1)

$$(5)^2 = 25.$$

It would appear then that by squaring both members of a radical equation we can eliminate the radical sign and thus obtain the root of the radical equation.

10.12/Radical Equations

Unfortunately, this statement is not always true. Consider, for instance, the equation

$$\sqrt{x} = -7. \qquad (3)$$

By examining equation (3) we know immediately that this equation has *no* roots for there is no number whose *principal square root* is a **negative** 7. As stated earlier, the principal square root must be a **positive** number.

However, let us proceed by squaring both members of (3).

$$\sqrt{x} = -7 \qquad (3)$$
$$x = +49$$

It would seem that 49 should be a root of equation (3). When we test to see if this is so we find that 49 does not satisfy this equation.

$$\sqrt{x} = -7$$
$$\sqrt{49} \stackrel{?}{=} -7$$
$$7 \neq -7 \qquad \text{Therefore 49 is not a root.}$$

The number 49 in this situation is called an **extraneous root** of the equation

$$\sqrt{x} = -7.$$

Actually, it is in no way a root of this equation. It is though, a root of the equation

$$x = 49,$$

which is the equation determined by squaring both members of the original equation.

In view of what we have just shown, squaring members of an equation may give us numbers that satisfy the **new** equation but *do not satisfy the original equation*. Hence, it is extremely important that we check all answers to determine whether they are actually roots of the original equation or merely extraneous roots.

Regardless of this weakness, we must square both sides of the equation when solving a radical equation. We do this to eliminate the radical.

Example 1

Solve and check the following equation.

$$\sqrt{x} + 5 = 11$$

Explanation If we square the left member of this equation, the following will occur:

$$(\sqrt{x} + 5)(\sqrt{x} + 5) = (\sqrt{x})^2 + 10\sqrt{x} + 25$$
$$= x + 10\sqrt{x} + 25$$

Thus, although we squared the left member, we did not eliminate the radical for the term $10\sqrt{x}$ will now appear on the left side of equation (1). In order to overcome this difficulty we must first *isolate* the radical. To do this, -5 is added to both members of the equation.

Solution
$$\sqrt{x} + 5 = 11$$
$$\sqrt{x} = 11 - 5 \quad \text{(Isolating the radical)}$$
$$\sqrt{x} = 6$$
$$x = 36 \quad \text{(Squaring both members)}$$

Check for $x = 36$:
$$\sqrt{x} + 5 = 11$$
$$\sqrt{36} + 5 \stackrel{?}{=} 11$$
$$6 + 5 \stackrel{?}{=} 11$$
$$11 = 11 \quad \text{Therefore, 36 is a root.}$$

Example 2

Solve and check the following equation.

$$\sqrt{2x - 3} + 4 = 9$$

Solution
$$\sqrt{2x - 3} + 4 = 9$$
$$\sqrt{2x - 3} = 9 - 4 \quad \text{(Isolating the radical)}$$
$$\sqrt{2x - 3} = 5$$
$$2x - 3 = 25 \quad \text{(Squaring both members)}$$
$$2x = 25 + 3$$
$$2x = 28$$
$$x = 14$$

10.12/Radical Equations

Check for $x = 14$:

$$\sqrt{2x - 3} + 4 = 9$$
$$\sqrt{2 \cdot 14 - 3} + 4 \stackrel{?}{=} 9$$
$$\sqrt{28 - 3} + 4 \stackrel{?}{=} 9$$
$$\sqrt{25} + 4 \stackrel{?}{=} 9$$
$$5 + 4 \stackrel{?}{=} 9$$
$$9 = 9 \qquad \text{Therefore, 14 is a root.}$$

Example 3

Solve and check the following equation.

$$\sqrt{x - 5} + 3 = x - 4$$

Solution

$$\sqrt{x - 5} + 3 = x - 4$$
$$\sqrt{x - 5} = x - 4 - 3 \qquad \text{(Isolating the radical)}$$
$$\sqrt{x - 5} = x - 7$$
$$x - 5 = (x - 7)^2 \qquad \text{(Squaring both members)}$$
$$x - 5 = x^2 - 14x + 49$$
$$0 = x^2 - 14x - x + 49 + 5$$
$$0 = x^2 - 15x + 54$$
$$0 = (x - 9)(x - 6)$$

$x - 9 = 0$ and $x - 6 = 0$

$x = 9 \qquad\qquad x = 6$

Check for $x = 9$:

$$\sqrt{x - 5} + 3 = x - 4$$
$$\sqrt{9 - 5} + 3 \stackrel{?}{=} 9 - 4$$
$$\sqrt{4} + 3 \stackrel{?}{=} 5$$
$$2 + 3 \stackrel{?}{=} 5$$
$$5 = 5$$

Therefore, 9 is a root.

Check for $x = 6$:

$$\sqrt{x - 5} + 3 = x - 4$$
$$\sqrt{6 - 5} + 3 \stackrel{?}{=} 6 - 4$$
$$\sqrt{1} + 3 \stackrel{?}{=} 2$$
$$1 + 3 \stackrel{?}{=} 2$$
$$4 \neq 2$$

Therefore, 6 is an extraneous root.

EXERCISES 10.12

A

Solve and check each of the following radical equations.

1. $\sqrt{x} = 8$
2. $\sqrt{3x} = 6$
3. $\sqrt{5x} = 5$
4. $\sqrt{y+4} = 7$
5. $\sqrt{y-3} = 4$
6. $\sqrt{2w-5} = 3$
7. $\sqrt{5w+1} = 6$
8. $\sqrt{w+3} = 8$
9. $\sqrt{s} + 10 = 7$
10. $2\sqrt{s} = 6$
11. $3\sqrt{a} = 9$
12. $5\sqrt{x} - 1 = 14$
13. $4\sqrt{2x} - 1 = 11$
14. $5 - \sqrt{y} = 4$
15. $11 - \sqrt{3a} = 2$
16. $\sqrt{x-5} + 7 = 10$
17. $\sqrt{3x-2} - 5 = 2$
18. $\sqrt{6-z} + 5 = 4$
19. $\sqrt{5-2z} - 3 = 4$
20. $2\sqrt{3w-1} + 1 = 7$
21. $3\sqrt{4w+3} - 1 = 11$
22. $\sqrt{\dfrac{x}{2}} = 3$
23. $\sqrt{\dfrac{3a}{4}} - 1 = 2$
24. $2 + \sqrt{\dfrac{2x+1}{3}} = 5$

B

Solve and check each of the following radical equations.

1. $\sqrt{x^2 - 9} = 4$
2. $\sqrt{x^2 - 13} + 1 = 7$
3. $3 + \sqrt{2x^2 - 7} = 8$
4. $4 = 11 - \sqrt{3x^2 + 1}$
5. $9 - \sqrt{25 - 4x^2} = 6$
6. $\sqrt{x^2 + 11} = x + 1$
7. $\sqrt{x} = x - 6$
8. $\sqrt{2x} + 4 = x$
9. $\sqrt{3x+1} = x - 1$
10. $\sqrt{5x-4} + 5 = x + 3$

C

Solve and check each of the following radical equations.

1. $\sqrt{3x} = \sqrt{x+10}$
2. $\sqrt{5x-16} = \sqrt{2x-1}$
3. $\sqrt{6x} - \sqrt{x+20} = 0$
4. $\sqrt{x^2 - x - 20} - \sqrt{3x+12} = 0$
5. $3\sqrt{x} = \sqrt{x} + 6$
6. $7\sqrt{2x} = 2(5 + \sqrt{2x})$
7.* $\sqrt{2x+1} = \sqrt{x+1}$
8.* $\sqrt{4x+1} = \sqrt{6x} - 1$
9.* $\sqrt{3x-2} = \sqrt{x-2} + 2$
10.* $\sqrt{4x+1} = \sqrt{x+2} + 1$

*The solution requires squaring both members in two different steps.

CHAPTER REVIEW

1. Solve and check each of the following quadratic equations. If the equation has no roots, indicate this by writing the words "no roots."
   a. $x^2 = 64$
   b. $3x^2 - 26 = 1$
   c. $5x^2 + 10 = 0$
   d. $2(x^2 - 2) = 46$

2. Solve and check each of the following equations.
   a. $(x - 5)^2 = 49$
   b. $x^2 - 6x + 9 = 16$

3. Solve each of the following equations.
   a. $(x + 4)^2 = 3$
   b. $(2x - 3)^2 = 5$

4. What will the third term of each of the following trinomials have to be so that each trinomial will be a perfect square?
   a. $x^2 + 6x +$ ___
   b. $x^2 - 16x +$ ___
   c. $y^2 + 7y +$ ___
   d. $y^2 + \frac{4}{5}y +$ ___

5. Solve each of the following equations by the completing the square method. Check your answers.
   a. $x^2 - 2x = 15$
   b. $x^2 - 4x - 2 = 30$

6. Solve each of the following equations by the completing the square method.
   a. $x^2 + \frac{5}{2}x = \frac{3}{2}$
   b. $2x^2 - 30 = -4x$

7. Write each of the following quadratic equations in standard form.
   a. $3x^2 - 4 = 5x$
   b. $2x(x - 3) = 5$

8. Determine the values of a, b, and c for each of the following quadratic equations.
   a. $5x^2 = 3x - 7$
   b. $2(5 - x) = 3x^2$

9. Determine the roots of each of the following quadratic equations by using the quadratic formula.
   a. $x^2 + 5x - 6 = 0$
   b. $3x^2 + 15 = 14x$
   c. $x(x + 3) = 18$
   d. $5(x^2 + 4x) = x - 12$

10. Use the square root table to find the approximate square root of the following numbers.
    a. 59
    b. 137

11. Find the approximate values of the roots of each of the following equations. Use the square root table.
    a. $x^2 - 3x = 5$
    b. $2x^2 + 3 = 6x$

12. The values of a, b, and c below represent the lengths of the legs and the hypotenuse respectively of a right triangle. Determine the side that is not given in each exercise. Approximate your answer if necessary.

    a. $a = 12$, $b = ?$, $c = 13$ c. $a = ?$, $b = 9$, $c = 14$
    b. $a = 6$, $b = 7$, $c = ?$ d. $a = x$, $b = x+3$, $c = 5$

13. Using a/b as the general form of the rational number where a and b are integers, state the values of a and b for each of the following numbers.

    a. $\frac{5}{7}$ b. 9

14. Express each of the following terminating decimals as the rational number a/b where a and b are integers.

    a. .57   c. .06
    b. .834   d. 3.1

15. Express each of the following repeating decimals as the rational number a/b where a and b are integers.

    a. .444...   c. 5.7272...
    b. .343434...   d. .8363636...

16. Find the product in each of the following exercises and simplify the answer if necessary.

    a. $(-\sqrt{2})(+\sqrt{5})$ d. $2a\sqrt{b^3} \cdot 3a\sqrt{bc}$
    b. $\sqrt{3x} \cdot \sqrt{5y}$ e. $5x\sqrt{6y^3} \cdot x\sqrt{2y^2}$
    c. $\sqrt{6xy} \cdot \sqrt{3x}$ f. $\sqrt{ab}\sqrt{3a}\sqrt{15b^3}$

17. Find the quotient in each of the following exercises.

    a. $\dfrac{\sqrt{8}}{\sqrt{2}}$ c. $\dfrac{\sqrt{24x^7}}{\sqrt{6x}}$

    b. $\dfrac{\sqrt{b^5}}{\sqrt{b^3}}$ d. $\dfrac{\sqrt{15x^3y^2}}{\sqrt{3x^3y}}$

18. Simplify each of the following radicals.

    a. $\sqrt{\dfrac{4}{25}}$ c. $\sqrt{\dfrac{8c^2}{9b^8}}$

    b. $\sqrt{\dfrac{25x^6}{16}}$ d. $\sqrt{\dfrac{a^5b^3}{4c^4}}$

19. Collect like terms in each of the following exercises.

    a. $5\sqrt{2} + 3\sqrt{2}$ c. $4\sqrt{x} + 2\sqrt{y} - \sqrt{x} + 3\sqrt{y}$
    b. $2\sqrt{3} - \sqrt{3} - 4\sqrt{3}$ d. $5a\sqrt{b} - 3a\sqrt{b} + a\sqrt{b}$

20. Collect like terms in each of the following exercises.
    a. $\sqrt{16x} + \sqrt{25x}$
    b. $\sqrt{5} + \sqrt{20}$
    c. $5\sqrt{x^3} - 4x\sqrt{x}$
    d. $\sqrt{5} + \sqrt{20} - \sqrt{45}$

21. Rationalize the denominator of each of the following fractions.
    a. $\dfrac{7}{\sqrt{3}}$
    b. $\dfrac{6}{\sqrt{12}}$
    c. $\dfrac{\sqrt{a}}{\sqrt{5b^3}}$
    d. $\dfrac{6x}{\sqrt{2x}}$

22. Solve and check each of the following radical equations.
    a. $\sqrt{x} + 6 = 11$
    b. $\sqrt{3x} - 5 = 4$
    c. $\sqrt{2x-3} + 5 = 10$
    d. $\sqrt{3x-5} + 7 = x + 2$

23. The side of a square is 6 meters long. What is the approximate length of the diagonal of the square?

24. The length of a rectangle is 7 meters longer than the width. The diagonal of the rectangle is 9 meters long. What are the length and width of the rectangle?

Answers to Odd-Numbered Problems

CHAPTER 1

Exercises 1.1, page 4

1.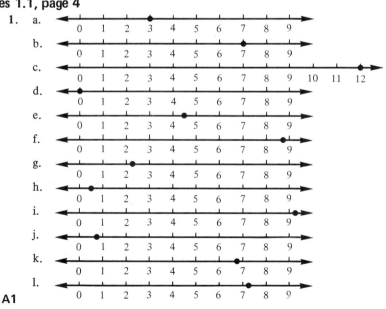

Answers to Odd-Numbered Problems

3. a. The numbers greater than 2 and less than 5.
    b. The numbers greater than 3 and less than 9.
    c. The numbers from 1 through 3 inclusive.
    d. The numbers from 4 through 10 inclusive.
    e. The numbers from 2 through 3 inclusive.
    f. The numbers from 0 to 3, including 0 but not 3.
    g. The numbers from 6 to 9, including 9 but not 6.
    h. The numbers from 4 to 5, including 4 but not 5.

Exercises 1.2, page 7

1. a. The temperature is 5 degrees below zero.
    b. The temperature is 10 degrees above zero.
    c. The temperature is 35 degrees above zero.
    d. The football team gained 5 yards on the last play.
    e. The quarterback lost 10 yards during the third down.
    f. The stock went up $2.50 in price during the day.
    g. Over the past month the stock fell $9.75 in price.
    h. The town is 1,520 feet above sea level.
    i. The lake is 240 feet below sea level.

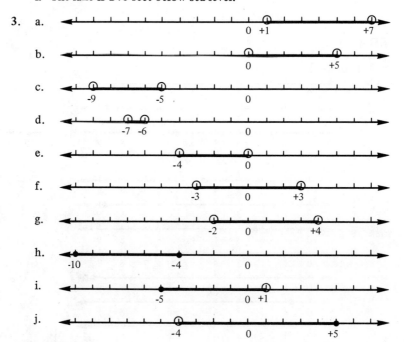

Exercises 1.3, page 12

1. a. $x > 12$
    b. $x < 14$
    c. $x < +11$
    d. $x > -5$
    e. $x \geqslant +4$
    f. $x \leqslant 0$
    g. $x \geqslant -10$
    h. $x = 6$

Chapter 1   A3

3.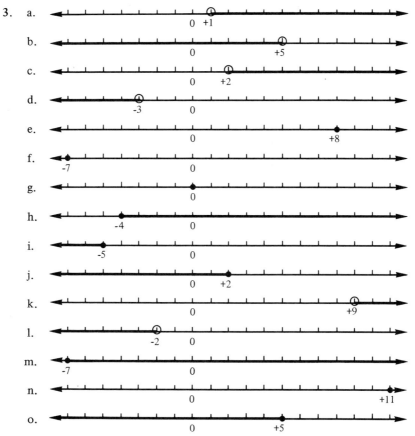

5. Answers may vary from those given below.
   a. +1, +2, +3, +6, +7, +8   d. −1, −2, −3   g. −5, −4, −3
   b. −3, −4, −5   e. 31, 32, 33   h. +8, +7, +6
   c. −9, −8, −7   f. 13, 14, 15   i. −7, −8, −9

Exercises 1.4, page 17

1. a. 6^3 d. 14^3 g. b^5
    b. 5^2 e. 11^5 h. $(\frac{2}{3})^2$
    c. 9^4 f. a^3 i. $(\frac{5}{6})^4$

3. a. 9   h. 529   o. 1
    b. 49   i. 216   p. .01
    c. 81   j. 1   q. .0004
    d. 1   k. 16   r. .000008
    e. 8   l. 625   s. .125
    f. 64   m. 243   t. $\frac{1}{8}$
    g. 225   n. 128   u. $\frac{16}{625}$

5. a. $\frac{4}{9}$ e. 1   i. .0409
    b. $\frac{25}{81}$ f. .0001   j. 8.064
    c. $\frac{27}{64}$ g. .000027   k. 1.9
    d. $\frac{1}{4}$ h. .13

Exercises 1.5, page 21

1. a. $5 + x = 7$ f. $36 = 52 - x$ k. $x^3 = 64$
    b. $x + 17 = 23$ g. $3x = 27$ l. $2x + 6 = 27$
    c. $x - 6 = 2$ h. $6x = 30$ m. $4x - 5 = 31$
    d. $12 - x = 1$ i. $x^2 = 81$
    e. $15 = x + 11$ j. $64 = x^2$

3. a. T   h. T   o. T
    b. T   i. T   p. F
    c. F   j. T   q. T
    d. T   k. F   r. T
    e. T   l. T   s. F
    f. F   m. T   t. T
    g. T   n. T

5. a. $\{2\frac{1}{2}\}$ e. $\{2\frac{1}{3}\}$
    b. $\{4\frac{1}{2}\}$ f. $\{2\frac{2}{5}\}$
    c. $\{9\frac{3}{4}\}$ g. $\{2\frac{2}{7}\}$
    d. $\{5\frac{1}{4}\}$ h. $\{1\frac{2}{9}\}$

Exercises 1.6, page 25

1. a. $x < 17$ f. $16 - x > 10$
    b. $x > 12$ g. $27 > 9x$
    c. $2x < 24$ h. $30 < x + 1$
    d. $4 + x > 10$ i. $8 < x + 2$
    e. $5x > 50$ j. $5 + 3x > 11$

3.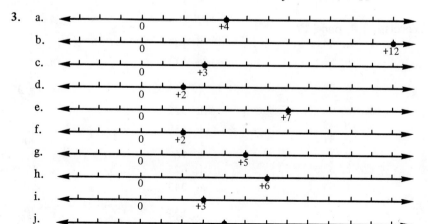

Chapter 1   A5

5.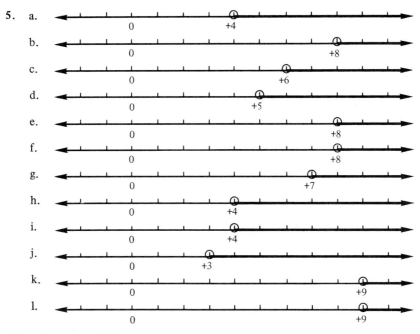

Chapter 1 Review, page 26

1.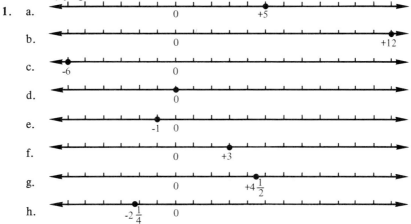

3. a. The numbers greater than 2 and less than 5.
   b. The numbers from −2 to +3, including −2 but not +3.
   c. The numbers from −5 to 0, including 0 but not −5.

5. a. The numbers are greater than five.
   b. The numbers are less than negative three.
   c. The numbers are equal to four.
   d. The numbers are greater than or equal to positive four.
   e. The numbers are less than or equal to negative six.
   f. The numbers are greater than or equal to positive two.

A6 Answers to Odd-Numbered Problems

7. a. 4^5 b. y^2 c. $(\frac{3}{5})^3$

9. a. 16 b. 9 c. 1 d. $\frac{64}{343}$

11. a. Six added to the number is equal to seventeen.
    b. The number subtracted from fifteen is equal to four.
    c. The number cubed is equal to one.
    d. The number added to nine is equal to twelve.

13. a. $x > -5$ b. $2x > 14$ c. $20 < 4x$

15. a. [number line with point at +5]
    b. [number line with open circle at +8]

CHAPTER 2

Exercises 2.1 (Part 1), page 32

A

1. +10	11. +7	21. -3
3. +12	13. -3	23. +2
5. +15	15. -6	25. 0
7. -10	17. +5	27. 0
9. -17	19. -9	29. -17

B

1. +6	5. +5	9. +3
3. -8	7. -2	

C

1. -1	3. +18	5. +2

Exercises 2.1 (Part 2), page 38

1. +7	11. -7	21. -14
3. +11	13. -8	23. -12
5. +4	15. 0	25. -16
7. -6	17. +10	27. +33
9. 0	19. +7	29. -12

Chapter 2 A7

B

1. +15
3. +9
5. -2
7. -13
9. 0
11. 0
13. -11
15. +2
17. +39
19. 0
21. +6
23. -4

C

1. +13
9. -14
17. +13
3. +4
11. -21
19. -15
5. +3
13. -3
7. +5
15. +4

D

1. +9
3. +2
5. +20
7. 0

Exercises 2.2 (Part 1), page 43

A

1. Polynomial
3. Not a polynomial
5. Polynomial

B

1. Binomial
5. Binomial
3. Trinomial
7. Monomial

C

1. $9a$
3. $9a$
5. $7ab$
7. $14cd$

Exercises 2.2 (Part 2), page 47

A

1. $9a$
11. Not Possible
21. 0
3. $7x$
13. 0
23. $+20xyz$
5. $-15w$
15. 0
25. $-6a^3b$
7. Not Possible
17. $+3ab$
27. 0
9. $-2x^3$
19. $-5x^2y$
29. $-35x^3y^3$

B

1. $+11x$
3. $-2a$
5. $+10xy$
7. $+4ab$
9. $+11x^2y$
11. Not Possible
13. $+20ab^3$
15. $+4ab^2c$

C

1. $8a$
3. $-16a$
5. $16x^2$
7. 0
9. $-2a^3b$
11. $-4x$
13. $-6ab$
15. 0
17. $-1xyz$
19. $-7x^2$

D

1. $2x$
3. $-23x^2$
5. $-3x^2$
7. $-3a$
9. $6x^4y$

Exercises 2.3, page 51

A

1. $-3x + 5y - 10w$
3. $9x^2 - 6x + 2$
5. $-2a + 2$
7. $-4x - 5y + 5w$
9. $-a - 6c$
11. $4a^2 + ab + b^2$
13. $9x + 9y + 6w$
15. $7x^2 - 5x + 6$
17. $3xy - y^2$
19. $7x + 3y + 6w$
21. 4

B

1. $8x + 2y - 7w$
3. $3x + y + 4$
5. $-2xy + 2y^2$
7. $12a - 7b - 1c$
9. $xy + y^2$

C

1. $9x + 5y - 4w$
3. $8a + 4b - 3c$
5. $8x^2 - 6xy + 3y^2$
7. $2x - y + 4w$
9. $11a + 3b + 4c$
11. $4x^2 - xy$

D

1. $3a + 4b$
3. $-7a + 3b$
5. $3x^2 - 2x$
7. $-a^2 - 5a$
9. $2x^2 + 3x$
11. $-4a^2 + 3b^2$
13. $x^2 + x$
15. b

Exercises 2.4 (Part 1), page 55

1. $+3$
3. $+8$
5. $+8$
7. $+5$
9. $+6$
11. $+5$
13. $+14$
15. -5
17. -9
19. -6
21. 0
23. -2
25. -3
27. -2
29. $+11$

Chapter 2 — A9

Exercises 2.4 (Part 2), page 58

A

1.	+6	11.	+8	21.	+12
3.	+9	13.	+14	23.	-8
5.	+8	15.	-2	25.	+5
7.	+2	17.	0	27.	-25
9.	+8	19.	+12	29.	+6

B

1.	+8	3.	+13	5.	-8
7.	-2	9.	-11	11.	-5
13.	-9	15.	-3	17.	-5
19.	-12	21.	+9	23.	-47
25.	+6	27.	+108		

Exercises 2.5, page 60

A

1.	$+3a$	13.	$-3a^5$	25.	$+2a$
3.	$+7c$	15.	$8b^2c$	27.	0
5.	$-2y$	17.	$1a^2$	29.	$+3xy$
7.	$+2ab$	19.	$+3x^2$	31.	0
9.	$+11a^3$	21.	$-a^3$		
11.	$-1a^2b$	23.	$-3x$		

B

1.	$2a - 8b + c$	3.	$8x^2 - 7$	5.	$-14b + 6c$
7.	$x^2 - 4x - 10$	9.	$-3x + 4y - 2$	11.	$-2a - 6b + 2c$
13.	$-2b + 3c$	15.	$-2xy$		

C

1.	$3x + 4y$	7.	$-5x^2 - 2x + 4$	13.	$-8x + 5y$
3.	$-2x^2 - 8x - 1$	9.	$-6x^2 - 4x + 2$	15.	$-1ab$
5.	$-2x^2 + 5$	11.	$-2x^2 + 3xy - 3y^2$		

D

1.	$-3a + b - c$	3.	$-3a^2 - ab - 3b^2$	5.	$4a - 6b + 6c$

Exercises 2.6, page 66

A

1. $7x + 4$
3. $9b - 8$
5. $3x + 3$
7. $8a - 2b$
9. $4x - 4y$
11. $-3a^2 + 12a + 4$
13. $-6x + 5y$
15. $-4x + 8y$
17. $-2x + 2y$
19. ab

B

1. $9x - 6$
3. $3a - 13c$
5. $16b + c$
7. $5a - 2b$
9. $3ab + b^2$

C

1. $9x - 1$
3. $-12z + 5$
5. $10x - 2$
7. $-19a - 2$
9. $3x^2 + 3x + 3$

Chapter 2 Review, page 67

1. a. $+9$
   b. -7
   c. -2
   d. $+1$

3. a. $+4$
   b. -2
   c. 0
   d. -4

5. a. $1x^2$
   b. $-10a$
   c. 0
   d. $-2x^2 y$

7. a. Not a polynomial
   b. Polynomial

9. $-2b + 3c$

11. a. $+4$
    b. -3
    c. -6
    d. -12
    e. $+9$
    f. 0

13. a. $a - 2b - 10c$
    b. $6x^2 - 2x$
    c. $a^2 + 10a - 6$

15. a. $4x + 4$
    b. $8x + 1$
    c. $-6x - 6$
    d. $6x - 3$
    e. $-2x + 1$

CHAPTER 3

Exercises 3.1 (Part 1), page 73

A

1. $+10$
3. -24
5. $+12$
7. -40
9. -60
11. -44
13. -35
15. -18
17. $+6$
19. 0
21. 0
23. -5
25. -108
27. $+24$
29. $-\frac{1}{4}$

Chapter 3   A11

B

1. +24	7. +8	13. +32
3. +72	9. −21	15. −72
5. −30	11. −30	

Exercises 3.1 (Part 2), page 75

A

1. +4	11. −64	21. +34
3. +16	13. +81	23. −48
5. +121	15. −27	25. +54
7. +1	17. +32	27. +9
9. +125	19. +1	29. −20

B

1. +25	7. −60	13. +30
3. −32	9. +98	15. −36
5. −72	11. +1	

Exercises 3.2, page 82

A

1. $+12a^7$	11. $-12a^5b^7$	21. $-24a^{12}$
3. $-18a^5$	13. $-42a^5b^2c^3$	23. $-16x^4y^4$
5. $-4a^8$	15. $-45abc$	25. $+6x^{10}$
7. $+14a^2$	17. $+18x^7y^2z^5$	27. $-2x^3y^3z^4$
9. $-a^2$	19. $5xyz$	29. $+x^3y^2$

B

1. $+4a^6$	9. $-x^{10}$	17. $-75a^9b^{10}$
3. $+a^2b^4$	11. $-18a^7$	19. $-250a^8$
5. $-64x^6y^3$	13. $2a^{13}$	
7. $81x^{12}$	15. $+20a^2b^9$	

Exercises 3.3 (Part 1), page 84

1. $+15a^5 + 6a^4$	17. $-24x^3 + 18x^4 - 42x^5$
3. $-18a^3 + 10a^4$	19. $-35a^3b - 28a^2b^2 + 21ab^3$
5. $-42x^2 - 18x$	21. $7x^2 - 21xy - 28y^2$
7. $-15a + 5$	23. $-15 + 35x + 50x^2$
9. $-2x^2y + 2xy^2$	25. $-3a^2 + 2ab + b^2$
11. $6x^4 + 10x^5$	27. $x^6 - x^5 + x^4 - x^3$
13. $-10a^3b + 12ab^3$	29. $-a + b + c - d$
15. $-a^2b + ab^2$	

Exercises 3.3 (Part 2), page 86

A
1. $6x^2 + 14x$
3. $-10x - 1$
5. $13x + 12$
7. $10x^2 + x$
9. $-10x^2 - 6xy$
11. $-7x^3y + 10xy^3$
13. $8a - 4$
15. $a + 10$
17. $a^2 + 5ab$
19. $-2x + 15$
21. $-2a^2 - 2a - 4$
23. $7a^2 - 3ab - b^2$

B
1. $12a - 4b$
3. $12a - 6$
5. $x^2 + 12x$
7. $6x - 34$
9. $19x + 6$

Exercises 3.3 (Part 3), page 90

A
1. $12x^2 + 17x + 6$
3. $6a^2 + 1a - 12$
5. $12y^2 - 38y + 30$
7. $10x^6 + 31x^3 + 24$
9. $-a^6 + 12a^3 - 35$
11. $20x^2 + 26xy + 8y^2$
13. $6x^2 - 1xy - y^2$
15. $4x^2 - 9y^2$
17. $15a^4 - 34a^2b + 15b^2$
19. $6a^3 - 9a^2 + 8a - 12$

B
1. $6x^2 + 19x + 15$
3. $10x^4 + 11x^2y - 6y^2$
5. $4a^2 + 12a + 9$
7. $2x^2y^4 - 3xy^2 - 20$
9. $a^2 + 2ab + b^2$

C
1. $4x^3 + 16x^2 + 19x + 10$
3. $x^3 + 2x^2y + 2xy^2 + y^3$
5. $x^3 - y^3$
7. $10a^4 + 9a^3 - 7a^2 - 18a - 8$
9. $a^4 - 2a^2b^2 + b^4$

D
1. $2x^3 + 11x^2 + 11x - 12$
3. $4a^2 + b^2 + c^2 + 4ab + 4ac + 2bc$
5. $a^3 + 6a^2 + 12a + 8$

Exercises 3.4 (Part 1), page 94
1. $6x^2 + 13x + 6$
3. $12x^2 + 13x - 4$
5. $14x^2 - x - 3$
7. $8x^2 - 22x + 15$
9. $x^2 + 5x + 6$
11. $x^2 - 2x - 63$
13. $x^2 - 11x + 30$
15. $15x^2 + 19xy + 6y^2$
17. $20x^2 - 13xy^2 + 2y^4$
19. $3a^2b^2 - 11ab + 6$
21. $9 + 10x + x^2$
23. $8 + 14x^2 - 15x^4$
25. $24 - 26xy + 5x^2y^2$
27. $6b^4 + 5b^2 - 25$
29. $20 + 3a^2 - 2a^4$
31. $2x^4 - 3x^2y^2 + y^4$
33. $20a^2b^2 + 19abc + 3c^2$
35. $x^2y^2 - 3xyz + 2z^2$
37. $2x^2y^4 + 3xy^2w - 9w^2$
39. $8x^2 - 14xy - 15y^2$

Chapter 3

Exercises 3.4 (Part 2), page 98

A

1. $4x^2 - 9$
3. $x^2 - 49$
5. $4x^2 - 1$
7. $4x^2 + 4x + 1$
9. $x^2 - 9y^2$
11. $64 - x^2$
13. $x^6 - 1$
15. $25x^2 - \frac{4}{49}$
17. $9x^4 - 12x^2 + 4$
19. $\frac{1}{4}x^2 - \frac{1}{9}y^2$
21. $16x^2y^2 - 9w^2$
23. $4x^2 - 25$
25. $25x^2 - 9y^2$
27. $16 - 9x^2$
29. $-25x^2 - 70x - 49$

B

1. $4a^2 + 12a + 9$
3. $a^2 - 6a + 9$
5. $16 - 24a + 9a^2$
7. $36 - 12a + a^2$
9. $a^2 - 2ab + b^2$
11. $9 + 6ab + a^2b^2$
13. $a^4 - 6a^2 + 9$
15. $16 + 8ab^2 + a^2b^4$
17. $\frac{1}{4}a^2 + 4a + 16$
19. $\frac{1}{9}a^6 - 4a^3 + 36$

C

1. $6x^2 + 6x + 9$
3. $25x^2 - 7x - 3$
5. $25x^2 - 6x - 1$
7. $10x^2 - 3x + 24$
9. $2x^2 + 5x - 18$
11. $-45x^2 + 3x + 5$
13. 25
15. $-6x^2 + 36x - 7$

Exercises 3.5, page 101

A

1. $+2$
3. $+2$
5. -4
7. -4
9. -6
11. $+2$
13. -1
15. $+6$
17. -8
19. 0

B

1. $+5$
3. -2
5. -4
7. $+4$
9. -4
11. $+\frac{1}{2}$
13. $-\frac{1}{3}$
15. 0
17. $-\frac{3}{5}$
19. $-\frac{1}{4}$
21. $+\frac{4}{5}$
23. $+9$

Exercises 3.6, page 106

A

1. x^4
3. x^5
5. 1
7. 1
9. $+2x$
11. $-5x^7$
13. $+2x^5$
15. $-15y$
17. $-8xy^6$
19. $+9x^9y$
21. $+yz$
23. $+12xy$
25. $+\frac{1}{4}$
27. $+1$
29. $+y$

A14 Answers to Odd-Numbered Problems

B

1. $2x^2$
3. $-3x^2$
5. $+2x^3y$
7. 0
9. -4
11. $-3x^2y$
13. -1
15. $-1xy$

Exercises 3.7, page 109

A

1. $x^4 + x^2$
3. $x^5 + 3x^3$
5. $x - 2x^2$
7. $-1 - 5x^2$
9. $2y + 3xy^2$
11. $-2x^3 + 3x^2 + 6x$
13. $-1 + 2x^2 + 3x^4$
15. $7x^2y^2z^2 - 9xyz - 11$
17. $x - y + z$
19. $a + b$

B

1. $x^2 - x$
3. $-1 + x^2$
5. $+2x^2y - 5x^3y^3$
7. $5y^3 - 7xz$
9. $-x - y - 1$

Exercises 3.8, page 115

A

1. $x + 3$
3. $x - 3$
5. $x - 5$
7. $3x + 4$
9. $x + 3$
11. $x + 2y$
13. $2x + 7y$
15. $6xy - 5$
17. $3xy - 1$
19. $2x + 7$

B

1. $x + 2$
3. $5x + 2$
5. $8x - 5y$
7. $3x^2 + 7$
9. $x^2 - 2x + 3$
11. $2x - 3$
13. $3x - 1$
15. $2x^2 - 3x - 1$
17. $9x^2 - 3x + 1$
19. $x^3 + 5x^2 + 2x + 3$

C

1. $x - 3$; Remainder = $+13$
3. $3x - 5y$; Remainder = $-25y^2$
5. $2x + 1$; Remainder = -4
7. $x^3 + x^2 + x + 1$; Remainder = $+2$
9. $x^2 + xy + y^2$; Remainder = $-4y^3$

Chapter 3 Review, page 117

1. a. $+20$
   b. -42
   c. 0
   d. $+30$
   e. $+36$
   f. -125
   g. $+18$
   h. -12
   i. $+1$
   j. $+144$

3. a. $-30a^7$
    b. $-6x^2y^5z$
    c. $+6x^3y^4z$
    d. $+16a^4b^6$
    e. $+x^{12}$
    f. $40a^7$

5. a. $12a - 10$
    b. $-x - 2y$
    c. $3x^2 + 3y^2$
    d. $-5a + 6b$

7. a. $6x^3 + 13x^2 + 8x + 3$
    b. $x^3 - 2x^2y - xy^2 + 2y^3$

9. a. $4x^2 - 25$
    b. $49x^2 - 4y^2$
    c. $-9a^2b^2 + 24ab - 16$
    d. $9a^2 + 12a + 4$
    e. $25x^2 - 60x + 36$
    f. $16 - 8xy + x^2y^2$

11. a. -2
    b. $+4$
    c. $+8$
    d. x^2
    e. $-8xy^4$
    f. $-3yz$
    g. $+1$
    h. $+5xy$

13. a. $x + 4$
    b. $2x + 5y$
    c. $x - 4$
    d. $x - 3$

CHAPTER 4

Exercises 4.1, page 123

A

1. -5	7. -12	13. $+5$
3. $+3$	9. -6	15. -2
5. -11	11. $+4$	

B

1. 4	15. -5	29. 9
3. 3	17. 6	31. -2
5. -7	19. -2	33. 11
7. -1	21. 0	35. 7
9. 24	23. -11	37. 17
11. 6	25. 15	39. 0
13. 2	27. 32	

C

1. 8	9. -14	17. -6
3. 6	11. 12	19. 4
5. -18	13. -13	
7. -3	15. 0	

Exercises 4.2, page 128

A

1. $\frac{1}{7}$
3. $\frac{1}{6}$
5. $-\frac{1}{2}$
7. $-\frac{1}{8}$
9. $\frac{1}{4}$
11. $-\frac{1}{5}$
13. $\frac{1}{4}$
15. $-\frac{1}{2}$
17. $\frac{3}{2}$
19. $-\frac{5}{4}$
21. $-\frac{8}{3}$
23. $-\frac{9}{4}$

B

1. 4
3. −5
5. 6
7. 14
9. −7
11. −5
13. 6
15. $5\frac{1}{5}$
17. $3\frac{1}{3}$
19. $-23\frac{3}{4}$
21. 9
23. −8
25. −4
27. 6
29. −12
31. −16
33. −9
35. $-3\frac{1}{6}$
37. −26
39. $-4\frac{4}{9}$

C

1. 9
3. −16
5. 42
7. 162
9. 6
11. −49
13. −6
15. 1

Exercises 4.3 (Part 1), page 132

A

1. 4
3. 8
5. 7
7. 9
9. 3
11. −2
13. −3
15. −3
17. 9
19. 14
21. 0
23. 7
25. 3
27. −1
29. $6\frac{2}{3}$
31. 0
33. $2\frac{1}{2}$
35. $\frac{3}{5}$
37. $-\frac{1}{2}$
39. $-\frac{1}{4}$

B

1. 9
3. 8
5. −20
7. −36
9. −64
11. $-1\frac{1}{3}$

Chapter 4 A17

Exercises 4.3 (Part 2), page 135

1. 3	11. −12	21. 9
3. 2	13. 9	23. 1
5. 8	15. 7	25. $\frac{1}{2}$
7. −4	17. −7	27. $-4\frac{1}{2}$
9. −7	19. 7	29. 4

Exercises 4.3 (Part 3), page 139

1. 1	9. −6	17. −3
3. 4	11. 0	19. −3
5. 8	13. 0	21. −6
7. 4	15. −1	23. All numbers

Exercises 4.3 (Part 4), page 142

A

1. 4	5. 1	9. 0
3. 2	7. 3	11. 1

B

1. 7	9. 0	17. 3
3. 3	11. 3	19. −3
5. −3	13. 3	21. −5
7. 9	15. −16	23. 0

C

1. 2	5. $-19\frac{1}{2}$	9. 1
3. −1	7. −2	

Exercises 4.4, page 148

1. 7, −11	9. 6, $-3\frac{3}{5}$	17. $7\frac{1}{3}$, $-2\frac{2}{3}$
3. 4, −4	11. 2, −2	19. 17, −9
5. 5, −5	13. 3, −3	21. No roots
7. 0, −6	15. No roots	23. −3, 5

Exercises 4.5, page 153

A

1. $x > 7$	9. $x < 4$	17. $x > 7$
3. $x > 2$	11. $x > -4$	19. $x < 2$
5. $x < -6$	13. $x > 6$	21. $x > 7$
7. $x > -6$	15. $x < 6$	23. $x > 2$

Answers to Odd-Numbered Problems

B

1. $x > 2$
3. $x < -6$
5. $x < -4$
7. $x > -2$
9. $x > 5$
11. $x > 9$
13. $x > -1$
15. $x > -2$

Exercises 4.6, page 158

1. $x > +1$ or $x < -1$

   -1 0 +1

3. $x > -7$ and $x < +7$

   -7 0 +7

5. $x > +1$ or $x < -5$

   -5 0 +1

7. $x > +5$ or $x < -4$

   -4 0 +5

9. $x > -6$ and $x < +2$

   -6 0 +2

11. $x > +4$ or $x < -6$

    -6 0 +4

13. $x > +4$ and $x < +6$

    0 +4 +6

15. $x > +10$ or $x < -4$

    -4 0 +10

17. $x > +9$ or $x < -1$

    -1 0 +9

19. $x > -3$ and $x < +2$

    -3 0 +2

Chapter 4 Review, page 158

1. a. 4
   b. 10
   c. 3
   d. –7

3. a. $4\frac{1}{2}$
   b. 5
   c. 5
   d. $\frac{1}{2}$
   e. 3
   f. 3
   g. 12
   h. 6

5. a. 12
   b. 6
   c. –6
   d. 7
   e. 14
   f. 4

7. a. $x > 7$
   b. $x > -7$
   c. $x > 7$
   d. $x > 4$
   e. $x > 4$
   f. $x < 5$
   g. $x > 6$
   h. $x > -1$
   i. $x > 4$
   j. $x < -3$

CHAPTER 5

Exercises 5.1, page 163

A

1. $2 + 5$
3. $9 + 4$
5. $10 - 7$
7. $9 - 2$
9. $12 + 3$
11. $9 - 7$
13. $7 \cdot 8$
15. $2 \cdot 9$
17. $20 \div 5$
19. $7 \div n$
21. $n \div 3$
23. $7 - 4$
25. $n - 20$
27. $3n + 2$
29. $3n + n$

B

1. $2n$
3. $n - 7$
5. $n - 5$
7. $n \div 4$
9. $\frac{1}{2}n$
11. $3n - 2$
13. $5(n + 1)$
15. $2n - (n + 1)$
17. $n \div 3$
19. $(2n - 3) \div 4$

Exercises 5.2 (Part 1), page 166

1. $n + 7 = 24$
3. $n + 3n = 28$
5. $2n = n + 5$
7. $n - 6 = 11$
9. $4n - 7 = 35$
11. $6n - 4n = n + 10$
13. $n + (n + 1) = 47$
15. $2n + 4n = 60$
17. $n + (n + 4) = 56$

Exercises 5.2 (Part 2), page 170

1. $3n + n = 56; n = 14$
3. $5n = 3n + 10; n = 5$
5. $n + (n + 1) = 83; n = 41$
7. $2n = 5n - 57; n = 19$
9. $3n - 19 = n + 7; n = 13$
11. $9n + 2n = 44; n = 4$
13. $45 + 6n = n; n = -9$
15. $n + 16 = 2(n + 1); n = 14$
17. $2(n + 5) = 38; n = 14$
19. $\frac{1}{2}n - 17 = 3; n = 40$

Exercises 5.2 (Part 3), page 174

1. $n + (n + 12) = 26; n = 7, n + 12 = 19$
3. $n + (2n + 7) = 61; n = 18, 2n + 7 = 43$
5. $n - (2n - 20) = 8; n = 12, 2n - 20 = 4$
7. $(3n + 2) - n = 12; n = 5$ (Sister), $3n + 2 = 17$ (Tom)
9. $n + (2n - 3) = 30; n = 11$ (Rocco), $2n - 3 = 19$ (Steve)
11. $(2n) + (n) + (n + 6) = 62$; 2nd = 14, 1st = 28, 3rd = 20
13. $2(3n + 1) + 5n = 2(6n)$; 1st = 2, 2nd = 12, 3rd = 7
15. $n - \frac{2}{3}n = 48$; Clarence 144, Mona, 96
17. $x + (x + 1) + (x + 1) + 1 = 21; x = 6, x + 1 = 7, x + 1 + 1 = 8$

Exercises 5.3, page 180

1. $5(x + 3) + 10x = 90$; 5 dimes, 8 nickels
3. $10(2x) + 25x = 270$; 6 quarters, 12 dimes
5. $25x + 10(x + 7) + 5(4x) = 510$; 8 quarters, 15 dimes, 32 nickels
7. $10x + 5(x + 4) = 245$; 15 dimes, 19 nickels
9. $11x + 13(2x) = 1665$; 45 (11¢ stamps), 90 (13¢ stamps)
11. $25x + 10(30 - x) = 615$; 21 quarters, 9 dimes
13. $12x + 9(33 - x) = 351$; 18 (12¢ oranges); 15 (9¢ oranges)
15. $325x + 293(-20x) = 6244$; 12 @ $3.25; 8 @ $2.93

Exercises 5.4 (Part 1), page 186

1. $n = .56(84); n = 47.04$
3. $n = 1.23(\$420); n = \$516.60$
5. $n = .045(\$2,000); n = \$90$
7. $73 = .38n; n = 192.11$ (approx.)
9. $.055n = \$33; n = \$600$
11. $.0627(\$134) = n; n = \$8.4018$
13. $16 = 80n; n = 20\%$
15. $225 = 50n; n = 450\%$
17. $356n = 124; n = 34.8\%$ (approx.)
19. $15n = 45; n = 300\%$

Exercises 5.4 (Part 2), page 189

Each of the starred answers has been rounded.

1. $\$10 = \$36.50n; n = 27.4\%*$
3. $\$4 = \$19.50n; n = 20.5\%*$
5. $\$5.45 = \$9.50n; n = 57.4\%*$
7. $\$70.50 = \$259.50n; n = 27.2\%*$
9. $.70(\$958) = n; n = \$670.60$
11. $n + .12n = \$184; n = \$164.29*$
13. $n + .51n = 59¢; n = 39¢*$
15. $n - .35n = \$23.95; n = \$36.85*$
17. $n + .23n = \$120,000; n = \$97,560.98*$

Chapter 5 Review, page 191

A

1. a. $n + 5$ c. $2n - 10$
   b. $2n$ d. $17 - n$

B

The starred answer has been rounded.

1. $5x + 12 = 47; x = 7$
3. $2(x + 5) = 24; x = 7$
5. $(2x + 1) - x = 14; x = 13$ (Sally), $2x + 1 = 27$ (Fred)
7. $x + (x + 7) + (3x) = 52;$ 1st $= 9$, 2nd $= 16$, 3rd $= 27$
9. $5(x + 11) + 10x = 295;$ 16 dimes, 27 nickels
11. $15x + 8(x + 10) = 1000;$ 40 (15¢ stamps), 50 (8¢ stamps)
13. $n = 1.53(\$260); n = \$397.80$
15. $2,000n = 180; n = 9\%$
17. $6 = 39n; n = 15.4\%*$

CHAPTER 6

Exercises 6.1, page 197

A

1. 1, 2, 3, 6
3. 1, 3, 5, 15
5. 1, 2, 3, 4, 6, 8, 12, 24
7. 1, 2, 4, 8, 16, 32
9. 1, 2, 3, 4, 6, 8, 12, 16, 24, 48
11. 1, 2, 3, 4, 6, 8, 9, 12, 18, 24, 36, 72

B

1. 1 and 4, 2 and 2
3. 1 and 6, 2 and 3
5. 1 and 16, 2 and 8, 4 and 4
7. 1 and 24, 2 and 12, 3 and 8, 4 and 6
9. 1 and 30, 2 and 15, 3 and 10, 5 and 6
11. 1 and 45, 3 and 15, 5 and 9
13. 1 and 50, 2 and 25, 5 and 10
15. 1 and 60, 2 and 30, 3 and 20, 4 and 15, 5 and 12, 6 and 10

C

1. 2
3. 1
5. 10
7. 11
9. 13
11. 16

D

1. 3
3. 3
5. 3, 5
7. 3
9. 2, 3
11. 2, 3

E

1. $2 \cdot 3$
3. $2^2 \cdot 3$
5. $2^3 \cdot 3$
7. $2^2 \cdot 3^2$
9. 2^6
11. $2^5 \cdot 3$
13. $2^2 \cdot 3^3$
15. 2^8

Exercises 6.2 (Part 1), page 201

A

1. $1, x, x^2$
3. $1, a, a^2, b, ab, a^2b$
5. $1, 2, a, 2a$
7. $1, 3, a, a^2, 3a, 3a^2$
9. $1, 2, 4, 8, a, a^2, 2a, 4a, 8a, 2a^2, 4a^2, 8a^2$

B

1. 2
3. 2, 4

5. 3, 9
7. 2, 3, 6

9. 2, 4, 8

C

1. x^2
3. b^2

5. $3a^2$
7. a

9. x^2y^3
11. x^2y^2

Exercises 6.2 (Part 2), page 205

1. $a(x+y)$
3. $5(a-3c)$
5. $a(x+1)$
7. $x^2(x^2+1)$
9. $3(2x+1)$
11. $4(3-x)$
13. $b(a+c)$
15. $ab(a+b)$
17. $b(b^4-b^2+1)$
19. $2a(a^2-3a+5)$
21. $10(2a^2-3a-1)$
23. $ab(c-a^2b^2)$
25. $xw^2(x^2y^2w^2+xy^4w-1)$
27. $3xy^2(3xy+2y^2-4x^2)$
29. $5(8-5x^3y^3-6x^6y^6)$

Exercises 6.2 (Part 3), page 209

A

1. $(x+y)(a+b)$
3. $(x+2)(a+b)$

5. $(a+4)(c-5)$
7. $(c-2)(3a+4b)$

9. $(x-5)(b+1)$
11. $(a+b)(1+x)$

B

1. $(x+y)(c+d)$
3. $(x+y)(a+5)$
5. $(x+1)(a+b)$

7. $(a+1)(a+b)$
9. $(x+1)(a+1)$
11. $(x-1)(5-a)$

13. $(x-2)(x^2+3)$

Exercises 6.3 (Part 1), page 213

1. $(x+3)(x-3)$
3. $(a+8)(a-8)$
5. Prime
7. $(2x+5)(2x-5)$
9. $(8+a)(8-a)$
11. $(2x+1)(2x-1)$
13. Prime
15. $(3xy+8)(3xy-8)$
17. $(y+.7)(y-.7)$
19. $(b+.2)(b-.2)$
21. $(x+2)(x-2)$
23. $(4xy+5a)(4xy-5a)$
25. $(y^4+2)(y^4-2)$
27. Prime
29. $(7x^5+9y^3)(7x^5-9y^3)$
31. $(6x^2+5y)(6x^2-5y)$
33. $(x^3y+8w^2)(x^3y-8w^2)$
35. Prime
37. $(a+b)(a-b)$
39. $(a+x+y)(a-x-y)$

Exercises 6.3 (Part 2), page 215

1. $a(x+y)(x-y)$
3. $3(x+3)(x-3)$
5. $7(1+2x)(1-2x)$
7. $ab(b+a)(b-a)$
9. $2a(x+2)(x-2)$
11. $(a^2+1)(a+1)(a-1)$
13. $(x^2+4)(x+2)(x-2)$
15. $(4x^2+5)(4x^2-5)$
17. $5(x^4+25)$
19. $(4+9y^2)(2+3y)(2-3y)$

Exercises 6.4 (Part 1), page 220

A

3. $(\ +\ )(\ +\ )$
5. $(\ +\ )(\ -\ )$ Larger
7. $(\ +\ )(\ -\ )$ Larger
9. $(\ +\ )(\ -\ )$ Larger
11. $(\ +\ )(\ -\ )$ Larger
13. $(\ -\ )(\ -\ )$

B

1. $(x+5)(x+1)$
3. $(x-3)(x-1)$
5. $(x+7)(x-1)$
7. $(x+2)(x+4)$
9. $(x-6)(x-2)$
11. $(x-8)(x+2)$
13. $(x-8)(x-3)$
15. $(x+4)(x-3)$
17. $(x-10)(x+3)$
19. $(x-2y)(x-y)$
21. $(a-2b)(a-7b)$
23. $(bc+8)(bc-1)$
25. $(wz-9)(wz+4)$
27. $(c^2-8)(c^2-5)$
29. $(x+6)(x-5)$
31. $(x^2-9y)(x^2+6y)$
33. $(c-16d^2)(c+3d^2)$
35. $(a-2b)(a-30b)$

Exercises 6.4 (Part 2), page 224

A

1. $(3x+2)(x+1)$
3. $(3x+1)(2x+1)$
5. $(4x+1)(2x-1)$
7. $(7x-2)(x-1)$
9. $(3x+7)(x-1)$
11. $(5a-11)(a+1)$
13. $(6a+5)(a-1)$
15. $(5a-8)(a-1)$
17. $(2a-3)(2a-1)$
19. $(3a+1)(2a+5)$
21. $(3x-4y)(x+2y)$
23. $(3y-2z)(y+2z)$
25. $(3y-2z)(y+4z)$
27. $(6x-1)(x-4)$
29. $(9x+4)(x-1)$
31. $(3x+1)(4x-3)$
33. $(9x-4)(x+2)$
35. $(3x+1)(5x-4)$
37. $(2x-3)(3x-5)$
39. $(2x+3y)(4x-5y)$

B

1. $(3x - 2y)(2x + 3y)$
3. Prime
5. $(2a - b)(3a - 5b)$

7. Prime
9. Prime

Exercises 6.4 (Part 3), page 227

A

1. Perfect square
3. Not a perfect square
5. Perfect square

7. Not a perfect square
9. Not a perfect square
11. Not a perfect square

B

1. $(x + 4)^2$
3. $(a - 1)^2$
5. $(4x + 7)^2$
7. $(4x - 5)^2$

9. $(a + 2b)^2$
11. $(a - 3b)^2$
13. $(3y + 8z)^2$
15. $(5x - 6y)^2$

Exercises 6.4 (Part 4), page 229

1. $a(x + 2)(x + 3)$
3. $c(x - 5)(x - 3)$
5. $5(x + 5)(x - 2)$
7. $4(a + 3)(a + 7)$
9. $2(5 - x)(2 - x)$
11. $a^2(5 - a)(3 - a)$
13. $2(a - 5)^2$
15. $b(a - 6)(a + 5)$
17. $a(a + 9b)(a - 3b)$

19. $(x^2 + 1)(x + 2)(x - 2)$
21. $2(3x + 5)(x + 1)$
23. $5(3a - 1)(a - 2)$
25. $xy(x - 5y)(x + 3y)$
27. $a(a - 1)^2$
29. $a^2(3a - 1)(a - 5)$
31. $9(x - y)^2$
33. $2x(2x - 3)(x + 1)$
35. $(x + 3)(x - 3)(x + 2)(x - 2)$

Exercises 6.5, page 230

1. $2(a + b)$
3. $5x(x - 2)$
5. $(x + 2)(a + b)$
7. $(1 - 4a)(1 - a)$
9. $(3x + 5)(3x - 5)$
11. $a^2b(3a - 11b^3)$
13. $(3a + 1)(a + 2)$
15. $(x - y)(a - b)$
17. $x^2(x - 1)$
19. $(4a - 1)(3a + 1)$
21. $(3a - 1)(a + 7)$
23. $(a + b)(x + 1)$
25. $(11x + 6y)(x - y)$

27. $2(x - 5)(x - 4)$
29. $(6x - 7y)(x + y)$
31. $(x - 7)(a - 1)$
33. Prime
35. $(x + y)(a + 3)$
37. $(3x - 7y)(2x - y)$
39. $2(2a - 1)(4a + 3)$
41. $(a + b)(x - y)$
43. $4(2x + 1)(3x - 5)$
45. $2a(x - 7)(x - 5)$
47. $x^2(3x - 1)(x - 9)$
49. $(x + 3)(x - 3)(x + 1)(x - 1)$

Chapter 6   A25

Exercises 6.6, page 235

A

1. 5, 4
3. 7, −7
5. 3, 3
7. 0, −2
9. 2, −3, 4

B

1. 5, 2
3. 8, −1
5. $\frac{7}{2}, -\frac{7}{2}$
7. 0, 5
9. $-1, -\frac{7}{5}$
11. $\frac{5}{3}, -\frac{5}{3}$
13. 0, 1
15. −3, −3
17. $\frac{1}{2}, \frac{1}{2}$
19. 0, 2, −2

C

1. 5, −8
3. 12, −2
5. 11, −7
7. 0, 9
9. 4, −8
11. $3, \frac{5}{2}$
13. 0, 3
15. 3, −8
17. $2, \frac{7}{3}$
19. 2, −3

Exercises 6.7 (Part 1), page 237

1. $x^2 + 3 = 39$; $x = 6$ or −6
3. $x^2 + x = 56$; $x = -8$ or 7
5. $x(x + 3) = 40$; 5 and 8 or −8 and −5
7. $x(2x) = 32$; 4 and 8 or −4 and −8
9. $(x + 4)(x + 1 + 2) = 30$; 2 and 3 or −9 and −8
11. $x(x + 2) = 63$; 7 and 9 or −9 and −7
13. $x^2 + 2(x + 2) = 28$; 4 and 6 or −6 and −4

Exercises 6.7 (Part 2), page 245

1. $x(x + 3) = 54$; width = 6″, length = 9″
3. $2x(x) = 32$; width = 4 yards, length = 8 yards
5. $x(15 - x) = 56$; length = 8″, width = 7″
7. $x(20 - x) = 99$; length = 11′, width = 9′
9. $x(x + 4) = 45$; side = 5″
11. $(x + 2)(x + 3 + 2) = 70$; width = 5 yards, length = 8 yards
13. $(2x + 8)(x + 8) = 330$; width = 7′, length = 14′

Chapter 6 Review, page 246

A

1. a. 1, 3, 9
   b. 1, 2, 4, 5, 10, 20
   c. 1, 2, 4, 7, 14, 28
   d. 1, a, b, ab
3. a. 5
   b. 2
   c. 2, 3, 7

Answers to Odd-Numbered Problems

5. a. $a(b+c)$
    b. $a(x-1)$
    c. $x^2(x^3-1)$
    d. $xy(x+y)$
    e. $4(2a^2-6a+1)$
    f. $7a^2(1-2a-5a^5)$
    g. $(x-y)(b+a)$
    h. $(a+b)(x-1)$

7. a. $(x+2)(x+1)$
    b. $(x-1)(x-7)$
    c. $(x-5)(x+1)$
    d. $(x+11)(x-1)$
    e. $(4-x)(3-x)$
    f. $(9+x)(2-x)$
    g. $(x-8y)(x+3y)$
    h. $(2x-1)(x-5)$
    i. $(2x+1)(2x-7)$
    j. $(3x-2)^2$
    k. $(3x+5)(2x-3)$
    l. $(9b-8y)(x+y)$
    m. $(3x+2y)(2x-5y)$
    n. $(4x-3y)(2x+5y)$

9. a. 4, –3
    b. 0, –7
    c. 7, 7
    d. 1, 3, –5

B

1. $x^2+5=41;\ x=6$ or -6
3. $x(x+2)=63;\ 7$ and 9 or -9 and -7
5. $x^2+(x+2)^2=52;\ 4$ and 6 or -6 and -4
7. $x(9-x)=20;$ width $=4$ yards, length $=5$ yards
9. $x^2+(x+2)(x+3)=58;$ side $=4'$

CHAPTER 7

Exercises 7.1 (Part 1), page 253

1. $\dfrac{a}{2b}$
3. $\dfrac{4}{5a}$
5. $\dfrac{3x}{2z}$
7. $\dfrac{z}{y}$
9. $\dfrac{1}{b}$
11. $\dfrac{3}{4}$
13. $\dfrac{2}{x}$
15. $\dfrac{x+5}{x-1}$

Exercises 7.1 (Part 2), page 256

A

1. -1
3. -1
5. $-\dfrac{6}{7}$
7. $-(x+2)$
9. $\dfrac{-5}{2b+a}$

B

1. $\dfrac{2}{5}$
3. $\dfrac{x-2}{2}$
5. $\dfrac{x+6}{x+2}$
7. $\dfrac{2(x+3)}{x-2}$
9. $\dfrac{x+3}{x-2}$
11. $\dfrac{x}{x+2}$
13. $\dfrac{3a+4b}{5}$
15. $\dfrac{3}{a+b}$
17. $\dfrac{x+1}{3}$
19. $\dfrac{-3}{x+5}$
21. $\dfrac{-1}{2a-3b}$
23. $\dfrac{(a-3)(a^2+1)}{a-1}$
25. $(x-2)(x+1)$
27. $\dfrac{3a+1}{y}$
29. $\dfrac{-(x+4)}{2+x}$

Chapter 7

Exercises 7.2, page 261

A

1. $\dfrac{b^2}{a}$
3. $\dfrac{3}{a}$
5. x
7. $\dfrac{a}{b}$
9. $\dfrac{x(a+2)}{a+1}$
11. $10x$

B

1. $x-1$
3. $\dfrac{x-3}{x+5}$
5. $\dfrac{7x}{x+2}$
7. $\dfrac{7}{2}$
9. $\dfrac{7}{5x}$
11. $\dfrac{3}{2}$
13. $\dfrac{x+7}{x+2}$
15. $\dfrac{4}{5b}$
17. $\dfrac{x+2}{x-7}$
19. $\dfrac{x+3}{x-4}$

Exercises 7.3, page 266

A

1. $\dfrac{2b}{9a^2}$
3. $\dfrac{6}{x}$
5. $\dfrac{2}{5x}$
7. $\dfrac{15x^3 y}{4}$
9. $\dfrac{a^4}{c}$

B

1. $\dfrac{x+5}{x+2}$
3. $\dfrac{3a(x+2)}{2(x+1)}$
5. $\dfrac{2x}{y(2x+1)}$
7. 1
9. $\dfrac{1}{x(x-2)}$
11. $\dfrac{x+1}{2x+1}$

C

1. $\dfrac{4x}{3}$
3. $\dfrac{x+3}{2(x+5)}$

Exercises 7.4 (Part 1), page 271

A

1. $\dfrac{5x+4y}{2}$
3. $\dfrac{9x-4y}{a}$
5. $\dfrac{2x^2}{y}$
7. $\dfrac{-5x^2}{2y}$
9. $\dfrac{9a}{2b}$
11. $\dfrac{10a-5c}{b}$

B

1. $\dfrac{2x+7}{3}$
3. $\dfrac{7y-9}{x}$
5. $\dfrac{12a-4}{6x}$ or $\dfrac{6a-2}{3x}$
7. $\dfrac{2x-2}{5a}$
9. $\dfrac{7x-1}{ab}$
11. $\dfrac{11a-c}{b}$

C

1. $x+2$
3. $2x+3$
5. $x+4$
7. $a-2$
9. $-(x+3y)$

Exercises 7.4 (Part 2), page 276

A

1. $\dfrac{17a}{6}$
3. $\dfrac{31b}{8}$
5. $\dfrac{-a}{20}$
7. $\dfrac{13ab}{36}$
9. $\dfrac{a^2b}{20}$
11. $\dfrac{7x}{10}$

B

1. $\dfrac{3x+1}{4}$
3. $\dfrac{10x+1}{9}$
5. $\dfrac{14x-19}{6}$
7. $\dfrac{-5x-21}{16}$
9. $\dfrac{-4x-5y}{8}$

Exercises 7.4 (Part 3), page 280

1. x^5
3. $x^5 y^3$
5. xy
7. $12x^2$
9. 60
11. 120
13. $x^4 y^3 w^3$
15. $36x^5 y^2 w^2$

Exercises 7.4 (Part 4), page 282

A

1. $\dfrac{29x}{24}$
3. $\dfrac{2x+4}{x^3}$
5. $\dfrac{3y+4x}{xy}$
7. $\dfrac{25y+18x}{30xy}$
9. $\dfrac{9b^2+10a}{12a^2 b^2}$
11. $\dfrac{az-by+cx}{xyz}$

B

1. $\dfrac{7x+26}{12}$
3. $\dfrac{19x-22}{18}$
5. $\dfrac{7x-57}{36}$
7. $\dfrac{-7x-2}{24x}$
9. $\dfrac{x+3}{x^2 y^2}$
11. $\dfrac{2xy-y^2-x^2}{xy}$
13. $\dfrac{xy^2+y^3+2xy+x^3-x^2 y}{x^2 y^2}$

Exercises 7.4 (Part 5), page 284

A

1. $\dfrac{7x - y}{(x + y)(x - y)}$
3. $\dfrac{5x - 27}{(x - 5)(x - 6)}$
5. $\dfrac{x + 12}{(5x - 3)(4x - 1)}$

7. $\dfrac{-6x}{(x + 3)(x - 3)}$
9. $\dfrac{2}{5(x - 2)}$

B

1. $\dfrac{11x + 17}{(x + 1)(x + 2)(x + 3)}$
3. $\dfrac{7x^2 - 4x}{x(x + 2)(x - 2)}$
5. $\dfrac{7x - 6}{(2x - 1)(x + 1)(x - 1)}$

7. $\dfrac{-x^2 - x}{(2x - 1)(3x - 2)(x - 1)}$
9. $\dfrac{-x - 10}{x(x - 5)^2}$
11. $\dfrac{6x^2 - 13x - 5}{2x(x - 1)(x + 1)}$

C

1. $\dfrac{10x + 13}{(x - 1)(x + 2)(x + 3)}$
3. $\dfrac{-8x - 13}{x(x - 3)(x + 5)}$

5. $\dfrac{-5x - 3}{3(x - 2)(x + 3)}$

Exercises 7.5 (Part 1), page 287

1. $2x$
3. $-3a$
5. $12a$
7. $-8a$
9. $x + 3$
11. $12x - 8$
13. $-12x - 28$
15. $49 - 21a$

Exercises 7.5 (Part 2), page 290

A

1. 6
3. 6
5. −10
7. 9
9. $-\dfrac{3}{2}$

B

1. 6
3. 3
5. 6
7. 4
9. −3
11. −4

C

1. 12
3. 1

Exercises 7.6, page 293
1. $x > 3$
3. $x < 9$
5. $x > \frac{1}{3}$
7. $x > 5$
9. $x < -\frac{2}{3}$
11. $x < \frac{1}{2}$

Exercises 7.7, page 296

A

1. 4
3. 6
5. −2
7. 1
9. 8
11. −11

B

1. 3, 4
3. 4, 5
5. 3
7. 3, 6
9. 2, 5
11. 3, −4
13. 7, −2
15. 5, 6

C

1. 2
3. −3
5. 0, 7

Exercises 7.8, page 302

A

1. a
3. rs
5. rt
7. $5n$
9. a^2
11. ab^2
13. $c + d$
15. $2a - b$
17. $2 - a$
19. $b - 1$

B

1. $\dfrac{b}{a}$
3. $b + a$
5. $b - c$
7. $\dfrac{c + b}{a}$
9. $\dfrac{cd - b}{ac}$
11. $\dfrac{2cd - 2b}{ac}$
13. $\dfrac{ab - c}{a}$
15. $\dfrac{2a + 3c}{8}$
17. $\dfrac{-5a - 3b}{2}$
19. $a^2 - b^2$
21. $\dfrac{c}{a + b}$
23. $\dfrac{m}{rs}$
25. $\dfrac{a}{b - c}$
27. $\dfrac{c}{a - b}$
29. $\dfrac{b}{a - 1}$
31. $\dfrac{ab}{2b + 3}$
33. $\dfrac{7}{ab + cb}$
35. $\dfrac{a}{a - b - c}$
37. b
39. $a + b$

Chapter 7 A31

C

1. $\dfrac{A}{l}$
3. $\dfrac{P}{4}$
5. $\dfrac{C}{\pi}$
7. $\dfrac{A-P}{Pr}$
9. $\dfrac{A}{1+rt}$
11. $\dfrac{3V}{h}$
13. $\dfrac{A-\pi R^2}{\pi R}$
15. $\dfrac{aB}{A}$
17. $\dfrac{2S-an}{n}$
19. $\dfrac{A}{1+b}$

Exercises 7.9 (Part 1), page 307

1. $\dfrac{x+15}{x} = 4$; smaller = 5, larger = 20
3. $\dfrac{x+2}{x+5+3} = \dfrac{3}{5}$; numerator = 7, denominator = 12
5. $\dfrac{x+30}{x} = 7$; Tommy's age = 5, Mr. Johnson's age = 35
7. $\dfrac{x}{x-1} - \dfrac{3}{4} = \dfrac{3}{8}$; numerator = 9, denominator = 8
9. $\dfrac{x-3}{2x-15} = 2$; larger = 9, smaller = 6
11. $\dfrac{12-x}{x} = x$
    When smaller is 3, larger is 9.
    When smaller is −4, larger is 16.
13. $\dfrac{1}{x} + \dfrac{1}{x+1} = \dfrac{5}{6}$; numbers are 2 and 3.
15. $\dfrac{1}{x} + \dfrac{1}{9-x} = \dfrac{1}{2}$; Mary's age is 6, Sarah's age is 3.

Exercises 7.9 (Part 2), page 312

A

1. $\dfrac{1}{12}$
3. $\dfrac{1}{3}$
5. $\dfrac{h}{16}$.
7. a. $\dfrac{1}{5}$ b. $\dfrac{2}{25}$ c. $\dfrac{1}{5} - \dfrac{2}{25}$

B

1. $\dfrac{x}{15} + \dfrac{x}{10} = 1$; 6 days
3. $\dfrac{x}{12} + \dfrac{x}{8} = 1$; $4\tfrac{4}{5}$ hours or 4 hours, 48 minutes
5. $\dfrac{x}{5} - \dfrac{x}{15} = 1$; $7\tfrac{1}{2}$ hours or 7 hours, 30 minutes

7. $\dfrac{4}{10} + \dfrac{x}{10} - \dfrac{x}{25} = 1$; 10 hours

9. $\dfrac{6}{18} + \dfrac{x}{12} = 1$; 8 hours

11. $\dfrac{3}{12} + \dfrac{5}{12} + \dfrac{5}{x} = 1$; 15 days

Chapter 7 Review, page 314

A

1. a. $\dfrac{a}{2b}$ c. 10

   b. $\dfrac{3x}{2y^2}$ d. $\dfrac{3a^3}{2}$

3. a. $\dfrac{y^2}{x^2}$ d. $\dfrac{a-5}{a+1}$

   b. $\dfrac{x+1}{x-1}$ e. $\dfrac{1}{x}$

   c. $\dfrac{4a}{b}$ f. $\dfrac{2}{y-5}$

5. a. $\dfrac{11x}{7}$ c. $\dfrac{4x-y}{a}$

   b. $\dfrac{-5a^2}{2x}$ d. $\dfrac{-6b}{x-y}$

7. a. $\dfrac{5b+4a}{ab}$ d. $\dfrac{3y+4}{xy}$

   b. $\dfrac{6b+3+2a}{ab}$ e. $\dfrac{2ab-2b^2-2a^2}{ab}$

   c. $\dfrac{8b-5a}{6a^2b^2}$ f. $\dfrac{3ab-2b^2-a^2-6}{4ab}$

9. a. 6   c. -2
   b. 4   d. 5

11. a. $x > 4$ c. $x < -25$
    b. $x < -2$ d. $x < 6$

B

1. $\dfrac{x-3}{x+8+1} = \dfrac{2}{3}$; numerator = 27, denominator = 35

3. $\dfrac{x}{15} + \dfrac{x}{9} = 1$; $5\dfrac{5}{8}$ minutes

CHAPTER 8

Exercises 8.1 (Part 1), page 321

A

1. a. The point is 4 units to the right of the vertical axis.
   b. The point is 5 units above the horizontal axis.
3. a. The point is 6 units to the right of the vertical axis.
   b. The point is 2 units above the horizontal axis.
5. a. The point is 4 units to the left of the vertical axis.
   b. The point is 8 units above the horizontal axis.
7. a. The point is 3 units to the right of the vertical axis.
   b. The point is 4 units below the horizontal axis.
9. a. The point is 10 units to the right of the vertical axis.
   b. The point is 4 units below the horizontal axis.
11. a. The point is 6 units to the left of the vertical axis.
    b. The point is 11 units below the horizontal axis.
13. a. The point is 7 units to the right of the vertical axis.
    b. The point is 0 units from the horizontal axis.

B

1. (+3, +2)
3. (-7, +6)
5. (-15, -20)
7. (-8, 0)

Exercises 8.1 (Part 2), page 326

A

1. $x = 5, y = 2$
3. $x = -2, y = +7$
5. $x = -9, y = -7$

B

1. (5, 7)
3. (-2, -4)
5. (-6, 0)

C

1. (5, -4)
3. (0, -7)
5. (2, 6)

D

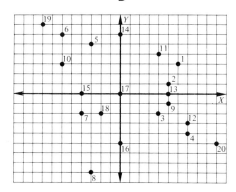

E

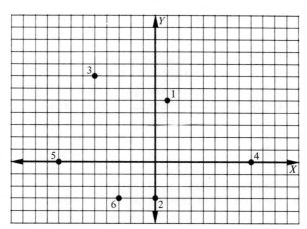

F

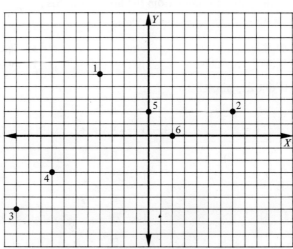

G

1. (−4, 0)   7. (4, −4)   13. (0, −3)
3. (2, 4)   9. (−5, −2)   15. (−5, 3)
5. (−3, 2)   11. (7, 0)

H

1. $x = -3, y = 1$ 7. $x = 7, y = -2$ 13. $x = -6, y = 3$
3. $x = 3, y = 1$ 9. $x = 0, y = 3$
5. $x = -5, y = -4$ 11. $x = 1, y = -1$

Chapter 8   A35

I

1. abscissa = −5, ordinate = −2   5. abscissa = −6, ordinate = 4
3. abscissa = 3, ordinate = 4   7. abscissa = −2, ordinate = −5

J

1. Positive   7. Seven units right of vertical axis
3. Negative   9. On the horizontal axis
5. First

K

1. 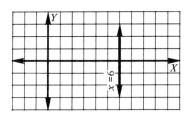   7.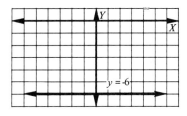

3. 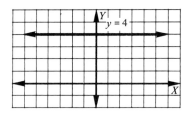   9.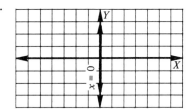

5.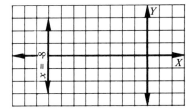

Exercises 8.2 (Part 1), page 334

1. (0, 4); (1, 3); (2, 2); (3, 1). There are many others.
3. (0, 12); (1, 11); (2, 10); (3, 9). There are many others.
5. (2, 0); (3, 1); (4, 2); (5, 3). There are many others.

Exercises 8.2 (Part 2), page 336

A

1. Yes—satisfies equation
3. No—does not satisfy equation
5. No—does not satisfy equation
7. No—does not satisfy equation
9. Yes—satisfies equation
11. Yes—satisfies equation

B

1. No—not on graph
3. Yes—on graph

C

1. Will satisfy equation since point is on graph.
3. Will not satisfy equation since point is not on graph.

Exercises 8.2 (Part 3), page 343

A

1. $y = 5 - 3x$
3. $y = 8 + 7x$
5. $y = 5x - 9$
7. $4x - 5 = y$
9. $3 - 8x = y$
11. $5 - x = y$

B

1.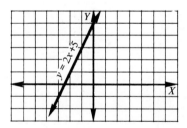

5.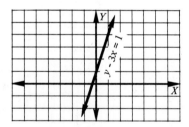

3.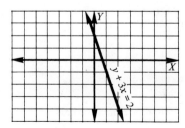

7.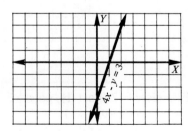

9.

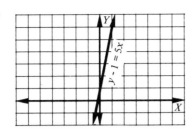

Exercises 8.2 (Part 4), page 345

A

1. $x = 7 - 4y$
3. $x = 4 - 5y$
5. $x = -3y - 9$
7. $3y - 5 = x$
9. $y + 9 = x$
11. $6 + 2y = x$

B

1.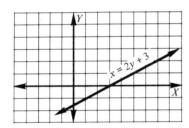

7.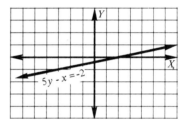

3.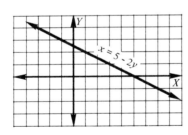

9.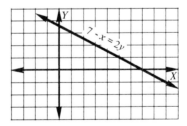

5.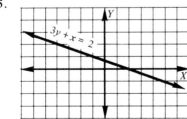

A38 Answers to Odd-Numbered Problems

Exercises 8.2 (Part 5), page 350

A

1. $y = 3 - \dfrac{3x}{2}$
3. $\dfrac{5x}{2} - 3 = y$
5. $\dfrac{3x}{2} - 4 = y$

B

1. $x = 4 - \dfrac{5y}{2}$
3. $\dfrac{3y}{2} - 3 = x$
5. $3 + \dfrac{5y}{2} = x$

C

1. $y = 10 - \dfrac{5x}{2}$
3. $8 - \dfrac{4y}{3} = x$

D

1. $(-4, -5), (-2, -2), (0, 1), (6, 10)$
3. $(-6, 9), (-3, 7), (0, 5), (12, -3)$

E

1. $(13, 9), (9, 6), (1, 0), (-3, -3)$
3. $(-13, 15), (-6, 10), (15, -5), (36, -20)$

F

1.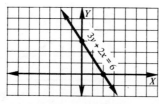

7.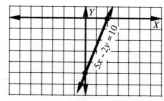

3.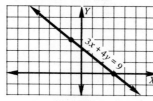

9.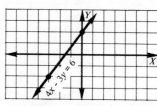

5.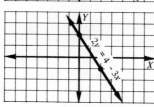

11.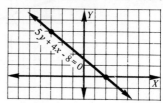

Chapter 8

Exercises 8.3 (Part 1), page 357

A

1. $\frac{5}{4}$
3. $\frac{1}{1}$
5. $\frac{9}{4}$
7. $\frac{1}{6}$
9. $-\frac{2}{1}$
11. $\frac{2}{3}$
13. $-\frac{3}{4}$
15. $\frac{11}{2}$

B

1. $\frac{1}{1}$
3. $\frac{5}{2}$
5. $-\frac{3}{5}$
7. $-\frac{2}{3}$
9. $-\frac{1}{3}$

C

1. $\frac{3}{1}$
3. $\frac{1}{4}$
5. $-\frac{1}{2}$
7. $\frac{5}{1}$
9. $-\frac{2}{3}$
11. $-\frac{5}{2}$
13. $\frac{3}{5}$
15. $\frac{7}{5}$

Exercises 8.3 (Part 2), page 364

A

1. $\frac{2}{3}$
3. $\frac{4}{1}$
5. $\frac{6}{7}$
7. $\frac{8}{1}$
9. $\frac{1}{1}$
11. $-\frac{3}{1}$
13. $-\frac{7}{2}$
15. $\frac{3}{4}$
17. $\frac{2}{3}$
19. $\frac{1}{5}$

B

1.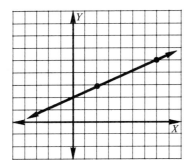
3.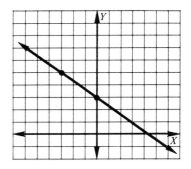

A40 Answers to Odd-Numbered Problems

5.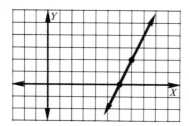

9.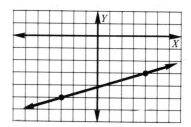

7.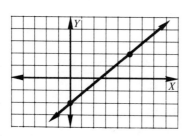

11.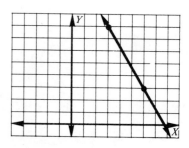

C

1.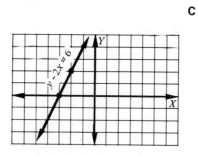

7.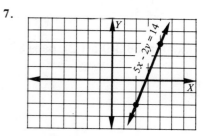

3.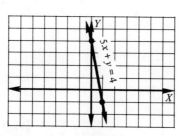

9.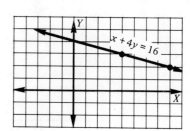

5.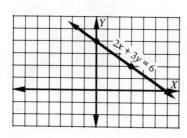

Chapter 8   A41

D

1. Parallel—lines have the same slope, 3.
3. Parallel—lines have the same slope, $-\frac{3}{2}$.
5. Not parallel—lines have different slopes, $\frac{3}{4}$ and $-\frac{5}{2}$.
7. Not parallel—lines have different slopes, $\frac{3}{2}$ and $-\frac{3}{2}$.

Exercises 8.4, page 370

1.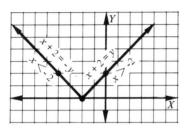

7.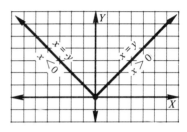

3.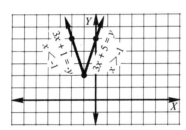

9.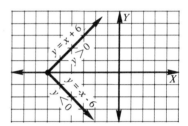

5.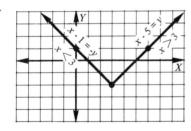

Exercises 8.5, page 372

1.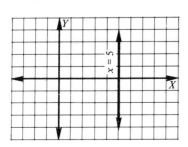

3.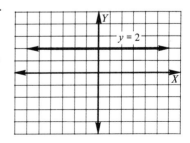

5.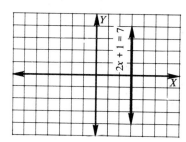

11.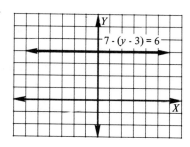

7.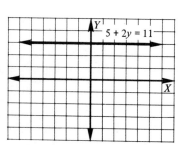

13.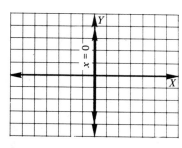

9.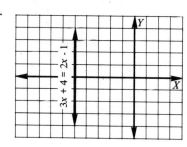

Exercises 8.6 (Part 1), page 376

A

Each of the following answers begins with the words, "The set of points in the half-plane..."

1. to the right of the line $x = 7$.
3. to the left of the line $x = 9$.
5. above the line $y = 6$.
7. below the line $y = 2$.
9. to the right of the line $x = -6$.
11. to the left of the line $x = -4$.
13. above the line $y = -1$.
15. to the right of the line $x = 0$ (right of vertical axis).
17. to the right of the line $x = 10$ and also including the line $x = 10$.
19. to the left of the line $x = -2$ and also including the line $x = -2$.

Chapter 8

B

1.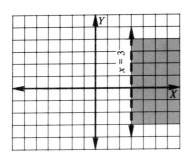

7.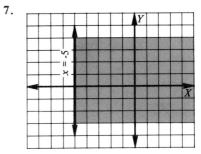

3.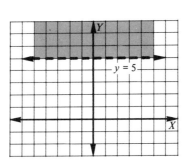

9.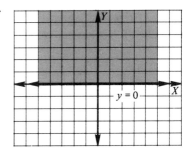

5.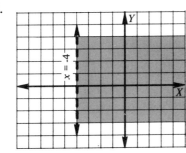

C

1.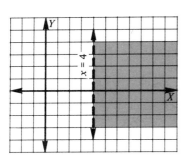

3.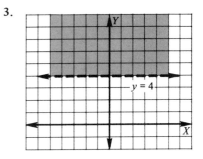

5.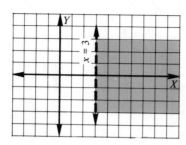

9.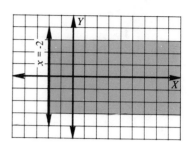

7.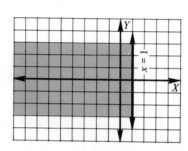

Exercises 8.6 (Part 2), page 380

A

Each of the following answers begins with the words, "The set of points in the half-plane..."

1. to the right of the line $x = 5y - 3$.
3. above the line $y = 4x - 7$.
5. to the left of the line $x = 5 - 2y$.
7. below the line $y = 2 - \frac{5x}{2}$.
9. to the right of the line $x = 3y - 6$.
11. above the line $y = 2 - \frac{6x}{5}$.

B

1.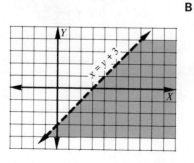

3.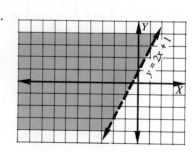

5.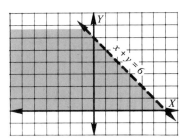

11.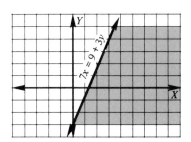

7.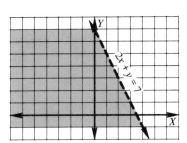

13.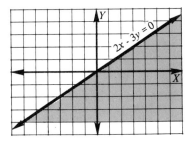

9.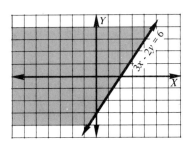

Chapter 8 Review, page 381

1. a. The point is 3 units to the right of the vertical axis.
      The point is 7 units above the horizontal axis.
   b. The point is 4 units to the left of the vertical axis.
      The point is 2 units above the horizontal axis.
   c. The point is 5 units to the right of the vertical axis.
      The point is 3 units below the horizontal axis.
   d. The point is 0 units from the vertical axis.
      The point is 4 units above the horizontal axis.

3. a. (−4, +3)   b. (−1, 5)

5. a. Fourth
   b. Third
   c. The coordinates will satisfy the equation.
   d. The point named by these coordinates will not line on the graph of the equation.

A46 Answers to Odd-Numbered Problems

7. a. $y = 5 - 2x$
   b. $x = 3 - \dfrac{5y}{2}$
   c. $\dfrac{3x}{2} - \dfrac{7}{2} = y$
   d. $7y - 3 = x$

9. a. $\dfrac{7}{3}$
   b. $-\dfrac{10}{3}$

11. a. $\dfrac{2}{3}$
    b. $\dfrac{1}{2}$

13.

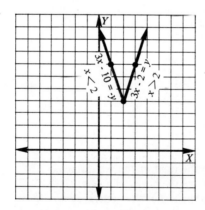

15.

17. a. The half-plane to the right of the line $x = 9 - 2y$
    b. The half-plane below the line $y = 3x + 4$

CHAPTER 9

Exercises 9.1 (Part 1), page 390

1. Consistent
3. Inconsistent
5. Consistent
7. Inconsistent
9. Dependent
11. Consistent

Chapter 9   A47

Exercises 9.1 (Part 2), page 394

A

1. (3, 4)
3. (6, 1)
5. (6, 0)
7. (2, −5)
9. (−2, −3)
11. (1, 3)
13. (4, −5)
15. (6, −3)
17. (3, 6)
19. (4, 2)

B

1. Inconsistent
3. (−2, −3)
5. Dependent
7. (3, −1)
9. (0, 0)

Exercises 9.2 (Part 1), page 398

1. (1, 3)
3. (4, −2)
5. (−3, 6)
7. (10, 2)
9. (−14, 2)
11. (3, −12)
13. (9, 3)
15. (−4, −1)
17. (12, 37)
19. $(6, \frac{1}{2})$

Exercises 9.2 (Part 2), page 401

1. (2, 3)
3. (1, −1)
5. (−2, 5)
7. (10, 8)
9. (−12, −12)
11. (−4, −6)
13. $(6, -\frac{1}{2})$

Exercises 9.3 (Part 1), page 406

1. (9, 2)
3. (−3, 7)
5. (7, −6)
7. (12, 9)
9. $(-\frac{7}{4}, -3)$
11. $(-5, \frac{5}{4})$
13. $(\frac{1}{4}, -\frac{3}{4})$

Exercises 9.3 (Part 2), page 409

1. (1, 2)
3. (−2, 5)
5. (−5, −3)
7. (−3, 0)
9. $(-2, \frac{7}{4})$

Exercises 9.3 (Part 3), page 412

1. (5, 7)
3. (3, −5)
5. (−5, 1)
7. (4, −5)
9. (3, 4)

Exercises 9.3 (Part 4), page 416

A

1. $5x - 6y = -3$
3. $3a - 5b = 2$
5. $6c - 2d = -5$
7. $2b - 6c = 5$
9. $3a + 2b = 30$
11. $15w - 9z = 4$
13. $-8x + 3y = -8$
15. $10y - 15z = 40$

	B	
1. (2, 4)	5. (6, −2)	9. (3, −10)
3. (−2, −3)	7. (12, −12)	

Exercises 9.4, page 419

1.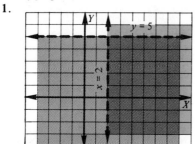

7.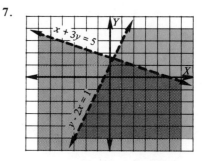

3.

9.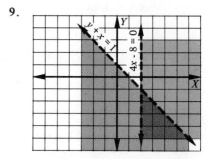

5.

11.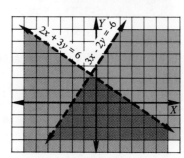

Exercises 9.5 (Part 1), page 423

1. $x + y = 43$, $x − y = 5$; 24 and 19
3. $h + l = \$75{,}000$, $h = 4l$; land: \$15,000, house: \$60,000
5. $f + s = 670$, $f = s − 20$; second day: 345, first day: 325

Chapter 9   A49

7. $r = \frac{1}{4}s$, $s + r = \$275$; salary: $\$220$, raise: $\$55$
9. $l = 3w$, $2l + 2w = 64$; width: 8 inches, length: 24 inches
11. $\frac{1}{2} + s = 3$, $\frac{s}{5} + l = 24$; smaller: -10, larger: 26
13. $x = y + 6$, $10x + 25y = 655$; dimes: 23, quarters: 17
15. $x = y + 4$, $89x + 95y = 2748$; 89¢: 17 pounds, 95¢: 13 pounds
17. $5r + 4g = 313$, $4r + 5g = 308$; green: 32¢, red: 37¢

Exercises 9.5 (Part 2), page 428

1. $u = t + 2$, $10t + u = 6u + 5$; 35
3. $u = 4t$, $10t + u = 3u + 4$; 28
5. $u = 3t - 2$, $10t + u - 2 = 5u$; 37
7. $u = t + 1$, $10u + t = 7(t + u)$; 12
9. $u = t + 4$, $10u + t + 10t + u = 154$; 59
11. $u = t + 1$, $10t + u = 8u + 2$; 34

Exercises 9.5 (Part 3), page 435

A

1. $x + y = 6$
3. $5x - y = 7$
5. $2x - 3y = 2$
7. $x + y = 4$
9. $5x + 4y = 2$

B

1. $6(x - y) = 60$, $5(x + y) = 60$
   Boat speed: 11 mph, current speed: 1 mph
3. $6(x - y) = 1,500$, $5(x + y) = 1,500$
   Plane speed: 275 mph, wind speed: 25 mph
5. $6.5(x - y) = 1,170$; $4.5(x + y) = 1,170$
   Plane speed: 220 mph, wind speed: 40 mph
7. $1.25(x - y) = 7.5$; $.75(x + y) = 7.5$
   Boat speed: 8 mph, current speed: 2 mph
9. $75(x - y) = 150$, $60(x + y) = 150$
   a. Wind velocity: .25 mpm
   b. Wind velocity: 15 mph

Exercises 9.5 (Part 4), page 441

A

1. $5x + 6y = 240,000$
3. $65x - 75y = 684,000$
5. $30x + 45y = 5,600,000$
7. $75x + 40y = 12,500,000$

B

1. $x + y = \$8,000$; $.07x + .05y = \$520$
   7% account: \$6,000; 5% account: \$2,000
3. $x + y = \$14,000$; $.07x + .055y = \$882.50$
   7% account: \$7,500; 5.5% account: \$6,500
5. $x + y = \$18,000$; $.045x = .065y + \$40$
   4.5% bonds: \$11,000; 6.5% bonds: \$7,000
7. $x = y + \$3,000$; $.06x + .04y = \$1,380$
   6% account: \$15,000; 4% account: \$12,000
9. $x = y - \$8,600$; $.06x + .09y = \$3,114$
   6% investment: \$15,600; 9% investment: \$24,200
11. $x = y + \$4,100$; $.05x = .08y - \$140$
    5% investment: \$15,600; 8% investment: \$11,500

Chapter 9 Review, page 442

A

1. a. Consistent—lines not parallel
   b. Inconsistent—lines parallel
   c. Consistent—lines not parallel
   d. Dependent—lines coincide

3. a. (8, 2)
   b. (2, -3)
   c. (5, 4)
   d. (-3, -7)

5. a. $3x - 5y = 4$
   b. $2a - 3b = 7$
   c. $4x + 3y = 60$
   d. $6a - 20b = 14$

7.
   a.

   c.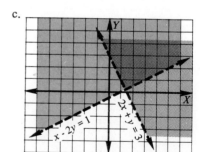

   b.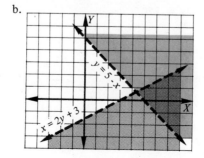

   d.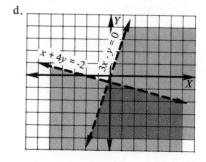

B

1. $x + 2y = -13$, $2x + y = 25$; 21 and -17
3. $x + y = 81$, $5x + 10y = 585$
   Nickels: 45, dimes: 36
5. $u = t + 2$, $10t + u = 7u + 4$; 46
7. $8(x - y) = 24$, $6(x + y) = 24$
   Boat speed: $3\frac{1}{2}$ mph; current speed: $\frac{1}{2}$ mph
9. $x + y = \$18,000$; $.06x + .07y = \$1,150$
   6% account: $11,000; 7% account: $7,000

CHAPTER 10

Exercises 10.1, page 449

1. $+4, -4$
3. $+\sqrt{11}, -\sqrt{11}$
5. No roots
7. $+4, -4$
9. $+\sqrt{7}, -\sqrt{7}$
11. $+6, -6$
13. $+\sqrt{11}, -\sqrt{11}$
15. $+\sqrt{2}, -\sqrt{2}$
17. $+4, -4$
19. $+2, -2$

Exercises 10.2, page 452

A

1. $3, -5$
3. $-4, -6$
5. $4, -1$
7. $4, -\frac{4}{3}$
9. $-1, -5$
11. $10, 2$
13. $0, -2$
15. $11, 9$

B

1. $-3 + \sqrt{5}, -3 - \sqrt{5}$
3. $-7 + \sqrt{10}, -7 - \sqrt{10}$
5. $\dfrac{-3 + \sqrt{7}}{2}, \dfrac{-3 - \sqrt{7}}{2}$

Exercises 10.3 (Part 1), page 455

1. 16
3. 81
5. 49
7. 225
9. $\dfrac{25}{4}$
11. $\dfrac{225}{4}$
13. $\dfrac{1}{25}$
15. $\dfrac{1}{9}$
17. $\dfrac{4}{25}$
19. $\dfrac{25}{64}$

Exercises 10.3 (Part 2), page 460

A

1. $1, -5$
3. $4, 6$
5. $2, -4$
7. $6, 8$
9. $7, -5$
11. $1, 3$
13. $1, 9$
15. $9, -7$
17. $5, 7$
19. $7, -4$

Answers to Odd-Numbered Problems

B

1. $\frac{5}{2}, 1$
3. No Roots
5. $\frac{1}{3}, 4$
7. $\frac{2}{5}, -2$

C

1. $4, -2$
3. $-1 + \sqrt{5}, -1 - \sqrt{5}$
5. $3 + \sqrt{14}, 3 - \sqrt{14}$
7. $\frac{1}{2}, -3$
9. $\frac{1}{3}, 2$
11. $2, -\frac{3}{5}$

Exercises 10.4 (Part 1), page 466

A

1. $3x^2 + 5x + 6 = 0$
3. $6x^2 + 3x - 5 = 0$
5. $6x^2 - 3x - 4 = 0$
7. $4x^2 - 2x - 3 = 0$
9. $3x^2 + 2x - 4 = 0$
11. $5x^2 - 6x + 0 = 0$

B

1. $a = 3, b = 5, c = 6$
3. $a = 2, b = -3, c = -5$
5. $a = 1, b = -4, c = 7$
7. $a = 5, b = -1, c = 3$
9. $a = 3, b = 5, c = -6$
11. $a = 1, b = -5, c = 4$
13. $a = 4, b = 2, c = -7$
15. $a = 4, b = -9, c = 0$
17. $a = 2, b = 0, c = -11$
19. $a = 2, b = -3, c = -4$

C

1. $-2, -5$
3. $\frac{3}{2}, 1$
5. $4, -\frac{1}{2}$
7. $0, \frac{8}{5}$
9. $2, -2$
11. $\frac{1}{4}, \frac{2}{3}$
13. $\frac{5}{2}, -\frac{5}{2}$
15. $\frac{5}{3}, -\frac{1}{2}$
17. $-3, -\frac{1}{3}$
19. $\frac{1}{3}, -\frac{1}{2}$

Exercises 10.4 (Part 2), page 471

A

1. 3.162
3. 7.483
5. 9.327
7. 9.747
9. 10.344
11. 4.359

Chapter 10   A53

B

1. $\dfrac{-5 \pm \sqrt{57}}{4}$ 7. $\dfrac{-5 \pm \sqrt{73}}{4}$

3. $\dfrac{2 \pm \sqrt{52}}{6}$ 9. No roots

5. $\dfrac{-7 \pm \sqrt{145}}{12}$ 11. $\dfrac{5 \pm \sqrt{193}}{14}$

C

1. .5485, −1.2152   9. 1.211, −.3539
3. .443, −1.693   11. .7746, −.7746
5. 6.5415, .4585   13. 2.633, −.633
7. .8792, −.3792

Exercises 10.5, page 475

A

1. 10   9. 10
3. 12   11. 7.28
5. 12.207   13. $a = 3.123$, $b = 5.123$
7. 5.291   15. $a = 6.472$, $b = 9.472$, $c = 11.472$

B

1. 11.314 cm   7. 1.7015 cm, 4.7015 cm
3. 7.071 m   9. 100 km
5. 9.22 in.   11. 10.198 ft

Exercises 10.6, page 481

A

1. $a = 2$, $b = 3$ 7. $a = 6$, $b = 1$
3. $a = 7$, $b = 10$ 9. $a = 0$, $b = 1$
5. $a = 3$, $b = -7$

B

1. $\dfrac{43}{100}$ 9. $\dfrac{307}{100}$

3. $\dfrac{8}{10} = \dfrac{4}{5}$ 11. $\dfrac{9{,}371}{1{,}000}$

5. $\dfrac{7}{1{,}000}$ 13. $\dfrac{32{,}492}{1{,}000} = \dfrac{8{,}123}{250}$

7. $\dfrac{526}{10} = \dfrac{263}{5}$ 15. $\dfrac{56{,}794}{10{,}000} = \dfrac{28{,}397}{5{,}000}$

C

1. $\frac{3}{11}$
3. $\frac{476}{999}$
5. $\frac{740}{999}$
7. $\frac{5}{9}$
9. $\frac{647}{99}$
11. $\frac{25{,}588}{99}$
13. $\frac{1{,}762}{9}$
15. $\frac{67{,}464}{99{,}900} = \frac{1{,}874}{2{,}775}$

Exercises 10.7, page 485

A

1. $+4, -4$
3. $+1, -1$
5. $+12, -12$

B

1. 8
3. -2
5. -10
7. y^3
9. -1

Exercises 10.8 (Part 1), page 489

A

1. $\sqrt{35}$
3. $-\sqrt{10}$
5. $-\sqrt{42}$
7. $\sqrt{6ab}$
9. Does not exist

B

1. $2\sqrt{2}$
3. $2\sqrt{5}$
5. $b^2\sqrt{b}$
7. $a\sqrt{b}$
9. $2\sqrt{a}$
11. $2\sqrt{3a}$
13. $ab\sqrt{a}$
15. $3ab^2\sqrt{b}$
17. $5x^3y^2\sqrt{xy}$
19. $4ab\sqrt{ac}$

Exercises 10.8 (Part 2), page 491

A

1. x^4
3. $x\sqrt{y}$
5. a^2b^3
7. $4\sqrt{xy}$
9. $3b\sqrt{2a}$
11. $a^2bc\sqrt{c}$
13. $15\sqrt{xy}$
15. $10xy$
17. $10x^2y\sqrt{y}$
19. $12a\sqrt{3a}$
21. $bc^5\sqrt{5b}$
23. $x^2y\sqrt{x}$

Chapter 10

1. $\sqrt{6} + \sqrt{10}$
3. $\sqrt{ab} - \sqrt{ac}$
5. $6a + 10\sqrt{a}$

B

7. $y\sqrt{5x} - x^2\sqrt{5}$
9. $x\sqrt{y} - xy + y\sqrt{x}$

Exercises 10.9, page 495

A

1. 3
3. 2
5. 4
7. x
9. xy^2
11. $2a$
13. $\sqrt{11}$
15. $8ac^2$

B

1. $\dfrac{2}{3}$
3. $\dfrac{x^3}{3y}$
5. $\dfrac{a^4\sqrt{3}}{4}$
7. $\dfrac{x\sqrt{y}}{z^2}$
9. $\dfrac{3\sqrt{a}}{5b}$
11. $\dfrac{2\sqrt{2b}}{c}$
13. $\dfrac{3x\sqrt{3xy}}{a}$
15. $\dfrac{2b\sqrt{2ab}}{3c^3}$

C

1. $2\sqrt{3}$
3. $\dfrac{1}{x^2}$
5. $\dfrac{2\sqrt{2}}{a}$
7. $\dfrac{y}{3w^3}$
9. $\dfrac{2x\sqrt{x}}{3y^2}$

Exercises 10.10, page 499

A

1. $9\sqrt{5}$
3. $5\sqrt{7}$
5. $2\sqrt{6}$
7. $\sqrt{10}$
9. $-2\sqrt{11}$
11. $8\sqrt{2} - 4\sqrt{3}$
13. $4\sqrt{5} + 7\sqrt{3}$
15. $8\sqrt{a} - 2\sqrt{b}$
17. 0
19. $5x\sqrt{y}$

B

1. $5\sqrt{a}$
3. $-\sqrt{x}$
5. $-\sqrt{3}$
7. $19\sqrt{x}$
9. $9\sqrt{2}$
11. $5\sqrt{2} + 5\sqrt{3}$
13. $-x\sqrt{x}$
15. $3x\sqrt{y} - y\sqrt{x}$
17. $31a\sqrt{a}$
19. $6x\sqrt{xy} - 4y\sqrt{xy}$

Answers to Odd-Numbered Problems

Exercises 10.11, page 502

A

1. $\dfrac{5\sqrt{2}}{2}$
3. $\dfrac{7\sqrt{2}}{4}$
5. $\dfrac{\sqrt{15}}{6}$
7. $\sqrt{b}$
9. $\dfrac{\sqrt{6x}}{3x}$
11. $\dfrac{\sqrt{ax}}{x^2}$
13. $\dfrac{3\sqrt{2ab}}{2b}$
15. $\dfrac{\sqrt{3b}}{3b}$
17. $\dfrac{\sqrt{2}}{a}$
19. $\dfrac{\sqrt{ab}}{b}$

B

1. .707
3. 2.236
5. .8163
7. .7638
9. .8451

Exercises 10.12, page 508

A

1. 64
3. 5
5. 19
7. 7
9. No root
11. 9
13. $\dfrac{9}{2}$
15. 27
17. 17
19. −22
21. $\dfrac{13}{4}$
23. 12

B

1. 5, −5
3. 4, −4
5. 2, −2
7. 9, 4 (ext.)
9. 5, 0 (ext.)

C

1. 5
3. 4
5. 9
7. 0, 4
9. 2, 6

Chapter 10 Review, page 509

1. a. 8, −8
   b. 3, −3
   c. No roots
   d. 5, −5

3. a. $-4+\sqrt{3},\ -4-\sqrt{3}$
   b. $\dfrac{3+\sqrt{5}}{2},\ \dfrac{3-\sqrt{5}}{2}$

5. a. 5, −3
    b. 8, −4
7. a. $3x^2 - 5x - 4 = 0$
    b. $2x^2 - 6x - 5 = 0$
9. a. 1, −6
    b. 3, $\frac{5}{3}$
    c. 3, −6
    d. $-3, -\frac{4}{5}$
11. a. 4.1925, −1.1925
    b. 2.366, .634
13. a. $a = 5, b = 7$
    b. $a = 9, b = 1$
15. a. $\frac{4}{9}$
    b. $\frac{34}{99}$
    c. $\frac{63}{11}$
    d. $\frac{46}{55}$
17. a. 2
    b. b
    c. $2x^3$
    d. $\sqrt{5y}$
19. a. $8\sqrt{2}$
    b. $-3\sqrt{3}$
    c. $3\sqrt{x} + 5\sqrt{y}$
    d. $3a\sqrt{b}$
21. a. $\frac{7\sqrt{3}}{3}$
    b. $\sqrt{3}$
    c. $\frac{\sqrt{5ab}}{5b^2}$
    d. $3\sqrt{2x}$
23. 8.485 m

Index

A

Abscissa, 322
Absolute value, 36, 37, 143
   linear equation in one variable, 143-149
   linear equation in two variables, 365-370
Addition
   addition principle, 36
   associative principle, 38
   fractions, 267-285
   like terms, 45
   monomials, 40-48
   polynomials, 49-53
   real numbers, 29-39
Addition principle, 36
Additive inverse, 57
Area problems, 238-246
Axes, 318, 323
   horizontal, 318
   vertical, 318

B

Binomial, 40
Braces, 20
Brackets, 14

C

Coefficient, 43, 299-302
Coin problems, 175-181
Collecting like terms, 50
Common factor, 198-210
   binomial, 205-210
   monomial, 201-205
Completing the square, 454
Coordinate
   axes, 323
   plane, 323
Coordinates, 322

D

Difference of two squares, 210-216
Digit problems, 425-429

Distributive principle, 42
Division
   fractions, 262-267
   monomials, 102-107
   polynomial divided by monomial, 107-110
   polynomials, 110-117
   real numbers, 99-102

E

Elements of solution set, 19
English sentence, 11
Equation
   addition principle, 119-124
   containing fractions, 285-291
   fractional equations, 293-297
   linear, 119, 231
   literal, 298-304
   multiplication principle, 125-130
   quadratic, 231-236
   radical, 504-508
   root, 121
   solution set, 18-22
   solving an equation, 121
Equations
   consistent, 388
   dependent, 388
   equivalent, 388
   inconsistent, 388
Exponent, 13-18
Extended number line, 5-8
Extraneous root, 505

F

Factoring, 193-248
   common factor, 198-210
   difference of two squares, 210-216
   factors of a number, 194-198
   pairs of factors, 194
   perfect square trinomial, 225-228
   prime factors, 196
   summary, 230
   trinomials, 216-230
FOIL method, 92-95
Fractional equations, 286-291
Fractions, 249-316
   addition and subtraction, 267-285
   division, 262-267

   equations, 286-291, 293-297
   inequalities, 291-293
   lowest terms, 251-257
   multiplication, 258-262
   multiplication principle, 250
   principle of 1, 250 ,273
   problem solving, 304-314

G

Graph
   inequality, 23
   linear equation, absolute values, 365-370
   linear equation, two variables, 330-352
   linear inequality, 373-380
   naming points, 318-330
   open sentence, two variables, 317-383
   parallel lines, 363
   point-slope method, 359-365
   relation to equation, 334
   solving a linear system, 386-395
Greatest common factor, 198

I

Inequality, 22-36
   with absolute values, 153-158
   addition principle, 150
   with fractions, 291-293
   graph, one variable, 157
   graph, two variables, 373-380
   multiplication principle, 151
   solution in one variable, 149-153
Interest problems, 436-442
Inverse
   additive, 57
   multiplicative, 103, 262
Irrational number, 482-485

L

Linear system of equations
   problem solving, 420-442
   solution by addition or subtraction, 402-416
   solution by graphing, 386-395
   solution by substitution, 395-401

Index

Linear system of inequalities, solution by graphing, 417–419
Literal equations, 298–304
Lowest common multiple, 277

M

Mathematical phrase, 161–165
Mathematical sentence, 8–13
Monomial, 40
Motion problems, 429–436
Multiple of a number, 5, 274
    lowest common multiple, 277
Multiplication
    commutative principle, 70
    distributive principle, 42
    FOIL method, 94
    fractions, 258–262
    monomials, 76–82
    monomial times polynomial, 84
    multiplication principle, 72
    polynomials, 83–99
    real numbers, 69–76
    special products, 91–99
    square of a binomial, 96–97
Multiplication principle, fractions, 250
Multiplicative inverse, 103, 262

N

Number line, 1–5
    extended number line, 5–8
Number problems, 165–175, 236–238, 304–308, 420–425

O

Open sentence in one variable, 119–159
Ordered pair of numbers, 320
Ordinate, 322
Origin, 6

P

Pairs of factors, 194
Parallel lines, 363
Parentheses, 14
    simplifying expressions, 62–67

Percent
    renaming a decimal as a percent, 183
    renaming a percent as a decimal, 182
Percent problems, 181–190
Perfect square, 195, 211
Perfect square trinomial, 225–228, 452
Placeholder, 9
Point-slope graphing, 361
Polynomials
    addition, 49–53
    binomial, 40
    division, 110–117
    monomial, 40
    multiplication, 83–99
    subtraction, 59–62
    trinomial, 40
Prime factors, 196
Prime number, 196
Principal square root, 447, 504
Principle of 1, 250, 273
Principle of 0, 232
Problem solving, 161–192
    area, 238–246
    coin, 175–181
    digit, 425–429
    interest, 436–442
    motion, 429–436
    number, 165–175, 236–238, 304–308, 420–425
    percent, 181–190
    Theorem of Pythagoras, 476
    two variables, 420–442
    work, 308–314
Product of square roots, 485–491
    principle of multiplication of radicals, 486

Q

Quadrants, 325
Quadratic equation
    problem solving, 236–246
    pure, 445–449
    solution by completing the square, 452–461
    solution by factoring, 231–236

solution by quadratic formula, 461–471
   standard form, 231, 445
Quadratic formula, 465
Quotient of square roots, 491–496
   principle of division of radicals, 493

R

Radical equations, 504–508
Radicand, 447
Rational expression, 249
Rationalizing the denominator of a fraction, 499–503
Rational number, 477
Real numbers, 2, 445–511
   addition, 29–39
   division, 99–102
   multiplication, 69–76
   subtraction, 53–59
Reciprocal, 263
Relatively prime numbers, 251
Repeating decimal, 478
Right triangle, 471
Rise, 353
Root of an equation, 121
   extraneous root, 505
Run, 354

S

Signed numbers, 33
Sign of direction, 6, 30
Sign of operation, 6, 30
Simplifying a radical, 487
Slope of a line, 352–365
   rise, 353
   run, 354
Solution set
   equality, 18–22
   inequality, 22–26
Square of a binomial, 96–97
Square root, 195, 211, 226, 469
   table, 470
Standard form
   linear equation, two variables, 231
   quadratic equation, 231, 445, 461, 463
Subtraction
   additive inverse, 57
   additive method, 53
   polynomials, 59–62
   real numbers, 53–59
   subtraction principle, equality, 405
   subtraction principle, real numbers, 57
Sum of square roots, 496–499
Systems of open sentences, 385–444
   solving the system, 385

T

Terminating decimal, 478
Theorem of Pythagoras, 471–476
   Pythagorian society, 473
Trinomial, 40

V

Variable, 9

W

Work problems, 308–314